HISTORY OF THE BUREAU OF ECONOMIC GEOLOGY, 1909-1960

W. Keene Ferguson

Bureau of Economic Geology
W. L. Fisher, Director
The University of Texas at Austin
Austin, Texas 78712

1981

Dr. J. A. Udden knows a lot about the geology of [the Terlingua district] which has not been published.... He is a geologist, I am a mining man. We look at things from a different standpoint. This Bureau virtually takes the place of a State Geological Survey, but we have only a small amount for field work. The Board of Regents thought it was time to bring the University into closer touch with the material development of the State and that is what I am trying to do.

Dr. William Battle Phillips to H. D. McCaskey of the U. S. Geological Survey, Feb. 2, 1910.

This State is so large and the cost of covering it so great that it would be best to restrict any appropriation for field work to specific purposes. For instance, if we had an appropriation of $15,000 a year for each of the following two years devoted entirely to oil or artesian water or whatnot, something could be accomplished.... A general or mineralogical survey is not within the possibilities now and it is useless to go into this fight. But an oil, gas and water survey would appeal to a great many people.

W. B. Phillips to C. R. Troxel, Apr. 25, 1911.

It appears to me that the University of Texas and the State of Texas have now reached a point where the function which distinguishes a university from any school of lower rank—the function of research and advancement of knowledge—should be duly recognized and means for its proper activity in this direction should be secured.... Investigations conducted by the Bureau must be planned on a broad basis, and complete familiarity with the conditions in our own State is necessary to insure the best results.

Johan August Udden to the Investigating Committee of the State Legislature, Dec. 1, 1917.

In our State, we have so to speak put the cart before the horse. We have described the formations before we have described the fossils on which the classification of the formations is based! While I recognize that it is necessary for all federal and state surveys to give considerable attention to the directly profitable part of geology, I realize that there is a limit to everything, and we are beginning to feel the want of real scientific work that must be done in order that the economic geologist can afterwards do any real service.

J. A. Udden to J. P. Buchanan, House of Representatives of the United States, Mar. 24, 1919.

CONTENTS

Foreword v
Preface vii
1. A Firm Foundation 1
2. "The Public Servant is Worthy of His Hire" 9
3. The Publications of the Bureau 19
4. The Bureau and the Oil Geologist 41
5. The U.S.G.S. vs. the B.E.G. 51
6. Scientific Servant of the State 61
7. Emergency Years: Depression and War 77
8. Dr. Sellards and the Search for Early Man 87
9. Multifaceted Geological Research in Texas 95
10. The Bureau and Resource Management 115
11. From the Ouachita Foldbelt to *Eohippus* 125
12. The Maturation of Oil Company Research Attitudes 139
Epilogue 145
Photographs 152
Notes 167
Appendices 235
I. Biographies 235
Virgil E. Barnes 235
Mary Josephine Casey 238
Ross Allen Maxwell 239
II. Memorials 241
Walter Scott Adkins 241
Esther Richards Applin 248
Charles Laurence Baker 251
Joshua William Beede 258
Emil Böse 260
Halbert Pleasant Bybee 260
Alexander Deussen 265
Alva Christine Ellisor 270

CONTENTS ***continued***

Julia Anna Gardner 274
Ralph Alexander Liddle 278
John Tipton Lonsdale 282
Leroy Thompson Patton 286
William Battle Phillips 289
Frederick Byron Plummer 290
Helen Jeanne Plummer 292
Edwin Leight Porch, Jr. 295
Thomas Mann Prettyman 296
Elias Howard Sellards 297
Johan August Udden 301
III. The Red River Investigations 307
IV. Udden's Correspondence Regarding Bureau Investigations of University Lands 309
V. The Association of American State Geologists 315
VI. A Proposed Institute of Geologic Research at The University of Texas 317
VII. Some Events in the Geologic History of Texas 323

FOREWORD

In 1969 The University of Texas Press published *Geology and Politics in Frontier Texas, 1845-1909*, by Dr. W. Keene Ferguson. This perceptive scholarly work examined the relationship of politics and public-sector geological investigations in the first half century of Texas statehood.

Texas is a state richly endowed with energy and mineral resources. This endowment was perceived in the days of the early state geological surveys, but the extent and magnitude of the state's resources were not fully recognized until later.

The several, legislatively created geologic surveys of the 1800's, chronicled by Ferguson, were the forerunners of the present Bureau of Economic Geology. The Bureau was established as a research unit of the University of Texas in 1909 by the Board of Regents. Since that time the Bureau has filled the dual role of state geological survey, a quasi-state agency, and of a research organization with a mission directed toward Texas resources.

State geological surveys or their equivalents exist in 49 states of the Nation. Many operate, as the Bureau, as administrative units of a major state university.

In 1971, I invited Dr. Ferguson, as an independent historian, to undertake a historical analysis of the Bureau of Economic Geology through the first 50 years of its existence (1909-1960), as a parallel volume to his previous history. The manuscript of this work, *The History of the Bureau of Economic Geology, 1909-1960*, was completed by Dr. Ferguson in 1977. We have made available to him Bureau files and records and have supported him in his research. But his conclusions and observations are, as they should be, those of an independent historian.

W. L. FISHER

Director
1981

PREFACE

The Bureau of Economic Geology. This bureaucratic title belies the essence of this organization — an ongoing group of professionals with a definite personality. In fact, the history of public geological research in Texas is the story of dedicated, inquisitive scientists shunning the higher salaries and opportunities offered by private enterprise to serve their science and the State of Texas. The Bureau has survived the emergence of extensive private geological research departments, the creation of geological departments in the various State agencies, and the threatened encroachment of the Federal Geological Survey. Not only has the Bureau survived, it has gained strength and stature throughout its history and now stands in the vanguard of basic geological research, energy investigations, and ecological studies in Texas. Throughout, the geologists in this story have made the Bureau a receptive, personal organization welcoming all inquiries and visitors in search of help in solving the many geological problems in relation to both mineral production and public resource demands. The people of the Bureau have served both the public interest and private enterprise for 50 years with never a hint of scandal to cloud the horizon.

This history covers the first half century of the Bureau's existence up to the death of the fourth director, Dr. John T. Lonsdale, in 1960. The Board of Regents of the University of Texas officially activated the Bureau of Economic Geology on October 21, 1909, in response to the urgings of William B. Phillips and University President Sidney E. Mezes. Phillips, who became the first director of the Bureau, had also headed the short-lived predecessor of the Bureau, the University of Texas Mineral Survey, from 1901 to 1905 (the history of the Mineral Survey and all previous State geological surveys in Texas is presented in *Geology and Politics of Frontier Texas* by the same author; see note 3). From Director J. A. Udden's pioneer contributions to the developing science of oil geology to current Director William L. Fisher's leadership in the search for solutions to the state's ecological and energy problems, from the few offices in old "B Hall" housing a staff you could count on one hand to today's generously staffed and funded modern offices, the Bureau has been outstanding in its service to the people of the State of Texas.

This book is divided roughly into three sections—the years under the directorship of Professor Phillips and Dr. J. A. Udden (1909-1932), the directorship of Dr. E. H. Sellards (1932-1945), and the directorship of Dr. John T. Lonsdale (1945-1960). The geological research efforts of the U. S. Geological Survey and oil geologists in Texas have been followed along with the contributions of the geologists of the Bureau. The outstanding research work of all of these geologists in Texas has been surveyed in the more technical chapters entitled "The Publications of the Bureau" (1909-1932), "Multifaceted Geological Research in Texas" (1933-1945), and "From the Ouachita Foldbelt to *Eohippus*" (1945-1960). Also, the work of the Federal Geological Survey in public resource management has been followed to provide perspective on the Bureau's role in the public management of the natural resources of Texas. The burgeoning role of geological research by the oil industry has been traced in an effort to illustrate the cooperation between the Bureau and private enterprise.

Many vital and interested people have provided invaluable assistance in the preparation of this history. First, current Director William L. Fisher made it all possible by offering his encouragement, the Bureau's funding of the project, and the availability of the Bureau's outstanding facilities. All of the personnel of the Bureau offered every assistance. These efforts made the work not only possible but pleasurable. The author is especially indebted to Virgil E. Barnes, Josephine Casey, Peter T. Flawn, Roselle Girard, Edgar W. Owen, Wallace Pratt, and Keith Young, who read the manuscript and made the necessary suggestions, additions, and corrections that contributed so greatly toward the effectiveness of the final publication. Elizabeth T. Moore, who retired from the Bureau in 1976 after 18 years of service, guided this work through the crucial stages in preparation for publication. Her suggestions and editing were helpful.

The current Bureau editorial staff, under the direction of Susann Doenges, edited and proofread the final galleys. Typesetting was by Charlotte Frere, under the direction of Lucille Harrell. This publication was designed by Judy P. Culwell and assembled by Micheline Davis.

The American Association of Petroleum Geologists, the American Journal of Science, and the Geological Society of America kindly granted permission to reprint memorials that appear in Appendix II.

I have enjoyed this fulfilling labor. Now let the story speak for itself.

W. KEENE FERGUSON

Austin, Texas
September 1981

ONE

A Firm Foundation

By the beginning of the 1920's, the Bureau of Economic Geology, which had been created in 1909, was well established as a viable, permanent organization for research on the geology of Texas. The increased awareness of the value of scientific research that was brought to the fore by the First World War, and the successful defense fought by the University of Texas against anti-academic attacks of Governor James E. Ferguson, were necessary concomitants to the establishment of the scientific and fiscal stability of the Bureau (1). However, the real credit for the success and growth of the Bureau must go to the first two directors, Dr. William Battle Phillips and Dr. Johan August Udden, who led the Bureau through its crucial early years to the attainment of scientific and legislative respectability (2).

These two eminent gentlemen had become acquainted while associated with the short-lived University of Texas Mineral Survey (1901-1905) (3). Phillips, an instructor in the Geology Department at the University, had been promoted to the rank of Professor of Field and Economic Geology and made Director of the University of Texas Mineral Survey in 1901 (4). In 1903, he contracted Udden, who was Professor of Natural Sciences at

Augustana College in Rock Island, Illinois, to investigate the geology of the Shafter silver mine district in Presidio County. Eight years later, Phillips, then Director of the Bureau of Economic Geology, hired Udden as the first field geologist for the Bureau; Udden went on to succeed Phillips as Director in 1915.

A very warm friendship developed between the two prominent scientists. When hearing of Phillips' death in 1918, Udden mourned the loss of his closest friend. However, as is evident in the epigraphs, their approaches to the business of the Bureau were different, as were their personalities.

Phillips' pragmatic approach to the work of the Bureau was coupled with a strong dedication that bordered on being aggressive and abusive. In a letter to a State representative condemning the shortage of funds for the Bureau, Phillips blasted out: "I do not care what you do with this letter . . . you can placard the Capitol with it . . . as I have no axe to grind nor any favors to ask personally; but I want some money for carrying on this work" (5). In answering a request made by the statistician of the U. S. Geological Survey for data on the mineral resources of the state, Phillips asserted that the information must be used for the purpose of strengthening the Bureau as well as for the Federal Survey (6): "My first purpose and almost the only one is to build up this Bureau just as yours is to build up the Geological Survey; . . . with me, it is Texas first, last, and always and I do not allow anything to interfere."

William Battle Phillips was born in Chapel Hill, North Carolina, in 1857. Both his father and maternal grandfather were professors at the University of North Carolina, where Phillips received his doctorate in mining engineering in 1883. He also studied at the School of Mines at Freiburg in Saxony, Germany, then the most famous mining school in the world. With his extensive academic and business experience in mining engineering, it was natural that, as Director of the University of Texas Mineral Survey and later the Bureau, he focused his attention on the Trans-Pecos, the only region in the state potentially rich in precious minerals (7).

Phillips was a leader in the successful struggle to revise the State mining laws and to prevent potential loss of revenue from prospecting leases and sales of the State lands in the Trans-Pecos (8). He directed most of his energies to this region (9):

> What the [Trans-Pecos] needs most are good topographic maps and a geological and mineralogical survey based on these maps. This could be accomplished without material expense to the State by contributing the money derived from the sale or lease of mineral lands owned by the State to the purpose of preparing these maps and geological reports. In other words, the mineral lands would be used to develop themselves. . . . [And] I would expend this money in the preparation of topographic maps with special reference to the location of dams for impounding storm waters to be used both for grazing and for agricultural purposes.

Yet, though his interest in the minerals of the Trans-Pecos, in the coals and lignites, and in the low-grade iron ores of the central and eastern parts of the state demonstrates his concern with the economic applicability of Bureau work, he was quick to note the importance of basic geological research as a foundation for efficient mineral exploitation. He was undoubtedly influenced in this respect by Udden, whom he hired as the first field geologist of the Bureau on September 1, 1911. When Udden commenced stratigraphic studies of the Carboniferous beds in the Glass Mountains of Trans-Pecos Texas in 1914, Phillips described the purpose of this research in words characteristic of Udden (10, 11):

> There is no coal there, but the Carboniferous outcrops on a very extensive scale and as our oil fields in the northwest part of the State [which had been the object of Udden's first work for the Bureau] derived their oil from the Carboniferous, we have thought it would be well to prepare special reports on the Carboniferous outcrops wherever they can be examined.

Perhaps the many qualities in Dr. Phillips' personality and the range of his professional competence are best summarized in a memorial in the *Quarter-Centennial Memorial Volume of the Division of Natural Resources* (12):

> He was a proficient mining engineer, chemist, geologist, mineralogist, metallurgist, and technologist, and [he was] memorable for sincerity and integrity, fearlessness and independence in mind and character, a widely interested, versatile, well-stocked, and broadly liberal mentality, untiring industry, thoroughness and earnestness of purpose.

When Phillips resigned in the summer of 1914 to become President of the Colorado School of Mines and turned over the direction of the Bureau to his hand-picked successor, Udden, he had used this unique combination of talents to establish firmly the viability of the Bureau in the eyes of the President and the Regents of the University of Texas. This is clearly demonstrated in the increase in appropriations for the Bureau from $7,000 in 1910 to $25,000 in 1914. Even though it would be Udden and his team of stratigraphers and paleontologists who would bring scientific respectability to the Bureau in the next decade and establish the precedents for University research, Dr. Phillips can truly be called the father of the Bureau of Economic Geology.

Dr. Udden was preeminently a scientist—more particularly a geologist. An insight into the nature of this genius, and his dedication to his chosen field, is found in a reply he made to a request from the editor of *Petroleum* for biographical information (13):

> I had my early training as a geologist under a naturalist of the old school, known to many of the older naturalists in this country—Dr. Joshua Lindahl, who was my early teacher in college. You might also state that I left my former position as Oscar II Professor of Natural History at Augustana College, feeling that the vast opportunities offered to me as a geologist in Texas would more than repay the loss of pleasant associations maintained for nearly a quarter of a century in Illinois. I may add that my expectations have been fully realized. The study of the geology of the Southwest has only begun, and still it has revealed so many new and unexpected things that I am firmly of the opinion that there will be interesting geologic work to do in Texas for a whole century to come.

The insatiable scientific curiosity of this Swedish gentleman lured him away from the comfortable professor's life at a small Midwestern college into the scientific frontiers of Texas at the age of 52.

His purposefulness and self-discipline are further demonstrated in his 40-year battle with diabetes. An associate at Augustana wrote that Udden was able to order his life so carefully and scientifically that he was able to live a full and active life in spite of this handicap (14).

Udden was typical of his generation in that his scientific intuitions were tempered with a respect for the practical applications of the results of scientific research. It was part of Udden's genius that he early recognized the need for basic geological information in any oil exploration program, and at the same time he foresaw the tremendous additions to the understanding of the stratigraphy and structure of the rocks of Texas that would come from drilling. Furthermore, this eminent scientist was one of the first men to recommend the use of the seismograph in oil exploration, and he was the leader in the development of subsurface geological studies—lithologic and paleontologic—in the Southwest. The tremendously important role played by Udden in the early decades of the oil industry is most aptly described in the following excerpt from a conversation with Edgar W. Owen, author of *The Trek of the Oil Finder: The History of Petroleum Exploration* (15, 16, 17, 18):

> Udden's little paper [on the potential value of seismograph exploration tools] that he delivered to the A.A.P.G. in 1920 was one of the real milestones of geological science.... A short time later, Udden was called in as a consultant for Louisiana Land and Exploration, who had enormous holdings in the Gulf Coast ... and Udden in his first report recommended that they undertake a seismic survey. Subsequently, they undertook one of the most fruitful seismograph campaigns that was ever conducted....
>
> Another major effect of Udden's work with the Bureau was a long paper in the A.A.P.G. in 1919 on the potential producing formations in Texas. Just to show you how far ahead he was of all the companies, it was thought generally at the time that they had found all the salt domes on the Gulf Coast—the Gulf Coast was washed up—and this he denied. Nobody had ever paid any attention to the Gulf Coast Eocene, but Udden said the Eocene from Mexico to the Mississippi River was one of the great potential producing formations. Nobody had paid any attention to the Ellenburger. He said that it was a major potential producing formation. All the way through he just gave the oil companies a whole new concept on the magnitude of the prospects in Texas at a time when Texas was doing damn little....
>
> Another major point in Udden's career was that microscopic sample examination he made of the Spur well in which he mentioned the occurrence of potash in the waters of some of the samples. While the influence of that was delayed, still that sample log stood as a great landmark for a good many years.

Udden constantly called for the systematic collection and use of the wealth of information made available in well cuttings. From his description of the Spur well on, Udden was the premier sample investigator; before the emergence of the company geological departments, there was a great demand for the interpretation of samples by Udden and his associates. In a 1921 letter bemoaning the legislative discontinuance of a $5,000 appropriation for subsurface work, Udden spoke of the backlog of samples sent in for examination and noted that he usually spent 2 or 3 hours in the laboratory going over samples that had been prepared by his assistants (19). In 1919, the Bureau published Handbook No. 1 compiled by Dr. Udden entitled *Aids to Identification of Geological Formations* (20). It was intended to encourage improvement in accuracy and uniformity of driller's logs.

Perhaps the most celebrated contribution that Dr. Udden made to the development of exploration methods in the Southwest was his pioneer work in micropaleontology (21). Udden had made his first microscopic examinations of cuttings from wells in Illinois from 1908 to 1911. According to F. B. Plummer, "Dr. Udden not only introduced micropaleontology into Texas from Illinois, but he generally sold the idea of its importance to the oil field geologist" (22).

In a 1912 study of the oil and gas fields of Wichita and Clay Counties, Udden presented a detailed lithologic description of the samples, and also included a list of *Fusulina* found in some of the cuttings (23). Then in the noted report on *The Deep Boring at Spur [Dickens County, Texas]* in 1914, Udden had a section entitled "Notes on Fossils" (24). Although he was unable to identify any species of Foraminifera, he sketched many that he found in the samples. Udden also coauthored a paper on the Thrall oil field in which he prepared a table showing the relative frequency and abundance of Foraminifera in the Taylor Marl in that field (25). In typical objective fashion, he added that no reliable determinations were attempted in the study as neither author had special knowledge in the study of foraminifers.

Udden also proceeded to train a number of geologists in micropaleontology. All of them went on to join oil companies. He first trained J. Wallace Bostick, the son-in-law of former State Geologist E. T. Dumble, who left to join Dumble's Rio Bravo Oil Company in 1918 (26). Miss Esther Richards (later Mrs. Paul Applin) continued the studies of Texas microfossils begun by Bostick. Miss Richards followed Bostick to join the ranks of the progressive Rio Bravo Oil Company in 1919 (at about the same time, Miss Alva C. Ellisor initiated micropaleontological work for the Humble Oil & Refining Company in their Fort Worth office).

V. V. Waite examined samples under Udden's direction from 1917 to 1918, at which time he left to join the Atlantic Company (27). In the summer of 1920, a private edition of *Microscopic Characteristics of the Bend and Ellenburger Limestones*, coauthored by Udden and Waite, appeared (28). This was an excellent paper, very advanced for those days, and included thin sections of cores. In June 1920, Miss Hedwig Kniker joined Dr. Udden's subsurface laboratory group at a time when samples from the Cretaceous and Tertiary wells in Central Texas had begun to flow into the Bureau's sample library (29). She and J. W.

Beede were to write the first Bureau publication dealing specifically with Texas microfossils: *Species of the Genus* Schwagerina *and Their Stratigraphic Significance* (30). This comprehensive study included correlations of Permian fusulines throughout the Northern Hemisphere. From the early 1920's on, the major oil companies began to hire their own micropaleontologists. Udden had awakened them to the value of such work and was also probably responsible for the establishment of the first micropaleontology course at the University (31).

Under Director Udden, the Bureau became an invaluable ally of the oil exploration companies. In subsequent chapters it will become quite evident that the Bureau, in spite of its limited funds and small staff, made the most vital research contributions in paleontology and stratigraphy. The Bureau's geologists were able to consolidate and coordinate some of the most important new geological information pouring out of oil exploration efforts and to interpret these data and feed them back to the companies. Thanks to Udden, the company geologists looked to the Bureau as a neutral research agency that could provide basic geological information which they could use as a guide to oil exploration. Udden described this working relationship most clearly (32):

> I find that in nearly every case, these private geological organizations are concerned with private local problems, and cannot give their attention to the general geologic questions which have the most important bearing on local problems; [and so] all look to the Bureau for information on such general geologic problems.

TWO

"The Public Servant Is Worthy of His Hire" (32)

The role of the University of Texas Mineral Survey, which the Legislature created in 1901 as a reform instrument to help the state realize the full economic potential of the public school and University lands, was weakened by legislation that provided a more practical approach to the problem of protecting public mineral lands from being exploited at the expense of the various school funds (33). At the same time, the tremendous impact of the post-Spindletop oil discoveries prompted some state leaders to consider establishing a permanent geological bureau as part of the University. This bureau would assist and encourage the development of the state's resources.

University of Texas President Sidney E. Mezes was very interested in extending the benefits of the University to as many Texas citizens as possible. This concern underlay his proposal to the Board of Regents in June of 1909 that the University establish a Bureau of Economic Geology (34):

> The Mineral Survey attached to the University was very useful to the people of the State, and did much to convince them of the University's ability to give useful and practical information. Its weakness was that it had to stand on its own feet instead of being part of the University, and for a variety of reasons it proved incapable of sustaining itself before the Legislature. A further weakness was the

insufficient specification of the services the Survey was prepared to render. This aroused expectations that could not be met, and led to disappointments and hostilities. I believe that a Bureau of Economic Geology, with carefully defined aims, which it would be possible to compose, would escape both these dangers and would in its usefulness justify its existence and bring support to the University.

The Board immediately approved the recommendation and placed $3,500 in the budget to cover the expenses and salary of a director.

Dr. Phillips, who had returned to private consulting work for the Chisos Mining Company after the termination of the Mineral Survey, may originally have suggested the idea of a bureau to Mezes (35). Whereas Mezes foresaw the increased public support for the University that such a bureau might create, Phillips emphasized the contributions that the bureau would make toward stimulating the Texas economy. He praised the Regents' action and noted that "no other institution of higher learning in the country has taken upon itself the duty of providing, at its own expense, an office to which one may apply for information" on the mineral resources of a state (36).

Wisely, the Board did not try to outline the functions of the bureau exactly as Mezes had recommended. A press release in July announced the appointment of Phillips as director and defined the general duties of the bureau. The Regents proposed that the bureau serve as an agency to provide information that could be freely consulted by geologists, engineers, and landowners alike — "a sort of clearing house of information." Mezes went on to stress the practical benefits that would be forthcoming and advocated the conversion of the old power house into a "testing laboratory of economic geology" (37). The Board officially activated the bureau on October 21, 1909 (38).

When Dr. Udden assumed the post of Director of the Bureau of Economic Geology and Technology in 1915, research at the University of Texas had its real beginning (39). He was an astute and capable State Geologist, well aware of the varying backgrounds and attitudes of State legislators, but he was first of all a scientist, alert to the fact that research is the lifeblood of any science. In a letter of recommendation for the noted paleontologist Emil Böse, Udden commented on how difficult it was to educate legislators to "the necessity of scientific work which must precede the economic application of geologic knowledge" (40). For example, he pointed out that the neglect of the study of

Texas fossils had been a serious hindrance to the development of petroleum resources in the northern part of the state.

It had taken the foresight and vigor of Dr. Phillips to give momentum to the Bureau. As a mining engineer he had naturally stressed studies of the mineral resources of the Trans-Pecos region and investigations into the utilization of the lignite, coal, petroleum, and natural gas resources of Texas. Dr. Phillips also stressed research in engineering and industrial chemistry under the auspices of the Bureau. Following his resignation and the appointment of Udden as director in 1915, the Bureau of Economic Geology and Technology was reorganized upon Udden's recommendation into the Division of Engineering under F. E. Giesecke, the Division of Chemistry under E. P. Schoch, and the Division of Economic Geology under Udden (41). Phillips had actively directed work in all three divisions.

With Udden as director, for the first time the correspondence of the Bureau included discussions of stratigraphy and paleontology. During the first year of 1915, Udden and Charles Laurence Baker continued their stratigraphic studies; Udden hired the world-renowned German invertebrate specialist Emil Bose, who undertook the first thorough faunal research in Texas. In 1916, Udden brought to completion the first compendium of Texas geology, University of Texas Bulletin 44, which included the first geological map of the state, an outstanding achievement (42). The new director also initiated a program of areal and stratigraphic geological investigations on the county level. Udden had indeed launched the research aspect of Bureau work.

Udden stepped down as active director of the Bureau in 1925 to be succeeded by Dr. E. H. Sellards. During the decade of his leadership he attracted outstanding geologists into the ranks of the Bureau. In the summer of 1921, F. B. Plummer wrote Udden inquiring about a possible position with the Bureau. In this letter Plummer touched on the attraction of Bureau service. Plummer commented that he had worked for Roxana Oil Company in north Texas for 6 years, and the time had come when he had "to decide whether [he wanted] to be moved around continually like a machine hunting for oil pools or settle down to a much less salary in some position where [he] can do more real scientific work" (43). For the record, Udden offered Plummer a post with the Bureau at a salary of $2,320, but because of the severe monetary sacrifice involved, Plummer cabled back, "Sorry, acceptance impossible" (44). However, in 1921 Plummer coauthored one of the finest research studies ever published by the Bureau, *The Stratigraphy of the Pennsylvanian Formations of North-Central Texas*, and in 1928, he did join the Bureau staff (45).

As a matter of fact, C. L. Baker, the first geologist to join the Bureau after Udden in 1914, left the Bureau in 1917 to join the oil exploration department of the Southern Pacific Railroad in San Francisco (46). Baker continued to communicate regularly with his close friend, Udden, discussing points of stratigraphy, especially of the Trans-Pecos; he helped re-edit Bulletin 44 in 1919, and returned to Bureau service in 1932 to assist in writing the two-volume compendium on Texas geology, *The Geology of Texas, Vol. I, Stratigraphy* (University of Texas Bulletin 3232) and *Vol. II, Structural and Economic Geology* (University of Texas Bulletin 3401). Baker described the structural geology of the Trans-Pecos in the second volume (47). His return to the Bureau had been hastened by the Rio Bravo Oil Company's depression layoffs.

Dr. Böse did not leave the Bureau because of salary. Böse, who was with the Mexican Geologic Survey from 1898 to 1915, at which date he joined the Bureau, was on vacation in Mexico in the summer of 1917 when the United States Congress passed a law tightening restrictions on the entry of German citizens into the United States. He was denied readmittance and felt that he had to resign from the Bureau. Udden wrote a letter to the State Department praising Böse's character and scientific reputation, whereupon Böse was allowed to return (48). Böse then accepted a position with the Richmond Oil Company, which allowed him to continue his investigations in South Texas and northern Mexico. He continued to keep in close contact with Udden, Baker, and also with Walter Scott Adkins, Cretaceous paleontologist who joined the Bureau in 1919, discussing points of Texas and Mexican paleontology and stratigraphy up to the time of Böse's death in a car wreck in 1927 (49). That very year the Bureau published his outstanding paper on the Cretaceous ammonites of northern Mexico and Texas (50).

Another enthusiastic and brilliant scientist that Udden brought to the ranks of the Bureau was Dr. J. W. Beede, probably on the recommendation of his former student, Professor H. P. Bybee of the University of Texas. Beede left his position as

Professor of Geology at Indiana University in 1917 to fill the vacancy created by the departure of Böse (51). He was a specialist in Pennsylvanian and Permian paleontology and made vital contributions to the stratigraphy of North-Central Texas. However, he likewise found the lure of higher salaries too strong and joined the Empire Company of Bartlesville, Oklahoma, in the fall of 1922 (52).

In 1918, the third director-to-be (in 1932) joined the staff of the Bureau. In that year, Dr. Udden offered Dr. Elias Howard Sellards, then Director of the Florida Geological Survey, a position with the Bureau. Sellards hesitated to leave Florida, but when Dr. E. H. Barbour, State Geologist of Nebraska, visited Sellards and spoke so glowingly of the geological research possibilities in Texas, Sellards decided to accept the offer (53). So Sellards, a noted paleobotanist, came to Texas primarily because of the promising research potential. He focused his early years of work on the study of the subsurface geology of the oil-producing regions in the state. His first paper for the Bureau, written in 1918, was *On the Underground Position of the Ellenburger Formation in North-Central Texas* (54). This report was a great aid to oil geologists. Sellards also directed the preparation of scientific evidence presented in the suit by the State of Texas over the ownership of some Red River Valley lands. This Texas-Oklahoma boundary dispute gained prominence in 1921, and the scientific results of the investigations were published by the Bureau in 1923 (55).

The departure of Adkins, who joined the Mexican subsidiary of Shell Petroleum Company in 1921, and of Beede had come during a period of retrenchment in Texas government that almost resulted in the elimination of the Bureau (56). In the summer of 1921, in a letter to University of Texas President Dr. R. E. Vinson, Udden reviewed the activities of the Bureau since its inception. He emphasized the success of Bulletin 44, *Review of the Geology of Texas*, and showed how it had greatly stimulated oil development in the state. He also praised the Permian work of Beede, the subsurface studies by Sellards, and the Cretaceous investigations by Adkins, and added that he could think of "no other profession where continuity of service is of such importance as in the case of the geologist" (57). In further support of the Bureau, Robert T. Hill asked his supervisor at the U.S. Geological Survey, Dr. David White, to urge President

Vinson and Governor Pat Neff to spare the Bureau. Hill observed that Neff's reason for cutting out the appropriations for the Bureau was that the Bureau was not engaged in teaching and therefore was not an essential part of the University. Hill declared that the splendid research work of the Bureau was basic to geological teaching at the University (58).

The Bureau was spared, but its budget for the fiscal year 1921-1922 was $29,430, down about $15,000 from the previous year. In 1925 the new State administration, elected on a platform of retrenchment, greeted Dr. Sellards' assumption of the active directorship of the Bureau with an even lower budget of $21,360 (59). As Associate Director of the Bureau, Sellards rallied oil company support for increased appropriations for the Bureau in 1927. Donald Barton, President of the Houston Geological Society, contributed a full-page editorial to the *Oil Weekly* entitled "Texas Niggardly with Funds for Geological Development" (60). Barton and other oil company geologists such as Dr. Frank DeWolf, Chief Geologist of the Humphreys Corporation, attended the legislative appropriation committee meeting in Austin in April 1927. At this meeting, Sellards' request for $50,000 was answered by the Board of Control's recommendation of $19,550. However, the lobbying campaign finally began to pay off with an appropriation of $40,950 for the fiscal year 1929-1930 (61).

Dr. Sellards also began trying to build up support for a geological research foundation for the state. He corresponded with University President H. Y. Benedict concerning the possibility of beginning an endowment for the collection and study of microfossils and vertebrate fossils (62). Sellards, and Udden before him, was quite successful in soliciting contributions from geological societies for funding Bureau publications. In 1927 the Fort Worth Geological Society made the first contribution toward a revolving publication fund. The Houston and San Antonio Societies soon followed (63).

Back in 1924, Udden had reported to the Acting President of the University, Dr. W. S. Sutton, that he had made attempts to raise money from oil companies for a similar publication fund (64). The Bureau correspondence contains no record of such contributions. In 1927 Sellards corresponded with Wallace Pratt of the Humble Oil & Refining Company about establishing a topographic mapping fund to which oil companies might contribute (65). Sellards felt that the State would match these funds, which could be used to foster more U. S. Geological Survey topographic work in the state. The Marland Oil Company, in conjunction with the Panhandle Geological Society, did contribute $1,253 to be used in setting bench marks in their region (66). However, Pratt's reply to Sellards pointed out the industry's reluctance to contribute to Bureau funds (67):

> Even at a more propitious time, it would be difficult for oil companies to appropriate money for topographic mapping, unless a definite proposal to map a definite area is made so that the advantage which the producing companies might expect could be presented to them in concrete form.

In June of 1925, Sellards suggested to University President W. M. W. Splawn that the scientific departments engaged in activities aside from or in addition to teaching be reorganized within a Division for the Development of Natural Resources as follows:

Bureau of Economic Geology and State Geologic Survey
Engineering Experiment Station
(changed to Bureau of Industrial Chemistry in 1927)
Industrial Chemistry Experiment Station
(changed to Bureau of Engineering Research in 1926)

Sellards declared that the function of these departments, under what became the Division of the Conservation and Development of the Natural Resources of Texas, "is to lay the foundation for the continued and enlarged development of the natural resources and the related economic interests of the state" (68).

In addition to this concern with budgetary matters, Sellards reassembled a talented geologic staff to replace the excellent staff decimated during the retrenchment years of the early 1920's. The Bureau publications during the first 7 years of Sellards' active direction of Bureau affairs continued the standard of quality established under Udden, prompting Dr. Levi Stanley Brown of Massachusetts Institute of Technology to describe the Bureau's work as the outstanding geological research work being done at that time (69).

After 4 years in commercial work, Adkins returned to the Bureau in the fall of 1925 to continue his "love affair" with Cretaceous paleontology. In 1928, Frederick Byron Plummer finally joined the Bureau staff and immediately took up work on the compendium of Texas geology that was being prepared. That same fall, Dr. John Tipton Lonsdale, destined to become the fourth director of the Bureau, resigned the position as geologist he had accepted in 1925 to become Chairman of the Department of Geology at Texas Agricultural and Mechanical College. However, Sellards retained him in the post of consulting geologist, as he did Dr. Gayle Scott of Texas Christian University the following year. The year 1926 also marks the beginning of Josephine Casey's 45-year career as a vital cog in Bureau affairs. C. L. Baker returned to Bureau service in 1932 at the request of Sellards, as noted above.

As another valuable Bureau contribution to the oil industry in the state, Sellards, who taught stratigraphy, also established a course in petroleum geology at the University. The lecturers in the pilot course, entitled "Petroleum Production from Salt Domes of the Coastal Plain," were the leading lights in Texas geological research: Donald C. Barton, Alexander Deussen, Plummer (who in 1930 added the title of Professor of Petroleum Engineering to his title of Geologist with the Bureau), W. E. Wrather, J. W. Beede, and C. L. Baker. This course marked the beginning of the Department of Petroleum Engineering at the University (70).

From this brief account of the administrative history of the Bureau from 1909 up to 1932, it becomes very evident that Phillips, Udden, and Sellards had to be administrators of rare ability to be able to bring any kind of continuity or coordination into the programs of the Bureau (71). In the forty-first annual report of the U. S. Geological Survey in 1920, Director George Otis Smith had editorialized on the resignation of 15 full geologists from his staff the previous year for financial reasons (72). He observed that the net result was a loss in efficiency out of all proportion to the false economy represented by the governmental pay scale. He added that the morale of his staff was very low. If the Texas legislators had recognized that economy can be achieved much better through the increased efficiency of scientific work that comes with tenure and familiarity with research—that continuity of service breeds efficiency—the Bureau might have become the leader in geologic research for the entire Gulf Coastal Plain and Great Plains regions. As it was, the quality of the research work was unexcelled in the country, and Bureau publications became the leading research works on the Permian stratigraphy and Cretaceous paleontology of the United States. Furthermore, the

Bureau was in the vanguard of the study of subsurface geology and had alerted the oil industry to the potential of micropaleontology. The next chapter includes a discussion of the leading contributions to the knowledge of the geology of Texas made by Bureau geologists or sponsored in part by the Bureau.

THREE

The Publications of the Bureau

In 1932, Dr. L. S. Brown, prominent crystallographer at Massachusetts Institute of Technology, wrote Dr. Sellards congratulating him on his new position as Director of the Bureau. In the closing of his letter, he added that "in an informal discussion in the Department here, the merits and histories of various state surveys came up and someone remarked that the Texas Survey publications stood at the very front in perception and genuine geologic value" (73). Indeed, the Bureau's geologists and their research studies gained the respect of all American geologists. Bureau staff members did work closely with Federal geologists and oil company researchers, and the Bureau published fine works by non-Bureau people, but in this chapter, the author will discuss only the outstanding research investigations made by the geologists of the Bureau.

The six major geographical-geological provinces of the state are (progressing inland from the coast) the Gulf Coastal Plain, the Cretaceous uplands, the Central Mineral Region, the Osage Plains, the High Plains, and the Trans-Pecos. Unconsolidated sands and clays ranging from Upper Cretaceous to Recent characterize the belt of Coastal Plain deposits that stretch from northeast Texas to the Rio Grande. The belt of Cretaceous highlands lies above this plain. The scenic limestone Hill Country of the Edwards Plateau forms the southwestern part of this region, and the ledges and valleys of the Black and Grand Prairies form the north arm. In the curve of this highland belt lie the uplifted igneous and metamorphic rocks of the Central Mineral Region. The Osage Plains lie to the north and west of the Cretaceous highlands. North-south-trending limestone ridges with scattered coal seams characterize the eastern part of the Osage Plains, whereas red beds and gypsum ridges predominate in the western part of this Permo-Carboniferous region. The High Plains of northwest Texas are set off from the Osage Plains by a bold escarpment. The Canadian, Red, and Brazos Rivers, and, to a lesser degree, the Colorado River have etched deep canyons into the Tertiary alluvial beds of this area. The unique basin-and-range country of the Trans-Pecos region contains rocks ranging from Precambrian to Recent. The discussions of geological research efforts will be divided on this regional basis (74).

In his concise survey of *Thirty-five Years of Progress in the Knowledge of the Geology of Texas,* Alexander Deussen emphasized the debt that geologists owed to the oil industry (75): "Much of the remarkable progress that has been achieved during this interval has been made possible by the activity of the driller." Deussen should have added that the tremendous increase in geological research energies expended in Texas that came with the increasing numbers of oil geologists was another major factor contributing to the remarkable advances made in geological research in Texas.

Heretofore, in nonindustrial Texas, deficient in precious minerals, the main thrust of geological research had been the work of government agencies such as state geological surveys and the U.S. Geological Survey. But in 1935, if an observer looked at the numbers of oil geologists and the vast amounts of research funds expended by oil companies, he might assume that the Bureau of Economic Geology, with its small staff and limited appropriations, had been relegated to the status of disseminator of information about the resources of the state. Never have figures been so misleading. In the first 23 years of its life, the Bureau played a vital and versatile research role alongside the burgeoning exploration departments of the oil companies.

One of the main reasons for the Bureau's agility and adaptiveness was the energetic and pragmatic scientific leadership of Dr. Udden. Not only did he have great geologic intuition, but he also had a talent for demonstrating the economic potential of basic research. As mentor of the leading oil company geologists of the early decades of the petroleum industry, Udden fostered the spirit of cooperative interchange of research information that has characterized the geologic frontier in Texas. He pioneered the use of what are today the basic exploration tools of the oil geologists: micropaleontology and detailed lithological interpretation of well samples. He foresaw the value of geophysical exploration methods and had a fine eye for talented geologists.

The nonpareil reconnaissance geologist C. L. Baker was a leader in the study of the geological intricacies of the structure and stratigraphy of the Trans-Pecos region, the High Plains, and the Permian red beds of the Osage Plains. Dr. Beede supported Baker's work with his excellent grasp of Permian paleontology. World-respected paleontologist Emil Böse made prodigious contributions to the Permian and Cretaceous paleontology of the state. W. S. Adkins took over where R. T. Hill left off and became the premier Cretaceous stratigrapher and paleontologist. The regional subsurface studies produced by E. H. Sellards were invaluable aids to the understanding of the geologic structure of the state.

In addition to these outstanding contributions by Bureau geologists, the Bureau supported other research efforts such as the brilliant study of the Glass Mountains by Philip B. King and Robert E. King, graduate students at Yale (76). The Bureau provided a publication medium for some of the outstanding early work by oil geologists, such as the work of F. B. Plummer and R. C. Moore on the Carboniferous strata of North-Central Texas and several other fine micropaleontological studies discussed below (77).

Drill records divulged the secrets of subsurface stratigraphy and aided the discovery of such subsurface structural features as the salt domes of the Coastal Plain, the Mexia-Luling fault trend, the Ouachita Foldbelt, and the Permian Basin complex. Bureau geologists played a major role in the interpretation of the latter two features and were leaders in basic stratigraphic research, as demonstrated in the review of their work which follows.

Prior to 1900, the geologists of the Dumble Survey had described the broader aspects of Gulf Coastal Plain geology, and G. D. Harris had initiated Tertiary paleontological studies in this region. During the first 25 years or so of the 20th century, the geologists of the oil companies and of the U. S. Geological Survey dominated investigations of the geology of the Coastal Plain. Obviously, the oil geologists were doing much important work, but they did not begin to publish much until the Bulletin of the American Association of Petroleum Geologists first appeared in 1916. With the exception of the fine studies by E. T. Dumble, of the Rio Bravo Oil Company, and R. A. F. Penrose, Jr., of the Dumble Survey, most of the publications on the Cenozoic of the Coastal Plain up to 1932 came from the pens of geologists of the U.S. Geological Survey, including W. C. Veatch, Alexander Deussen, L. W. Stephenson, and Julia Gardner (78).

Udden felt that the oil geologists and the Federal geologists were expending enough energies on the Coastal Plain, so the Bureau geologists concentrated their efforts in other regions (79). However, in the late 1920's they began to focus their attention on the Coastal Plain. In *The Geology and Resources of Colorado County,* T. L. Bailey made a real contribution to the study of coastal sediments (80). His observations include petrographic studies and sophisticated discussions of the origin of the post-Tertiary sediments. University of Texas Bulletin 2645, *The Gueydan, A New Middle Tertiary Formation from the Southwestern Coastal Plain of Texas,* by Bailey, is a pioneer study of the volcanic tuffs of South Texas (81).

Henryk B. Stenzel and B. Coleman Renick wrote a fine report—on the lower Claiborne section exposed in the Brazos River—that appeared in University of Texas Bulletin 3101 (82). Stenzel, who worked under Lonsdale at Texas A&M College at the time but who later joined the Bureau, did the paleontological work; Renick, a consultant, helped in the stratigraphic work (83). Theirs was the first detailed Tertiary stratigraphic work since the time of studies made by Federal geologists of the Coastal Plains Investigations Project of the U. S. Geological Survey.

The work of Bailey and Stenzel heralded the beginnings of the detailed stratigraphic studies that would finally lead to an understanding of the rapid lateral facies changes characteristic of these beds. Drilling provided the information on similar downdip facies changes such as those described in *The Pratt Well in Webb County* by R. A. Jones, a consultant from San Antonio (84).

Because the Cretaceous beds that formed the central uplands of Texas contained vast amounts of oil in the subsurface of the Coastal Plain, the demands for information on the surface stratigraphy and paleontology of these strata were great. In his review of geological research in Texas from 1909 to 1935, Deussen commented that it had not been possible to add much to R. T. Hill's classification, definition, and description of the Cretaceous rocks of the state as presented in his monumental work, *Geography and Geology of the Black and Grand Prairies, Texas* (85). The magnitude of Hill's reconnaissance work is indeed great, but the county-level stratigraphic and paleontological investigations carried out by the Bureau geologists from 1918 through the 1920's provided the necessary substructure for detailed subsurface stratigraphic understanding (86).

These county and regional studies published by the Bureau and authored by such geologists as Ellis W. Shuler, W. M. Winton, Gayle Scott, F. M. Bullard, and, of course, W. S. Adkins, represent the first small-scale geologic mapping done by the State Survey and include the first systematic areal stratigraphic studies (87). *The Geology of Tarrant County* by Adkins and Winton is outstanding among these reports.

In addition to the detailed paleontological discussions that accompanied his county reports, Adkins wrote several paleontological studies for the Bureau, including his pacesetting *Handbook of Texas Cretaceous Fossils* (88). Adkins' thorough knowledge of the Cretaceous paleontology of Texas enabled him to draw up excellent correlation charts for the Cretaceous strata across the state on the basis of paleontological zonation. He recognized early that good horizon marker fossils alone could solve the complexities of the lateral facies variations in the Cretaceous beds that extend from the Red River to San Antonio to the mountains of the Trans-Pecos.

Adkins' correlation of the Texas Cretaceous strata with the European stages was based on the perceptive ammonite studies in Emil Böse's fine paper on *Cretaceous Ammonites from Texas and Northern Mexico,* which demonstrated that ammonite zonation tends to reduce the magnitude of the Comanchean-Gulf unconformity in Texas (89).

Finally, when placed in the context of the stratigraphic or post-reconnaissance stage of geological exploration, Adkins' understanding of the Cretaceous as demonstrated in University of Texas Bulletin 3232 is just as outstanding as Hill's.

The coal beds and vertebrate fossils of the Pennsylvanian and Permian strata that stretch across the Osage Plains have attracted the attention of geologists from the time of Ferdinand Römer. The development of petroleum resources in this region (beginning in 1911) brought about active and intensive geological explorations, the likes of which the state had never known. Never before had there been so much information on the geology of one region. Most of the information was gathered by oil geologists, but the Bureau played a leading role in the interpretation of this information.

The Osage Plains extend from the foot of the High Plains east to the Cretaceous exposures of the Grand Prairie, north to the Red River, and south to the Llano Uplift and the Edwards Plateau. The Upper Pennsylvanian is exposed in the eastern part of this large region in the valleys of the Colorado and Brazos Rivers. The Callahan Divide, capped by Cretaceous remnants, separates these two areas of Pennsylvanian outcrop, and correlation of beds across this divide has been difficult.

W. F. Cummins, R. S. Tarr, and N. F. Drake, all of the Dumble Survey, investigated the Pennsylvanian beds of the Colorado River Basin, and their reports, especially those of Drake, remained the best available on these strata until F. B. Plummer and R. C. Moore published their exhaustive study of the Pennsylvanian strata of the Osage Plains in 1921 (90).

Prior to the work of Plummer and Moore in this area, Dr. Udden undertook one of the pioneer studies of the subsurface geology of an oil-producing area in Texas. Typically, Udden's work was clear and thorough, and the most exciting aspect was his detailed lithological description of the well samples, which included notations of fossil *Fusulina*. As noted in an earlier chapter, Udden pioneered the study of well samples and micropaleontology in the Mid-Continent and Texas, and alerted the geologists of oil companies to the value of these studies. In this 1912 precedent-setting Bureau publication, Bulletin 246, *A Reconnaissance Report on the Geology of the Oil and Gas Fields of Wichita and Clay Counties*, Udden correlated Drake's Colorado River Pennsylvanian section with the subsurface sections in wells in the Henrietta and Electra fields (91).

The work of Plummer and Moore remained the best study of the Pennsylvanian in Texas until the recent studies by L. F. Brown, Jr., and others (92). A large number of geologists worked on this study under the direction of Plummer, then with Roxana Oil Company. Moore, with the Kansas Geological Survey, did the paleontological work and also was the first to use polished sections of limestones. Plummer and Moore's publication is an outstanding example of sophisticated stratigraphic work. It includes a discussion of the conditions of sedimentation and an excellent geologic map. The mapping work had been begun by the geologists of Roxana in 1916. In the introduction Udden expresses "thanks to the liberal management of the Roxana Petroleum [Oil] Company for having donated this valuable contribution to the knowledge of the Pennsylvanian in North Central Texas" (93).

Permian formations cover the western one-half or more of the Osage Plains. On the Red River, the entire Permian is composed of red-bed facies—red or gray shales, sandstones and dolomites, and gypsum beds. Southward, a facies change occurs by which the lowermost group of the Permian, the Wichita, changes to marine shales and limestones. A similar change occurs downdip to the west. In the next group, the Clear Fork, the change to the south is less obvious. In the uppermost group, the Double Mountain, the red-bed facies continues throughout the exposed belt in the Osage Plains. However, the entire Permian changes to marine facies in the downdip. In the Glass Mountains of Trans-Pecos Texas, the entire Permian with the exception of the uppermost part is marine.

W. F. Cummins of the Dumble Survey initiated stratigraphic studies of these vertebrate fossil-bearing red beds that included excellent descriptions of the increasing age of the Pennsylvanian-Permian boundary (94). The first work of C. L. Baker for the Bureau was the investigation of the nonmarine deposits of the High Plains and the westernmost Osage Plains. In Baker's 1916 paper on *The Origin of Texas Red Beds* (95), he discussed the paleogeography and history of Permian sedimentation. His outstanding studies led him to the conclusion that these red beds were originally "maturely decomposed red residual soils formed under warm and moist climatic conditions" (96). Red beds were not indicators of arid conditions.

Dr. J. W. Beede, who had done pioneer work on the Permian invertebrate paleontology of Kansas, concentrated on these same beds in his work for the Bureau. In 1918 and 1921, the Bureau published studies on the geology of Runnels and Coke Counties

(97). Beede was the senior author of both reports, which included his descriptions of Permian sections exposed in the Colorado River bluffs. In the Coke County report, Beede remarked that the unconformity represented by the base of the San Angelo Formation was of great extent and was probably the most valuable horizon marker in the Permian of Texas (98). In his last year of service with the Bureau, 1922, Beede, with D. D. Christner, traced the San Angelo into Oklahoma where he found it to be continuous with the Duncan Sandstone.

In 1925 Beede presented a manuscript entitled *The San Angelo Formation, and The Geology of Foard County, Texas* to the Bureau (99). It included his observations on the stratigraphic importance of the San Angelo Formation and was accompanied by a map indicating the outcropping belt of this formation in Texas, which Beede felt would be of great service in any subsequent studies of the Texas Permian section.

In his well-written paper summarizing all knowledge on *The Correlation of the Permian of Kansas, Oklahoma, and Northern Texas*, Charles N. Gould, of the Oklahoma Geological Survey, clearly described the problems involved in trying to trace horizons across the line of color change marking the transition from marine to nonmarine facies (100). He noted that the paleontological work of Beede on invertebrates, of Samuel W. Williston and Ermine C. Case on vertebrates, and of Charles David White on plants, had contributed a great deal toward the solution of correlation problems (101). However, the keys to correlation were lithological marker beds that could be traced across the facies change. Beede traced one such key horizon, the San Angelo Formation, from the Colorado River north to Oklahoma (where it was equivalent to the Duncan Sandstone), and Gould and others were able to follow the Blaine Gypsum (the Texas equivalent of the Greer Formation) from Kansas south to the Colorado River.

In October 1925, a field conference of 25 geologists, including Dr. Sellards and Dr. Beede, was held in western Texas to study the Double Mountain Formation and correlating certain members of this formation with strata in Kansas and Oklahoma. The product of this conference was a chart showing the tentative correlation of the Permian red beds of Kansas, Oklahoma, and Texas. No attempt was made to correlate the non-red beds of Kansas and Texas (102). Field conferences similar to this one represent the most effective form of interstate cooperative research activities. In fact, the field conferences and meetings sponsored by the rapidly growing regional geological societies in Texas were becoming the instruments for interchange of geological knowledge among all geologic researchers working in the state—similar to the regional and national meetings of the American Association of Petroleum Geologists.

In 1929 the Bureau published two more papers on the Permo-Carboniferous rocks of the Osage Plains. In his *Depositional History of the Red Beds and Saline Residues of the Texas Permian* (103), Baker presented an excellent review of the geologic literature on the Permian of Texas. He credited Beede with accomplishing the most important paleontological and stratigraphic work on the Permian of the Texas Osage Plains and added that the Glass Mountains appeared destined to become the type for the marine Permian of the world. He also stated that the work of the King brothers confirmed very closely Udden's initial descriptions and divisions of these strata (see below). Baker's report also presented fine, detailed discussions of the paleogeography of the Permian period, in which he again declared that both red beds and saline residues were formed in nonarid climates.

M. G. Cheney's *Stratigraphic and Structural Studies in North-Central Texas* (104) was as practical as Baker's was theoretical. Cheney's work included no discussion of previous work; as he stated, it was a compilation of all available information of interest to the oil operator in North-Central Texas. The cross sections, the subsurface structure maps, the isopachous map, and the accompanying areal map were all the result of the cooperative sharing of information by all oil geologists working in the area.

Perhaps the most nearly complete marine Permian section in America is that of the Glass Mountains in Trans-Pecos Texas. Structurally, the Glass Mountains are a monocline of predominantly late Paleozoic formations, sloping to the northwest and forming the northwest flank of the Marathon Uplift. As Sellards noted in University of Texas Bulletin 3232, aside from a lead-silver prospect near Altuda, these Permian strata had no economic potential. However, he added: "Their underground connection (often in changed facies) with the petroleum-producing formations of the Permian Basin gives to them an exceptional interest economically as well as in their stratigraphic relations" (105).

In the first two decades or so of the Bureau's history, some of the most significant contributions to the knowledge of the state were studies of the Permian strata of the Trans-Pecos made by Bureau geologists and geologists sponsored by the Bureau.

The earliest significant reconnaissance work in the Trans-Pecos was done by W. H. von Streeruwitz, Joseph A. Taff, and Ralph S. Tarr of the Dumble Survey, and in 1900, R. T. Hill described the Glass Mountains and recognized the Paleozoic age of the strata (106). Dr. Udden did reconnaissance work for the University of Texas Mineral Survey in the Shafter silver mine district in Presidio County in 1904. In the concise and sophisticated report prepared for publication in 1905, Udden described the strata near Altuda, which is a station on the Southern Pacific Railroad at the west corner of the Marathon Paleozoic area, as "shales and limestones that clearly belong to the Upper Carboniferous series" (107), which were equivalent to his Cibolo Limestone section at Shafter.

In 1914 Udden began constructing a geologic map of the area of the Glass Mountains. In 1915 he sent C. L. Baker and field assistant W. F. Bowman to make a survey of parts of Brewster, Presidio, Jeff Davis, Culberson, and Reeves Counties (the region surrounding the Glass Mountains) for a map to be used in compiling the milestone first geologic map of the state. This map would be published with the *Review of the Geology of Texas* in 1916 (108). Early in September, Dr. Emil Böse relieved Baker and, with Bowman, measured the section of Permian rocks exposed in Gilliam Canyon of the Glass Mountains and made extensive fossil collections (109).

Udden was especially interested in the Guadalupian fauna of the Glass Mountains and had alerted the noted European-trained paleontologist Böse to the ammonites and *Richtofenia* found in these rocks. Böse's 1917 study of the paleontology of the Permian rocks exposed in Gilliam Canyon remains the most complete work on Permian ammonites in Texas (110). Böse was a thorough and painstaking researcher; he made the first attempts to correlate the Permian formations of the Trans-Pecos with those of the Osage Plains and also made the first satisfactory correlations with the Permian of Europe and Asia.

The Bureau also published Udden's *Notes on the Geology of the Glass Mountains* and *Geologic Exploration of the Southeastern Front Range of Trans-Pecos Texas* by Baker and Bowman in 1917

(111). Udden's paper is noted for its suggestion that there may be buried structures to the northeast, on trend with the Marathon fold structure, that might be oil bearing.

The report by Baker is a classic reconnaissance study from its romantic descriptions of "Scenic Features" (112):

> In the hazy sunlight of the late afternoon the canyon walls glow and assume shadowy outlines, the colors and contrasts of form merging into a harmonious whole. . . .

to its concise and well-written descriptions of the general structural characteristics of the area. Baker and Bowman mapped 4,500 mi^2 during their 4-month reconnaissance work; they included a prodigious amount of geologic information in their report. One especially perceptive observation was the suggestion of a marked similarity between the Marathon folding and the conditions in the Ouachita Uplift area in southeastern Oklahoma and the possibility of subsurface continuity between the two regions. This point is covered more fully below. Also, the physiographic descriptions of the Trans-Pecos are very sophisticated and continue the excellent precedent set by R. T. Hill.

In a manuscript prepared in 1922, C. L. Baker surveyed the geology of the Malone Mountains and the Quitman and Eagle Mountains (113). He was able to decipher the main structural features in these complexly overthrusted and overturned strata. His report included an excellent review of previous geological work in this area.

These were the foundations that preceded the brilliant Glass Mountains study carried out by Philip King and Robert King in the summers of 1925, 1926, and 1927 (114). In his introduction, "The Geology of the Glass Mountains, Stratigraphy," Sellards cites this study as the outgrowth of cooperation between Yale University, where the brothers were graduate students under Professor Charles Schuchert, and the Bureau; a part of the printing cost was paid out of a publication fund established by contributions from the Fort Worth Geological Society. Both Schuchert and the Bureau contributed to the expenses of this intensive study.

In his excellent summary of previous work, Philip King praised the efforts of Udden, Baker, and Böse in establishing a firm foundation for all subsequent studies of the Glass Mountains. He presented a very interesting discussion of the physiography of the region, including a critique of earlier physiographic descriptions in which he credits Udden in his Chisos Mountains paper with adding "to our knowledge of the physical features of this province with his description of the 'graded plains' which anticipated the better known works on desert pediments by many years" (115). The clarity and excellence of Philip King's writing and his geologic intuition are best demonstrated in his discussions of the stratigraphy and complex structure of the mountain strata. He found that Udden's formational units proved to be natural and practical in spite of the complexity of the stratigraphy and structure. King's superstructure, built on the foundations laid by Udden, Baker, and Böse, represents one of the outstanding pieces of geologic research work in the state.

Robert King, an excellent geologist, worked alongside his brother and wrote the *Faunal Summary and Correlation of the Permian Formations with Description of* Brachiopoda. He discussed the various Permian sections of the Trans-Pecos and compared them with other American and worldwide sections, as well as with the Permian of the Osage Plains. He also presented a very sophisticated description of the Permian fauna and paleogeography.

The Permian crops out at both the western (Trans-Pecos) and the eastern (Osage Plains) sides of the Permian Basin, which extends from the Pecos valley northward into New Mexico, Oklahoma, and Kansas. In this basin, the Permian is overlain by Triassic, Cretaceous, and Cenozoic formations. The problem of correlating across the basin had been attacked most successfully by following formations from either side toward the middle of the basin. Correlations of formations on opposite sides of the basin on the basis of fossils from surface exposures have been quite difficult because environments, hence fossils, are different.

The delineation of the Permian Basin and the understanding of the subsurface geology of the basin have come about mostly through the research work of oil company geologists. However, Dr. Udden was the first to become aware of the existence of a large basin in western Texas deposited during Permian time. His descriptions of the samples from the Spur well gave the first clues to the existence of this basin, and in 1915, his *Potash in the Texas Permian* included information on several wells on the High Plains that prompted him to announce the presence of a large Permian salt basin underlying the High Plains (116).

Udden foresaw that the discoveries of oil at Electra, Burkburnett, and Ranger in 1911, 1917, and 1919, respectively, would create a demand for information on the stratigraphy and paleontology of the surface exposures of the Pennsylvanian and Permian strata in the Osage Plains and in the Glass Mountains. Thus, he launched the investigations of the geology of the Glass Mountains in 1914; in 1917, he hired J. W. Beede, who was an expert on the Permian paleontology of Kansas. It was in the 1917 paper on the Glass Mountains that Udden first spoke of the possible subsurface extension (along a trend to the northeast of the structure of the Paleozoic rocks of the Glass Mountains) (117):

So, that there should exist, under the Comanchean and under the red beds, some places northeast of the Marathon Uplift where the Pennsylvanian and probably some of the Permo-Carboniferous lie folded under the relatively undisturbed red beds and the Comanchean limestones. The red beds are entirely impervious and would make an excellent cover for an oil pool. How far such covered places of tilted petroliferous formations of the Pennsylvanian may be found away from the exposures in the Marathon country no one can say, but it would be no surprise to find them at a distance of at least fifty or a hundred miles beyond the Brewster-Pecos county boundary. The trend of the Marathon Mountains would run through the southeast part of Pecos County into Upton and Reagan Counties, or even farther east than this.

Udden's suggestion that favorable structural conditions for oil accumulation might exist turned the attention of wildcatters to the region to be known as the Permian Basin, and led directly to the drilling of the Santa Rita No. 1 in May 1923, the discovery well for the Big Lake oil field and for the Permian Basin (118). The intensive period of geologic investigation and oil exploration that followed in the basin led to the discoveries of Yates, Winkler, Hobbs, and several other fields in rapid order. The obvious impact that the Santa Rita No. 1 had upon the value of the University of Texas lands will be discussed in a later chapter.

Actually, the Santa Rita No. 1 was drilled in the right place for the wrong reason. Udden's suggestion that the folding in the Marathon Uplift extended in the subsurface to the northeast was erroneous. The Paleozoic rocks of the Marathon Uplift and those of the Ouachita Mountains in Oklahoma are the surface exposures of a subsurface foldbelt that arcs through Texas northeast and almost due west from San Antonio to the aforementioned surface exposures. (This "Llanoria geosyncline" is discussed in detail below.) The Santa Rita No. 1 was actually drilled on a platform that extends southeastward from New Mexico into the Permian Basin.

In 1925, H. W. Hoots of the U.S. Geological Survey outlined the extent of the Permian Basin as best as could be done at the time and also attempted to correlate across the basin (119). His report includes several cross sections of the basin, a very good geological discussion of the strata in the basin, and one of the first published maps of the Permian Basin that shows "a major geosyncline extending due north and south and the deepest portion or axis lying roughly along the west sides of Midland, Martin, and Dawson Counties" (120).

Like the Permian Basin, the Ouachita System, or Llanoria geosyncline as it is called in Bulletin 3232, was discovered and delineated by the drill (121). In 1914 and 1915, Dr. Udden examined cuttings from wells at Georgetown and Leon Springs. He published his observations on the cuttings from these two wells in the 1919 Bulletin of the American Association of Petroleum Geologists (122). He found the cuttings to be similar and described them as pre-Paleozoic. Udden observed that both wells "lie in the belt of the Balcones escarpment, [which] proves that this belt coincides with an earlier uplift of considerable dimensions, most likely following the direction of the Balcones faults" (123).

It is hard to say who was the first to suggest that the facies of the Ouachita Mountains extended in the subsurface southwestward into Central Texas beneath the Balcones Fault Zone, and then westward from San Antonio to crop out again in the Marathon region. In 1917 C. L. Baker noted that the structure of the Marathon Basin area was the counterpart to that of the Ouachita Mountains of east-central Oklahoma. Both Baker and Udden failed to associate the similarity in the geology of these two areas with the subsurface foldbelt under the Balcones escarpment and thus did not suggest that all three geologic features were part of one continuous trend. Baker noted that the axes of the folds on the southwestern margins of the Ouachita Mountains strike southwest toward the Marathon Basin and have the same strike as the folds in the Marathon Basin. He felt that the deformation in both areas was contemporaneous, but he added (124):

> Whether a continuous folded mountain range once extended from Little Rock, Arkansas, southwestward to the Marathon Basin, or whether the two ranges now known only in the two widely separated regions were distinct cannot be told, for between the two lies a broad geosyncline (Llano Estacado geosyncline [i.e.,

Permian Basin]) formed of younger rocks deformed at a later date which entirely hide from view the older Paleozoic.

Additional wells drilled along the Balcones escarpment finally provided the additional clues that enabled Dr. Sellards to show on a map the continuity between the rocks of the Marathon Basin and those of the Ouachita Mountains, and those penetrated in the subsurface just southwest of the escarpment. Sellards himself best traced the order of investigations and publications on this geologic feature (125):

In January, 1929, Sellards mapped the probable continuation of sediments of the Ouachita facies southwestward in the general region of the Balcones fault zone to central Texas, and suggested their relationship to similar sediments in the Marathon region. In June and in August, 1929, [M. G.] Cheney postulated the extension of the Ouachita basin of Oklahoma southwestward into Texas and suggested its possible connection with the Marathon basin.

Sidney Powers, of the Amerada Petroleum Company, had forwarded a manuscript to the Geological Society of America in December 1927, in which he presented paleogeographic maps representative of Texas and Oklahoma during the Ordovician and four stages of the Pennsylvanian. The northern edge of the positive landmass during these times, Llanoria, roughly coincides with the Balcones scarp line on these maps. The location of this shoreline was based on well data. Powers stated (126):

The Ouachita Mountains, the Llano-Burnet Uplift, and the Marathon Uplift, and the Solitario Uplift are peripheral to Llanoria and are within fifty miles of schist struck by wells.... This edge of Llanoria was a Permian as well as Pennsylvanian shoreline and was afterward the hingeline of the Gulf Coast embayment, and as such became greatly fractured.

However, he never suggested that the northern shoreline of Llanoria was a geosyncline or that the Ouachita Mountains or the Marathon Uplift were the terminal surface expressions of a buried mountain range.

In a paper delivered at the Cleveland meeting of the Geological Society of America, also in December 1927, Raymond Moore postulated that the Ouachita trough was a continuation of the Appalachian trough or geosyncline and noted that in northeastern Texas, the Ouachita geosyncline can be traced under cover by means of well samples (127). Dr. Sellards delivered a paper on "The Underground Position of the Pre-Cambrian in Texas" at the New York meeting of the G.S.A. the following

December (128). Obviously Sellards, R. C. Moore, Sidney Powers, and others were gathering evidence and coming up with a picture of the position of the Ouachita geosyncline that became clearer with the drilling of each well along the Balcones escarpment and fault zone. By April 1929, there was enough evidence to prompt M. G. Cheney to comment (129):

> East Central Texas experienced much the same geologic history as did southeast Oklahoma . . . [and] the Marathon district of southwest Texas appears to be yet another salient of Appalachian basining and orogeny.

Finally, at the March 1930 meeting of the American Association of Petroleum Geologists in New Orleans, Sellards delivered a paper in which he described the slightly metamorphosed samples from wells drilled in the Balcones Fault Zone and noted that Hugh Miser had found them to be similar to the formations of the Ouachita series in Oklahoma. Sellards said (130) that the well samples were "accumulated in a trough or syncline in front of the land mass Llanoria, this trough being continuous toward the northeast with the Ouachita Mountains of Oklahoma and westward with the Marathon Uplift of Trans-Pecos Texas." Subsequent, more exact delineation of the location of the Ouachita geosyncline in Texas would come only with the drilling of more wells along the Balcones trend and the close examination of the Paleozoic samples from those wells.

In his discussion of geologic structure, Philip King discussed the regional relations of the Marathon folds and late Carboniferous diastrophism and sedimentation in the Southwest. He postulated (131) that "the Marathon basin is a small window through the younger strata, revealing only a fragment of the frontal part of a mountain system which must rival the Ouachitas in complexity of structure and magnitude of forces involved." He described both the Marathon and Ouachita facies as being geosynclinal and declared (132) that the folds in the Marathon Uplift and the Ouachita Mountains were "part of a general belt of deformation fringing the north border of Llanoria." He ended this discussion with the note that the current investigations being conducted by the Bureau of Economic Geology would provide the final answers to the regional relations of the Marathon folds.

The excellently done early stratigraphic investigations of the Triassic and Cenozoic beds of the High Plains of Northwest Texas by W. F. Cummins of the Dumble Survey and Charles N.

Gould of the Oklahoma Geological Survey attracted the attention of such noted vertebrate paleontologists as E. D. Cope and James W. Gidley, who collected specimens in this area around the turn of the century (133). Gidley collected for the American Museum of Natural History; R. S. Lull and Edward L. Troxell, his assistant, made collections for the Yale Peabody Museum. C. L. Baker's investigations of the *Geology and Underground Waters of the Northern Llano Estacado* marked the beginning of Bureau work in the High Plains (134). His work was followed by that of Leroy T. Patton, Dr. Udden, and Lyman C. Reed and Oscar M. Longnecker, whose work for the Rio Bravo Oil Company was published by the Bureau (135). W. D. Matthew, of the Paleontology Museum of the University of California, worked closely with the Bureau on the vertebrate paleontology and stratigraphy of these deposits—something that cannot be said for the aforementioned collectors, who never even consulted the Bureau on points of stratigraphy. None of them thought of leaving the fossil specimens with the Bureau (136).

Baker's paper on the Llano Estacado is another illustration of the rare geologic intuition that characterized all of his reconnaissance work. In his preface, Baker found it a "pleasure to be able to confirm in almost every particular the conclusions already reached by the pioneer investigators" (137). He gave special acknowledgment to the "great, classic paper, *The High Plains and Their Utilization,*" by W. D. Johnson, and praised his views on the agricultural prospects of the region (138).

Baker's history of Cenozoic deposition was the heart of his report, and it was in itself one of the fine early studies of the processes of nonmarine deposition (139). He described the later Cenozoic strata as alluvial deposits of rivers heading in the recently uplifted Rocky Mountains. He followed with a clearly written discussion on how stream meandering and gradation could result in a considerable thickness of stream deposits covering a rather large area such as the Llano Estacado; he pointed out how these sediments were then subjected to the constant eolian erosion and deposition typical of semiarid plains regions. His descriptions of the recent eolian redistribution of these Cenozoic sediments were based on extensive mechanical analyses of the sediments and were accompanied by concise discussions of eolian erosion and transportation that must have pleased Dr. Udden. Baker concluded that the chief processes in the building of the surface of the Llano Estacado were depositional rather than erosional. Later workers would conclude that the surficial deposits are in part wind eroded and redeposited lacustrine rather than alluvial deposits, but this does not dim the importance of Baker's studies.

From the summer of 1928 to the spring of 1929, Lyman Reed and Oscar Longnecker investigated the geology of Hemphill County for the Rio Bravo Oil Company. Although they were engaged primarily in plane table mapping in search of surface structures that might indicate subsurface traps favorable for the accumulation of oil, C. L. Baker, who was Chief Geologist of Rio Bravo at the time, encouraged them to explore all aspects of the geology of the area (140). This encouragement led them to collect vertebrate fossils, which Baker forwarded to W. D. Matthew, vertebrate paleontologist at the University of California. Matthew wrote a series of four articles dealing with the Pliocene formations and mammals of Hemphill County and the surrounding area of northern Texas (141). Dr. Charles Camp and V. L. Vanderhoof of the University of California subsequently joined Reed and Longnecker on collecting trips.

Baker was correct in taking credit for leading the search that uncovered one of "the most prolific fossil fields of Lower Pliocene age yet discovered" (142). In his introduction to the report of Reed and Longnecker, he added that it was indeed fortunate that Matthew studied the vertebrate fossils, for he was able to demonstrate that the fauna formerly considered to be Miocene were in fact lower Pliocene. With regard to Pliocene depositional history, Baker reported that both he and Matthew (143) "from somewhat different points of view, became convinced that the bulk of the lower Pliocene deposits are aeolian in origin, in part worked over and redeposited by the wind from original stream-laid deposits . . . and in part transported by the wind from more distant sources farther west." The results of the plane table mapping by Reed and Longnecker also indicated that the various layers of lower Pliocene sediments were originally laid down in beds of nearly uniform thickness over a hill-and-valley topography in much the same manner as the loess of the upper Mississippi Basin. Baker regards this as first-rate evidence of eolian origin (144).

The works of Baker and Reed and Longnecker represent significant studies in the branch of geological research called sedimentation, which was beginning to grow in importance at this time. Sedimentation, the systematic and statistical analysis of the various characteristics of sediments (grain size and shape and sorting, bedding, mineralogical content, and fossil content) for purposes of determining depositional environment, was just becoming a sophisticated geological research tool in the 1920's. However, the mechanical analyses of the lower Pliocene sediments undertaken by Reed and Longnecker and those by Harry X. Bay, which are discussed below, are the only such in-depth studies found in Bureau reports up to 1932 (145).

Bay's study of the Pennsylvanian conglomerates of the Osage Plains included analyses of size, sorting, shape, and megascopic mineral content. Bay was a graduate student under Dr. A. C. Trowbridge of Iowa State University, who is considered by many to be the father of sedimentation. Bay also conducted a series of studies on the reduction of angularity of rock pieces, similar to those making up the bulk of the conglomerates he investigated, with an abrasive mill. In this work, done under the direction of Dr. A. C. Tester, Bay was able to determine sedimentary features and patterns common to piedmont, littoral, floodplain, and deltaic environments of deposition.

The bibliography of this very interesting work contains all of the early works on sedimentation, including a 1914 paper on "The Mechanical Composition of Clastic Sediments" by Dr. Udden (146). Udden had intended a definitive work on the subject, but his other duties were too pressing, so he published this report on the data he had assembled up to that time. He collected an extensive variety of glacial tills, aqueous deposits, and eolian sediments, and through mechanical analyses came up with 32 tables for 32 categories of deposits. This excellent pioneer work demonstrates the scientific versatility and clarity of expression so characteristic of Udden's scientific thought and writing.

Udden was always searching for new keys for interpreting geological evidence and in 1928 wrote a challenging little paper entitled "A Neglected Field in Stratigraphy" (147). He began by discussing the development of sedimentary petrography, which enabled the geologist to understand better the environment of deposition in times past by subjecting sediments now being deposited to mechanical and mineralogical analyses. He then proceeded to say that perhaps one of the most important, as well as one of the most promising, lines of study of the stratified rocks, at least from a strictly scientific point of view, was the study of stratification itself (148):

> A most important element in history is time, and not only relative time, but time in cycles and periods of known duration. I am convinced that we make a mistake if we say that anything less than this should be looked for. [The variations in the rates of accumulation of sediments are so great that] who can wonder that there has grown into the minds of us the idea that we can deal only with relative age in geology? It has almost come to be a doctrine that such is the case. It is a doctrine that serves no good or stimulating purpose in our work.

Dr. Udden went on to discuss some of the more simple examples, such as annual laminae in glacial deposits. He added that we must find a unit of measurement of depositional periods lasting longer than a year or several decades. He noted that some workers had suggested a corresponding frequency in earthquakes and sunspots; he suggested that such a relationship might be the key to a yet unknown sedimentary cycle.

Perhaps one should not try to assign credit to individuals for particular contributions to the understanding of the geology of Texas. As is evident in earlier chapters, most contributions have been real team efforts, each researcher building on previous work, and all geologists sharing their knowledge. It is equally difficult to try to assess the advances in geological understanding that have been the direct result of Bureau research efforts. Yet there have been those few geologists who, because of their exceptional geological intuition or unusual organizational and analytical talents or simple strength of dedication to their chosen field, have made extraordinarily large contributions to the total knowledge of the geology of Texas. The Bureau has been especially blessed with men of this caliber: Udden, with his tremendous scientific curiosity and intuition; Böse, Adkins, and Beede, with their keen talent for interpreting paleontology; Baker, with his talent for perceiving the broad, basic geologic features of a relatively complex area; and Sellards, with his knack for organization and analysis. One can say with assurance that the investigations of the beautiful Glass Mountains by Udden, Baker, Böse, and finally the King brothers represent significant contributions to the understanding of North American geology, and were it not for the dedicated efforts of Böse and Adkins, especially, the understanding of Cretaceous stratigraphy of Texas and indeed the Southwest would be 20 years behind its present status.

These examples and the discussions included in this chapter demonstrate two things: one, the high quality of research work done by the geologists of the Bureau and geologists sponsored in part by the Bureau, and two, the ability of the Bureau to focus its research programs in directions that have greatly benefited the state's oil development. Not only has the quality of geological research in Texas been high, thanks in large part to the Bureau, but this research has greatly aided the state's economic development.

FOUR

The Bureau and the Oil Geologist

The contributions to the knowledge of the geology of Texas made by the geologists of the Bureau were largely the results of investigations made to facilitate the development of various oil-producing areas within the state. Dr. Udden was keenly aware of the importance of supporting oil exploration with sound stratigraphic information. The studies of the Permian stratigraphy of the Glass Mountains that he initiated were directed toward an understanding of the subsurface geology of the Osage Plains and the Permian Basin, both of which he felt were potential oil-producing regions. The Bureau coordinated its efforts with the needs of the oil industry.

Thus, the key influence on geological research in Texas in the late teens and twenties was the development of the oil industry, which was accompanied by the emergence of the commercial oil geologist. In a 1927 letter to R. A. F. Penrose, Jr., who had been a geologist with the Dumble Survey before becoming a very successful mining engineer, Dr. Sellards observed that "we must not overlook the fact that the character of the work which the oil geologists are doing is by no means superficial" (149). Sellards noted that on the contrary they were turning their attention more and more to very fundamental problems in geology.

Not only was the geological research work being done by the oil company geologists becoming more basic, but the numbers of commercial geologists were increasing in geometric fashion (150). The resultant plethora of information on the surface and subsurface geology of the state was unique in the annals of geological sciences. Never had so many scientists been engaged in research on one scientific frontier. However, Dr. Udden was quick to sense the dilemma in this situation. On one hand, oil exploration and drilling were making tremendous amounts of new geological information available; but on the other hand, this flood of new information was being gathered in an uncoordinated, piecemeal fashion, within an environment of competition. These new fragments of geological data could be lost forever, and those that were not lost might never be utilized to their fullest in solving the puzzles of Texas geology.

Udden realized that the Bureau, as a neutral agency, would have to take the lead in fostering the collection and preservation of the rock samples encountered in drilling. Hence he did his utmost to encourage accurate collection and description of well samples and to entice oil companies to contribute such records to a well sample pool to be administered by the Bureau. To Udden it was obvious that the key to the realization of the full potential of the vast amount of geological information, both surface and subsurface, which was being gathered by the commercial oil geologists, lay in the cooperative interchange of these data (151). One of the most significant contributions of the Bureau under Udden's directorship was the fostering of this spirit of cooperation—convincing oil company geologists and the scientific world as well of the advantages that could be gained through interchange of research information.

Everett L. De Golyer, in a short work on *The Development of the Art of Prospecting*, traced the gradual acceptance by the oil industry of geology as a vital oil exploration tool (152). He felt that the initial impulse along this line came with the publication of a series of oil field studies undertaken in the first decade of the 1900's by the U.S. Geological Survey. Consequently, Mid-Continent oil companies began to establish geological departments in the teens. Texas oil men showed little interest in geology at this time, probably because (1) most of the important oil fields were in the Coastal Plain, where surface mapping was undependable, and (2) little was known about salt domes. The discovery of oil along faults along the inner coastal belt and in the Osage Plains of North-Central Texas finally awoke company leaders to the value of geological investigations.

The exception to the pattern was the experienced geological department established by E. T. Dumble for the Southern Pacific Company back in 1897. The Southern Pacific created the Rio Bravo Oil Company in 1903, under Dumble's leadership, to operate oil properties on the Gulf Coast. Edgar W. Owen observed that Dumble recognized the potential value of precise subsurface stratigraphy more clearly than anybody else in a position of authority in the oil companies, but accessible information was too meager to permit accuracy (153). He also stated that in spite of Dumble's pioneer work in subsurface geology and micropaleontology, "knowledge of Gulf Coast subsurface stratigraphy was only rudimentary until after the oil companies established micropaleontology laboratories in 1920 and improved drilling methods yielded suitable well samples" (154).

The emergence of the geological exploration department in the Humble Oil & Refining Company was more typical than the early example set by Dumble (155). The rise of the geologist to a position of paramount importance in Humble's exploration organization was a direct outgrowth of oil discoveries outside of the lower Coastal Plain and of the astute leadership of Wallace Pratt, Humble's first chief geologist. In a paper entitled "Geological Structure and Producing Areas in North Texas Petroleum Fields" delivered at the fourth annual meeting of the American Association of Petroleum Geologists in 1919, Pratt commented (156):

> The problems of geology in North-Central Texas today are research problems, and a geological department in the petroleum industry today should be on the same basis as are research departments in other large progressive industries.

During the late teens and twenties, Pratt corresponded regularly with Dr. Udden about problems of stratigraphy and structure (157). Their interchange of information indicated the growing awareness on the part of company geologists of the value of sharing "strategic" geological research information (158). This spirit of cooperation between the Bureau and commercial geologists was initiated and pushed by Dr. Udden, who realized that they were all vitally concerned with increasing the knowledge of the geology of Texas (159).

The recognized need for a regional organization specifically devoted to stimulating cooperation and interchange of research data among the petroleum geologists of the southern Great Plains led to the birth of the Southwestern Association of Petroleum Geologists (S.W.A.P.G.) in 1917 (160). This organization was the brainchild of Charles H. Taylor, head of the Department of Geology at the University of Oklahoma, and E. L. De Golyer. Taylor realized that most of the geologists of the Southwest could not avail themselves of the benefits of existing geological societies, which as a rule held meetings in the East. Both he and De Golyer felt there was a real need for an organization for exchange of geologic thought among the burgeoning numbers of geologists in the Southwest. In addition to Taylor, Charles N. Gould, Oklahoma State Geologist, and Raymond C. Moore, State Geologist for Kansas, along with two commercial geologists, made up the committee in charge of the first meeting of the S.W.A.P.G. at Tulsa in February in 1917.

Udden was an early member of the organization and contributed an interesting little paper entitled "Hints to Prospective Geologists" to the first bulletin. He declared that American professional geologists had been subject to too little criticism. They, "like men of other professions, ought to mercilessly criticize each other within their own circle, when occasion arises" (161). He emphasized the benefits of forums for the discussion of geological research problems—a need partly satisfied by the creation of the S.W.A.P.G. (later the American Association of Petroleum Geologists).

The American Association of Petroleum Geologists (A.A.P.G.) meetings, both national and later regional, were true geological research forums, and the Bulletin of the Association provided a medium of publication for private and public geologists alike. Except for some select works, such as the Bureau's University of Texas Bulletin 2132 and micropaleontological studies, most of the results of the geological research work done by the numerous consulting geologists and the well-funded geological departments of the oil companies have been presented in the A.A.P.G. Bulletins (162). In spite of publication contributions made by local geological societies, the funds available to the Bureau for publication purposes could not begin to meet the demands made by the tremendous research energies expended in Texas alone. Furthermore, the Bulletins of the A.A.P.G. could handle regional or interstate research studies, whereas the Bureau would be hard pressed to justify the publication of such studies to the State Board of Control and legislators. So the Bureau directors heartily welcomed the growth of the A.A.P.G. and were consistent contributors to its Bulletin.

Another significant cooperative service provided by the A.A.P.G. was that of endorsing and encouraging geological research on problems too broad and basic to be handled by a company department or by the ever short-handed state surveys. In a 1928 article on "Research and the A.A.P.G.," K. C. Heald promoted the idea that the Association should lend both interest and active support to geological work that seems to fall into the realm of pure science (163):

> For the more practical minded we may urge that it is never possible to know what branch of science will leap from the high ground of pure academic interest into the thick of the economic battle.

The Committee on Research of the A.A.P.G. was subsequently created in March of 1923 and charged with fostering and coordinating basic studies relating to petroleum geology. This committee sent an outline of suggested geologic research studies relating to oil to John D. Rockefeller. He proved to be receptive to the idea of a coordinated research program and in November 1925 donated $150,000 for scientific research in the geology, physics, and chemistry of petroleum. A few weeks later the Universal Oil Products Company of Chicago donated an equal amount. Both funds were placed at the disposal of the newly created American Petroleum Institute. In turn, this Institute turned to the National Research Council (N.R.C.) as well as to the A.A.P.G. for advice in the selection, organization, and prosecution of research projects.

The N.R.C. had been created under the auspices of the National Academy of Sciences for the purpose of organizing the scientific men of America for the war effort. In 1919 it was reorganized and transformed from an instrument of war to one of peace (164). Its new function was to promote research in the mathematical, physical, and biological sciences. Professor Nevin M. Fenneman was the first Chairman of the Division of Geology and Geography. He appointed Dr. Udden Chairman of the Subcommittee on Sedimentation (165). The main business of this division was to gather information on "gaps" in the science of geology and possible research topics in these areas. In other words, the N.R.C. did not have funds to support research but rather served the purpose of suggesting and organizing research projects, emphasizing cooperation and coordination of efforts.

An example of the cooperative outlook of the Division of Geology and Geography was the undertaking of "plans looking toward the conservation and possible publication of the scientific results of geological work done for commercial companies." The Committee on Obtaining Scientific Data from Commercial Examinations and Explorations sent letters to the mining and oil companies of the United States asking that they place with the state geological surveys or with the U.S. Geological Survey information no longer considered confidential (166). Clearly, the rapidly growing A.A.P.G. served as the organization through which the results of privately supported geological research could be shared. The cooperative and collective spirit that the N.R.C. encouraged in scientific research stimulated similar concerns within other private and public geological organizations as well (167).

In essence, the N.R.C. was a tremendously effective organization for coordinating and guiding research (168). Through the N.R.C., the petroleum geologist could secure a more immediate and effective contact with universities, foundations, government bureaus, and other organizations that conduct research. Furthermore, though the National Research Council had no great endowment, it knew where to look for research funds. There was close cooperation between the Committee on Studies in Petroleum Geology of the N.R.C. and the Committee on Research of the A.A.P.G. These committees held joint meetings during which they discussed and selected research projects, such as those sent on to the American Petroleum Institute (169).

To some degree it may be said on the state level that the Bureau of Economic Geology served the same purpose as these research committees. Bureau directors Udden and Sellards often suggested, directed, or simply encouraged research projects by non-Bureau geologists. They put the resources of the Bureau at the disposal of such investigators and in some cases were able to pay field expenses in addition to publishing the results of the research work. The Bureau encouraged and actively supported such research works as Emil Böse's studies of the Cretaceous ammonoids of South Texas and northern Mexico, the stratigraphic studies of the Pennsylvanian formations of North-Central Texas by F. B. Plummer and R. C. Moore, the extensive investigations of the Glass Mountains by Philip and Robert King, the micropaleontological studies by Dorothy Ogden Carsey and Helen Jeanne Plummer, and Julia Gardner's excellent study of the Midway Group (170). It is difficult to imagine any geologist's undertaking any in-depth research work on the geology of Texas without seeking the advice and the cooperation of the Bureau, and the several series issued by the Bureau remain an essential medium for the publication of research studies on Texas geology.

The directors and geologists of the Bureau found the A.A.P.G. to be an invaluable ally in coordinating and making maximum use of all available research energies being expended on furthering the understanding of the complex geology of Texas. Through the years Bureau personnel have been consistent contributors of research articles published in the Bulletin of the A.A.P.G., and more importantly, the personnel of the Bureau have been influential in the research activities of the A.A.P.G. (171).

The role played by F. B. Plummer in the formation of the Society of Economic Paleontologists and Mineralogists (S.E.P.M.) clearly illustrates the effective participation of Bureau personnel in the activities of the A.A.P.G. The S.E.P.M., an organization under the auspices of the A.A.P.G., was established in March 1927 (172). After accepting a position as geologist with the Bureau in October 1928, Plummer was elected Secretary-Treasurer of the S.E.P.M. in 1928. That same year, Dr. Charles N. Gould, Director of the Oklahoma Geological Survey, called a meeting of concerned geologists to discuss regional studies of the stratigraphy and paleontology of the Pennsylvanian and Permian strata of the Great Plains. At this meeting, Plummer announced that the S.E.P.M. was created to bring about just this type of regional stratigraphic study on a cooperative basis between state survey geologists and oil company geologists (173).

Plummer was subsequently elected Chairman of the Subcommittee for Texas to Further Cooperative Investigations of the Permo-Carboniferous, which met at Eastland, Texas, in July 1928 (174). Dr. Sellards was the only other nonindustry man on this 27-man committee, whose members were chosen by local geological societies. At this meeting, the committee considered possible Permo-Carboniferous research projects. The following were eventually approved by the Research Committee of the A.A.P.G. for cooperation and financial support when possible:

1. The study of the sedimentation and correlation of typical Pennsylvanian sands of the Mid-Continent.

2. The compilation of county maps of the areal geological outcrops of formations in Texas.

3. The study of the flora of typical geological formations with a view to obtaining botanical criteria as an aid to the identification of formations and as an aid in geological mapping (175).

Of these three projects submitted by Plummer, as chairman, only the second was initiated immediately. The A.A.P.G. joined with the Bureau in a cooperative program of compiling county geologic maps in North-Central Texas. A mimeographed circular in the 1930 files of the Bureau on the progress of this cooperative mapping program includes the statement that 12 new county maps had been compiled, drafted, and distributed to the geologists of Texas, and 8 more were in the process of being drafted (176). The writer of this circular observed that it would have taken one geological party 12 years to have covered the area in the same detail as that of the cooperative maps, which were completed in 18 months.

The Bureau also carried out an extensive research project sponsored by the American Petroleum Institute (A.P.I.). In 1926, K. C. Heald, Chairman of the Central Petroleum Committee of the N.R.C., recommended to the A.P.I. that they sponsor investigations of the underground waters in oil fields (Project 25B). The project was begun in February 1927 under the direction of Dr. Sellards. E. M. Hawtof, a student assistant working for the Bureau, and Dr. C. E. Van Ostrand, whose time was donated by the U. S. Geological Survey, conducted extensive studies of increase in temperature with depth in wells. They found that the wells closer to the axis of a structure had higher temperatures at the same depth than those farther away from the axis (177).

Upon completion of this first part of the project, Hawtof was succeeded by E. C. Sargent, and at Sellard's request, Plummer was made director of the continued work under Project 25B. Plummer and Sargent undertook investigations of the geochemistry and geothermal character of the waters of the Woodbine sand in East Texas. This proved to be the most prolific oil sand in the state with the discovery of the East Texas field in October 1930. The results of their work were published by the Bureau in University of Texas Bulletin 3138. They found that the saline content and total solids in well waters close to faults and over salt domes were markedly higher than in areas of normal structure. Furthermore, connate water temperatures were hotter along fault zones (178). The value of this knowledge for oil exploration in the East Texas province is obvious (179).

The tremendous growth and success of the A.A.P.G. and the effectiveness of the aforementioned research organizations clearly demonstrated that commercial geologists were aware of the tremendous advantages gained through the interchange of the results of private geological research and the coordination of broad, basic research projects. As the leaders of the oil industry became more aware of the exploratory value of sophisticated geological research investigations, so did oil geologists recognize the advantages gained by contributing all research data of a strategic nature into the common pool of knowledge of the geology of Texas and other oil-producing states. In a letter to University of Texas President H. Y. Benedict, Dr. Sellards expressed the essence of this cooperative spirit (180):

> In Texas, there has developed fortunately a strong tendency towards cooperation among geologists in assembling and making usable the geological information accumulated in the State. The Bureau is, I think, in a small part responsible for this, although the trend is, I believe, general for the science.

In a letter endorsing the organization of the Texas Academy of Science, Sellards added that there had been no time in the history of the sciences in which there had existed a more general appreciation of the benefits of mutual cooperation in scientific endeavor. The swift strides made in the advancement of the knowledge of the geology of Texas that were associated with the search for oil were made possible by this spirit of cooperation. As a result, Texas became one of the most vital scientific frontiers in the world (181).

Dr. Sellards rightly observed that the Bureau of Economic Geology had played an important role in promoting this cooperative attitude among commercial geologists. Dr. Udden had not only served as mentor to the leading oil company geologists, but also placed all of the resources of the Bureau at their disposal. Before the emergence of the A.A.P.G., the Bureau had been the clearinghouse for interchange of strategic data on the geology of the state. The examples set by both Udden and Sellards in extending all possible assistance to the oil industry promoted a spirit of cooperation among all Texas geologists. The ability of the Bureau to assume a vital role in cooperation with the well-funded geological research departments of the oil companies demonstrates the tremendous catalytic influence that can be made by neutral (institutional) or public research agencies of limited resources. The catalytic role played by the Bureau in encouraging research coordination and interchange of research data represents perhaps the most valuable contribution made by the Bureau.

FIVE

The U.S.G.S. vs. the B.E.G.

Tremendous amounts of money and research energies have been expended in Texas by the U.S. Geological Survey, and Federal geologists have made invaluable contributions to the knowledge of the geology of the state (182). The most important contributions to the geology of Texas by the U.S. Geological Survey before Udden's directorship include the report on the geology of the Black and Grand Prairies by R. T. Hill; the Nueces and Austin folios by Hill in collaboration with T. W. Vaughan; the Uvalde folio by Vaughan; the El Paso and Van Horn folios by G. B. Richardson; the Llano-Burnet folio by Sidney Paige; two papers on the geology and ground waters of the Panhandle by Charles N. Gould; and two papers on the water resources of north Texas by C. H. Gordon (183). The U.S. Geological Survey publications on the Cenozoic of the Texas Coastal Plain in the first two decades of the 1900's will be discussed later.

During the years between the close of the Dumble Survey and the beginning of the University of Texas Mineral Survey (1896-1901), and during the interval between the University Mineral Survey and the creation of the Bureau (1905-1909), the U.S. Geological Survey was conducting the only professional geological work in the state. Since its inception, the U.S. Survey has been in continuous operation in Texas, conducting both topographic and geologic work. However, its record of cooperation with state surveys demonstrates the problems inherent in Federal-State cooperative programs.

State Geologist Dumble struggled continuously to promote U.S. Geological Survey topographic work in Texas. As long as John Wesley Powell's Irrigation Survey was in operation, extensive Federal funds were expended in Texas for topographic surveys. However, Dumble had trouble getting Powell to coordinate his survey program with the needs of the Texas survey. Consequently, for a couple of years, W. H. von Streeruwitz led the State survey's own topographical survey crew (184). Unfortunately, both Federal and State survey efforts were handicapped by the failure of the State Legislature to provide adequate support. Historically, the Legislature has been unable to appreciate the need for topographical surveys in support of geologic investigations.

The inability to obtain appropriations of matching State funds to support cooperative topographic and geologic work with the Federal survey obviously handicapped the Bureau. But there were other, more deep-seated problems that diminished the chances for successful Federal-State cooperative projects. The question of proper credits for the cooperative work was tied in with the issue of encouraging legislative recognition of the output of the respective surveys. This problem cropped up in connection with the cooperative investigations of potash in the subsurface of the High Plains and also in the cooperative preparation of a geologic map of Texas. In 1924, Dr. Udden wrote a letter to W. C. Mendenhall, Chief Geologist of the U.S. Geological Survey, in which he expressed his views on the question of proper credits in connection with the map project (185):

> If it could only be arranged so that the map could be distributed both from our office and from your office, and if it could be arranged so that the cooperation on our part would not appear to be a less amount than your own, . . . the [map] plan would appeal to us. I am not quite clear in my mind whichever is the best policy for us: Whether to jointly cooperate with you in the future as we have done in the past, or to decide upon a policy of working separate. I am afraid that we are not getting the credit we should have for the work we have done for the bringing to the front of our possibilities in potash. That work was absolutely initiated by the Bureau. . . . Furthermore, I made it known to the Survey two or three years ago that I intended to issue a new map of the geology of Texas and that I also favored

your work in the southeastern and southern portions of the state for this purpose.... I do not know from your point of the best policy for your survey, whether it is a good idea for you to issue state maps. But I would like to see the U.S. Geological Survey issue more folios, and do more scientific work; in this we have no rivalry at all, the field is so great....

There was also an element of professional jealousy that hampered cooperation between the State and the Federal surveys. In a letter to Udden in 1925, R. T. Hill, a veteran of scientific backbiting, described some of the infighting that went on (186):

I find that [G. H.] Girty's bitterness towards Beede, and [N. H.] Darton's hatred for Baker and a general dislike for Böse have caused these people [of the U.S.G.S.] to overlook the good work those people have done.... Darton had the effrontery to tell me the other day that he personally defeated the possibility of C. L. Baker getting a position in the University of Arizona on account of Baker's American Journal of Science paper. Now I have never seen or known Baker but I have seen his good work, and when I told [W. E.] Wrather of this Darton incident, he said that he [Darton] was taking the bread and butter out of Baker's mouth; that he, Wrather, had done his best to get Baker the job and that the act of Darton's was an awful outrage. But it was just another illustration of the fact that men in their ambitions, passions, and jealousies never hesitate at whom they strike. They have been doing it to me for forty years, and they are not going to like me any better for trying to get your Permian section recognized, because its recognition proves their missed opportunities.

All in all, most evidence seems to support Phillips' statement: "I suppose if the world continues to revolve the U.S.G.S. will after a while ascertain that there are some people in the world besides themselves" (187). E. W. Owen observed that the empire-building goals of the Geological Survey caused the Federal survey to operate as if the State surveys did not exist (188). As the old school of geologists, such as Darton, Girty, and Director G. O. Smith, faded from the ranks of the U.S. Geological Survey, a mutual spirit of cooperation similar to that developed between commercial geologists and the men of the Bureau began to be evident in U.S.G.S.-Bureau relations. This trend came to the fore when William E. Wrather became Director of the Federal survey in 1943 (189).

The following statement appears in the Twenty-Ninth Annual Report of the U. S. Geological Survey (1909) (190):

A geologic study of the Atlantic Coastal Plain is being carried on in cooperation with the official surveys of the various states concerned. Although the need for such work was long recognized, the plans for it were not perfected until about a year ago, when, in response to an invitation sent out by the Director of the National Survey, the state geologists from the various states came to Washington and met in conference with those members of the federal organization that were most directly interested in the proposed investigation. They agreed to undertake the cooperative survey of the geology and ground water resources of the Coastal Plain.

One of the real strengths of the U. S. Geological Survey lies in its ability to undertake broad, regional investigations such as this. The Federal survey has the funds and the wherewithal to conduct these studies, even without financial cooperation by the states (as in the case of Texas). The A.A.P.G. could not come up with the funds to undertake a similar extensive study of the Permo-Carboniferous strata of the southern Great Plains (191). The state surveys could not finance regional work or justify it to state legislatures.

Dr. T. Wayland Vaughan was named the supervising geologist in charge of these Coastal Plain studies. Initial work in Texas was undertaken by Alexander Deussen in 1908, 1 year before the creation of the Bureau. He investigated the Tertiary and Pleistocene deposits on the entire Texas Coastal Plain. A. C. Trowbridge did extensive work in South Texas. Dr. L. W. Stephenson, who later took charge of this project, examined the Eocene-Cretaceous contact in Texas and did quite extensive field work in the inner Coastal Plain Cretaceous deposits of northeast Texas and southwest Arkansas. Both Deussen and Stephenson were assisted by G. C. Matson and R. B. Dole and invertebrate paleontologist T. W. Stanton and paleobotanist E. W. Berry. Many individual papers were produced by these workers, but there was no overall compendium of the U. S. Geological Survey Coastal Plain investigations, although Deussen's outstanding Professional Paper did cover most of the Texas Coastal Plain (192).

After its creation, the Bureau was not consulted with regard to the scope and direction of these investigations (193). The Bureau did place its resources, including field notes and maps, at the disposal of the Federal geologists working in Texas (194).

Actually, the first real cooperative geologic work conducted by the U. S. Geological Survey and the Bureau of Economic Geology was in connection with the preparation of the geologic map of Texas that was finally published in 1932. The Federal survey had begun compiling material for a new geologic map of the United States in 1924. Chief Geologist W. C. Mendenhall wrote

Dr. Udden a letter describing how the Federal survey had initiated interest in compiling a new geologic map of Texas as an incident to the preparation of the new geologic map of the whole country (195). Mendenhall rather patronizingly stated that he believed he was safe in assuming that the Bureau would want to cooperate in this work. Udden fired back the reply quoted previously, in which he declared that the Bureau had initiated work on a new map of Texas a few years before. Udden went over the whole issue of cooperation and proper credits. The Federal survey went on with its project under the direction of G. W. Stose and later N. H. Darton, and, needless to say, ended up with top billing on the map legend (196).

Bureau field work devoted to the compilation of this map included L. W. Stephenson's work on the Upper Cretaceous of the Coastal Plain, T. W. Stanton's investigations of the Lower Cretaceous of the High Plains, N. H. Darton's reconnaissance work on the Permian of the Guadalupe Mountains and the red beds of western Texas and eastern New Mexico, and R. T. Hill's compilation of work in the Trans-Pecos region (197). The Federal geologists took charge of the compilation of the map, utilizing the work done by the Bureau toward this mapping project and also utilizing the areal work done by the oil companies (198).

Probably the most effective cooperative geological work between the Federal survey and the Bureau was Julia Gardner's stratigraphic and paleontological studies of the Midway Group of Texas. She began her work in 1925, with the Federal survey paying her salary and the Bureau paying her field expenses. Miss Gardner succeeded in untangling a mass of conflicting concepts about Texas Tertiary stratigraphy (199).

The U.S. Geological Survey and the Bureau of Economic Geology also cooperated in conducting investigations of the potash showings in the connate waters found in many wells in West Texas. The discovery of potash in subsurface waters was the result of Dr. Udden's studies of well samples from what was at the time the deepest boring in Texas near Spur in Dickens County, in the southeast part of the Panhandle. The prominent citizens of the area had contributed toward the drilling of this well in search of a water supply for the town of Spur. W. E. Wrather of the U.S. Geological Survey called Udden's attention to this well (200). Udden welcomed the opportunity of gaining information on the subsurface geology of this area in which there was much State land and arranged to have the samples sent to the Bureau. His careful examination of these samples set an excellent pattern for the fast-growing numbers of oil geologists.

Udden was mainly interested in clues to the subsurface stratigraphy of this region, but he was also well aware of the Legislature's concern with the economic contributions made by the Bureau. Following the fine descriptions of the microlithology and micropaleontology of the Spur well samples in his report on *The Deep Boring at Spur*, Udden announced that the most important economic result of the boring was that it "proved the existence in the dolomite beds of a stratum, or horizon, from which came a water sufficiently rich in potash to hold out inducement to the prospective search for this mineral" (201).

The U.S. Department of Agriculture and the U.S. Geological Survey in 1911 had begun conducting extensive investigations in California and Nevada in search of domestic sources of potash salts. In that year Congress appropriated $40,000 to be used by the U.S. Geological Survey in "determining geological conditions favorable to the preservation of potash salts" (202). The search intensified with the outbreak of the "European War," as Germany, the supplier of potash to the United States, curtailed exportation. In 1918, Dr. David White of the Federal survey asked the Bureau to cooperate in an investigation of "potash waters in the deep drilling now going on in the western part of Texas" (203). The Bureau did pay $1,800 of the salary of a field worker who was to watch drillings in Texas, New Mexico, and Oklahoma for shows of potash. From 1918 to 1920, Orbey C. Wheeler did the field work. D. D. Christner continued the investigation in 1921, until the cooperative program was halted owing to the inability of the Bureau to provide more funds (204).

The Federal survey continued to keep a watch on wells in this area and effectively lobbied for funds to conduct a core drilling exploration program in West Texas. In June 1926, Congress passed an appropriation of $500,000 to support a 5-year core drill exploration and development project. By the time the project had terminated in 1931, the Federal geologists had successfully outlined the most promising areas for potash exploration in Texas and New Mexico (205).

The Bureau and all other state geological surveys looked to the Federal survey to provide the topographic maps so vital to stratigraphic studies and field mapping. From its inception, the U.S. Geological Survey had conducted a continuous program of topographic surveying (206). Unfortunately, the amount of topographic coverage achieved yearly depended on the amount of funds a state could come up with to match the monies that the Federal survey was prepared to spend in that state. For example, for the biennium 1921-1922, the U.S. Geological Survey allocated $40,000 to be spent on topographic mapping in Texas. Udden's request for an equal amount of matching state funds was unsuccessful. Since the State Legislature appropriated $9,459.46, the Federal survey expended this amount in Texas instead of $40,000 (207).

This "matching-funds" program of the Topographic Branch of the U.S. Geological Survey was such an accepted procedure that in the 1931 Congressional appropriation for topographic mapping, 70 percent of the funds were restricted to being used only in states where matching funds were provided. This obvious favoritism to the more prosperous states who could buy more coverage had been one of the targets of the Temple Act passed in 1926. Dr. H. W. Temple, Congressman from Pennsylvania, sponsored this act, which authorized the completion of the standard topographic map of the United States in 20 years. The Appropriations Committee assumed that the spirit of the Temple Act was met if sufficient Federal appropriations were made annually to meet the matching funds offered by states. Temple felt that Federal appropriations should be made under a definite Federal program regardless of state financial cooperation, although such was invited. He felt that $10,000 should be expended annually in each state regardless of its ability to furnish matching funds (208). This debate over the interpretation of the Temple Act was pushed aside as the New Deal public works programs got underway in 1933.

The concern with home defense during World War I alerted the War Department to the need for accurate maps showing existing roads and for information on the location of emergency road-building materials. The lack of improved roads along the Mexican border and the south Texas coast was of special concern following the publication of the famous Zimmerman Telegram, which alluded to a German-Mexican alliance (209). The Council of National Defense requested the National Resources Committee to supervise the undertaking of investigations of roads and road-building materials in the "behind the front" areas in Texas. This Committee turned the responsibility over to the U.S. Geological Survey in the summer of 1917, and the Bureau of Economic Geology was asked to provide matching funds for making military maps of the coast of Texas and the border areas in Brewster and Terrell Counties (210). The State Legislature appropriated $10,000, and almost the entire staff of the Bureau went to work on the project, the scope of which was expanded to include mapping the entire state.

Three reports were prepared, one each on Terrell and Val Verde Counties and a statewide study of road-building materials (211). The authors of the county reports gave brief geological discussions, mapped the existing roads and trails on the Rio Grande border, and classified them according to their usefulness. Also, in the Val Verde County report, Geologist J. R. Roberts estimated the possible value of manganese ores — "such ores being urgently needed by industries during the war."

So, during the years 1909-1932, the U.S. Geological Survey did much more than extensive topographic work in Texas. The Federal survey cooperated in compiling a new geologic map of the state and in conducting investigations of the potash-bearing waters in the subsurface of the Texas High Plains (212). Through the Bureau, the State provided matching funds for the topographic work, and the Bureau contributed both funds and personnel to the potash investigations. The Federal survey continued those investigations after the Bureau found itself unable to contribute further amounts of its limited resources. The Bureau and the oil companies as well made all of their research information available for the compilation of the state geologic map. Finally, Federal geologists conducted extensive investigations of the geology of the Coastal Plain of Texas. Bureau geologists never hesitated to place all of their knowledge at the disposal of these Federal geologists (213).

Even though the Bureau supported Federal work in Texas in these several ways, the Director of the Bureau had little or no say in the Federal research projects, with the exception of the topographic work. This held especially true for geologic research. In no real sense of the word could the geological studies conducted in Texas by the U.S. Geological Survey relating to the Coastal Plain investigations be described as a cooperative undertaking. The Federal geologists determined the areas and procedures of investigation without consulting the Bureau. At the same time the Bureau scientists made no attempt to coordinate their work with that of the Federal geologists. There seems to have been a reluctance on the part of both organizations to place themselves in a subordinate position, especially on a geologic research project. Regional work such as the Coastal Plain project was outside the scope and pocketbook of a state survey, but several states could assign representatives to join Federal geologists in the coordination and direction of such projects to the mutual advantage of both the state surveys and the U.S. Geological Survey. Dr. C. N. Gould, of the Oklahoma Survey, in a letter to Dr. Sellards outlined such a scheme for the undertaking of a comprehensive study of the Tertiary deposits of the High Plains (214).

Yet the spirit of interchange of geologic research information was very evident. This informal exchange of knowledge among all geologists—Federal, State, and private—typified the research environment in Texas and represented cooperation at its finest. The leaders of the Federal and State organizations, concerned with the tie between production credits and appropriations and reluctant to subordinate the professional energies of their surveys in any way, found cooperation a ticklish affair, but the individual workers were not bound by such shackles. Perhaps the best description of the cooperative interchange of geological knowledge on the Texas research frontier is found in Julia Gardner's acknowledgments in the preface to her work on the Midway Group in Texas (215):

> The difficulty of adequate acknowledgment in a paper such as this is very great. There is in it comparatively little taken directly from an outside source. There is also little that does not bear the impress of the generosity of my fellow workers—of information gained in maps, in records and reports, in correspondence and conversation. In the pattern of the composite fabric it is impossible to trace the threads from which it is woven.
>
> I hold in grateful memory Dr. J. A. Udden, former Director for the Bureau of Economic Geology, for his interest and for the broader outlook which he gave to the entire problem. I particularly wish to thank those who have made the work possible, and who have awaited the results so patiently—Dr. E. H. Sellards, Director of the Bureau of Economic Geology, and Dr. L. W. Stephenson, in charge of the Section of Coastal Plain Investigations of the U. S. Geological Survey. Without their constant encouragement and advice this investigation would have been neither begun nor finished.
>
> To practically all of the petroleum companies operating in Texas I am indebted for information either in reports or on maps. I take pleasure in acknowledging an especial obligation to [the following] Messrs. . . .

SIX

Scientific Servant of the State

In an industrial society, man does not live in harmony with his environment, . . . he consumes and changes it.

P. T. Flawn, Nov. 1965 (216)

Conservation is, by definition, the wise use of natural resources, without waste.

Arthur Duggan, 1952 (217)

From 1909 to 1932, the Bureau of Economic Geology rendered outstanding service to the overall economy of Texas by virtue of its invaluable assistance to the oil industry. Furthermore, the value of its geological research contributions during these years was immense. However, apparently the geologists of the Bureau were so engrossed in their geological research work, especially with respect to the needs of oil exploration, that they did not recognize the importance of the early 20th century conservation or resource management movement spearheaded by President Theodore Roosevelt. This movement had its antecedents in the work on classification of public lands by Major John Wesley Powell and also by E. T. Dumble and W. H. von Streeruwitz in Texas. This movement had greatly expanded the public service role and thus the importance of the U.S. Geological Survey, although contemporary legislatures ignored these efforts for the most part.

In Geological Circulars published by the Bureau in 1965 and 1970, former director Dr. Peter T. Flawn described the constantly evolving public service role of the Bureau of Economic Geology (218). Geological surveys were the first attempts at organized research by the state, and their primary duty was to inventory the natural resources, "to discover and evaluate mineral resources on a reconnaissance scale and to develop new knowledge on the composition, structure, and geologic history of that part of the earth's crust called Texas" (219). Today the Bureau continues to fulfill this duty, though there has been a change in emphasis to include a much broader spectrum of natural resources including water, sand, gravel, and other nonmetallic minerals. However, it has assumed a new function of participating in resource use and planning programs.

One of the most important themes in the history of the Bureau has been its increasing involvement in programs for the planned utilization of natural resources. Such resource management was a major facet of the conservation ethic at the turn of the century (220). The concept of conservation itself has undergone a major transition in the 20th century, from concern with prevention of waste and preservation of unspoiled areas under the regime of Theodore Roosevelt and Gifford Pinchot to the current primary concern with the preservation of the quality of the environment, including cleanliness and beauty. Now the Bureau has assumed the role of assisting the state in the consideration of "all of the physical, social, and legal problems attendant on the use of land" (221).

Perhaps the most perceptive, concise description of the temper of the conservation movement launched under President Theodore Roosevelt is the following excerpt from Frank E. Smith in *The Politics of Conservation* (222):

> With a man in the White House who could be counted a friend, the champions of various conservation causes were ready to guide and advise him. All came with some background of special interest and special training in a single-purpose aspect of conservation or resource development, but most of them naturally supported adherence to a coordinated policy. Waste was something that they all instinctively reacted against, and waste was the major byproduct of uncoordinated development. Most of them, knowingly or not, were touched by the reformist philosophy of the day typified by Henry George. They believed that the nation's resources had to be protected from waste, and preserved and developed for the best use of the people. . . . They conceived of the conservation movement as a fight for economic justice.

The expression of this conservation ethic in the programs of the U.S. Geological Survey is of special interest to the analysis of the public service role of the Bureau of Economic Geology. In many areas the U.S. Geological Survey led the way in pursuing public service programs.

In the Annual Report of the U. S. Geological Survey for 1909, the year of the Bureau's birth, Director G. O. Smith discussed the issue of land classification, which he viewed as the necessary prerequisite to the most beneficial utilization of the public lands (223):

> The classification of the public domain and the investigation of its resources not only facilitates the work of fulfilling the requirements of existing law, but also . . . aids in the development in the land laws of the principle of relative worth, which is that the land must be so disposed of or so reserved as to insure utilization for the purpose for which it is most valuable. . . . No principle is more fundamental to real conservation and at the same time more beneficial to the mining and other industries than this of giving preference to the highest possible use of the public lands.

In the U.S. Geological Survey Bulletin on *The Classification of Public Lands* (1913), Director Smith reemphasized that "the utilization of lands for their greatest value necessitates the determination of that value, which is, briefly, land classification" (224). Smith noted the growing public concern with classification, which began in 1905 and 1906 with the realization that large areas of valuable coal lands in the West had been obtained from the government by means of agricultural entrees. The public demand for an end to these frauds led President Roosevelt in June 1906 to direct the Secretary of the Interior to withdraw from entry all valuable coal lands. It was at this time that the Geological Survey became actively engaged in land classification work and began assisting the General Land Office in the administration of the public lands (225). The act establishing the Federal survey in 1870 had given it the legal power to classify public lands, but it took the momentum of the conservation movement to prompt Congress to appropriate the necessary additional funds to carry out this program.

Further concern with the management of the natural resources on Federal lands came as the result of an investigation of the oil fields of California in the San Joaquin Valley, where there were extensive Federal lands. Investigating geologist Ralph Arnold assumed these studies in 1905, and his correspondence with his superiors in Washington undoubtedly prompted Director Smith to write Secretary of the Interior James Garfield in February 1908, calling attention to the superiority of oil as a naval fuel as well as to the oil potential of the Federal lands (226).

Subsequently, in 1908, President Roosevelt appointed a National Conservation Commission headed by Gifford Pinchot.

This Commission, which included prominent politicians, scientists, and industrialists, made the first comprehensive inventory of the nation's natural resources (227). In this report, based primarily on information provided by the Geological Survey, oil was declared a rare mineral that should be restricted in use and hoarded (228). On the basis of this recommendation, Roosevelt in September 1909 authorized the withdrawal of all public oil lands from entry because "it seemed desirable to reserve certain fuel oil deposits for the use of the American Navy" (229).

This conservation ethic of managed development of natural resources was further developed in the next decade. The Quality of Water Investigations by the U.S. Geological Survey represent the earliest concern with the control and management of natural resources not located on public lands. Following the passage of the Reclamation Act in June 1902, the Geological Survey established a separate Hydrographic Branch to administer the many water studies that would be the bases of all reclamation and irrigation projects. The major subdivisions of this Branch were the Reclamation Service, the Division of Hydrography (underground water investigations), and the Division of Hydro-Economics. This latter division was to handle investigations of the quality of water resources "with special reference to the effect of [pollution] upon industrial development, and the value of various sources of supply for domestic use or other industrial purposes" (230).

The head of the Quality of Water Investigations Branch, Marshall Ora Leighton, brought a scientist's objectivity to these pioneer studies of pollution. His main concern was the most efficient and economic usage of water as a natural resource. In his first report on *Sewage Pollution in the Metropolitan Area near New York City and Its Effect on Inland Water Resources* (1902) (231), Leighton called pollution "one of the most far-reaching drains upon human wealth that has yet been considered" (232). Later in the same report he declared that if the cost of purifying sewage before turning it into a river far exceeds the cost of providing a substitute pure water supply from highland or other sources, "it is manifestly better . . . to use a stream as a sewage course rather than a source of water supply" (233). Like other conservationists of his day, Leighton emphasized the economic advantages to be gained through controlled use of resources and disposal of wastes, as best exemplified by his title *Pecuniary Damage of Municipal Water Supply Occasioned by City and Industrial Wastes* (234).

Leighton recognized pollution of natural waters as the natural outgrowth of "social municipal progress." He felt that pollution should be accepted, but at the same time he felt that it had to be managed with the overall economic well-being of the people in mind. He viewed the problem as one of limiting the destruction caused by pollution to only one or two of the resources vested in the natural waters—health, drinking water, water for industrial use, ice crop, fishing crop, and realty values. In other words, Leighton was a scientific conservationist rather than an emotional preservationist, and his approach has been echoed in the words of present-day conservationists. Leighton wrote (235):

> The question may be asked: Is not the damage done to these streams by the sewage of the cities and by industrial plants merely a part of the compensation which we should willingly pay for the advantages that we have gained through the establishment of these polluting influences? Undoubtedly, to a certain degree; but the fact still remains that in most cases when such damage is carried beyond a certain point it must be regarded as needless and unwarranted.

In 1904, Edwin Goodell wrote a Quality of Water Investigations report entitled *A Review of the Laws Forbidding Pollution of Inland Waters in the United States* (236). In general, Goodell found that the jurisdiction over pollution of waters was established only in those states with high industrial development and resultant high pollution—the New England states with their mills and the industrial states of New York and New Jersey. Goodell observed that the courts had just begun to show themselves fully alive to the effect of pollution on public health and the existence and validity of public rights in respect to pollution (237).

At the close of 1910, the efforts of the Quality of Water Investigations section were combined with those of the Underground Water section. Because of the lack of public interest in pollution, water quality studies were confined almost exclusively to the mineral content of underground waters. The many Quality of Water Investigations reports published by the Geological Survey in the first decade of the 20th century remain as outstanding pioneer studies of the effects of pollution (238).

In 1920, another new dimension was added to the concept of resource management. In his introduction to the Annual Report for 1920, "Planning for the Future" (239), Director Smith explained that the major function of the Geological Survey was to collect and arrange data upon which the nation could base its plans for future development. He praised the precedent set by the pioneer planning work done by the irrigation survey from 1888 to 1891 (240). Engineering data such as those gathered by this survey were an invaluable aid in planning the storage of flood waters for power and irrigation in the West and for power development, improvement of navigation, and flood protection in the East.

Smith's article was in support of the most outstanding example of planned development of natural resources prior to the programs initiated under the New Deal: the Super Power Project, which was an outgrowth of a cooperative program between the Geological Survey and the Fuel Administration initiated in response to the wartime demand for fuel conservation. Actually, in 1910 M. O. Leighton and Rene Tavernier had written a paper for the Geological Survey on *The Public Utility of Water Powers and Their Governmental Regulation* (241). In his introduction, Leighton said that water power was a natural monopoly by reason of the natural laws of stream flow, and that long distance transmission of electricity made water power a public utility in every sense of the term. Leighton's arguments supported President Roosevelt's call for a national water power policy in his January 14, 1910, message to Congress. Leighton agreed with the President that the people must be awakened to the need for Federal regulation of water power development and operation.

However, it took the extreme wartime concern with fuel and power conservation to prod Congress into making a special appropriation for the continuation of the Geological Survey investigations of power resources following World War I. Industry added support in December 1918. In this month President E. G. Buckland of the New York, New Haven and Hartford Railroad Company, upon the advice of W. S. Murray, Consulting Engineer of New York City, urged the Secretary of the Interior to authorize a survey of the sources of energy in the industrial Northeast. An intensive lobbying campaign sponsored by the industries in this region was launched, and Congress appropriated $125,000 for the fiscal year 1921 to be used in making an intensive analysis of the ever-growing energy requirements of the industrial region between Boston and Washington (242).

The Geological Survey supervised this Super Power Project and published results of the studies in Professional Paper 123, *A*

Super Power System for the Region between Boston and Washington. W. S. Murray, the senior author, conceived of electrical utilities as the key to economic power and saw fit to label the planned power system as nothing but a super utility. Unlike the Pacific coastal region where water power abounds, the Super Power zone had small hydraulic resources that could support only 21 percent of the power demands of the railroads and industrial plants in the zone. The rest of the power would be supplied by steam electric generator plants fueled by coal. Thus, an analysis of the availability of coal reserves was included in the report.

The Super Power System was a coordinated plan for power production that made maximum use of existing generator and transmission equipment and recommended construction of large super power plants with attendant transmission lines. The authors of this report emphasized that this plan alone could bring order and coordination to the production of energy in this vital population and industrial region. The plan would produce the highest possible efficiency in power production and was therefore in tune with the conservation ethic of the day that demanded the maximum economic yield from natural resources, and could only come through managed resource development (243).

As a final note to this discussion of the public services provided by the U. S. Geological Survey, mention should be made of the natural history guidebooks produced by that organization. In 1915 the Geological Survey published a series of four guidebooks covering the geography, geology, mineral resources, industries, and history of the regions bordering the several transcontinental rail routes (244). These were to inform the many travelers going to the Panama-Pacific Exposition in San Francisco of the geology, resources, and history of the Great West. The fine reports included geologic-topographic index sheets of the routes. Also, in 1933 the Federal survey printed several regional geologic guidebooks for the International Geologic Congress held in Washington, D. C. One focused on the natural history and geology of the Trans-Pecos Texas area, and another covered the geology and typical oil fields of the eastern Oklahoma-Texas region (245).

These studies made by the Geological Survey dealing with the management of natural resources—public lands, minerals on public lands, water, and water power—reinforced and broadened the national conservation movement led by President Theodore Roosevelt and Gifford Pinchot. This movement, which achieved a great momentum during Roosevelt's presidency and again in the late teens as a result of the resource demands of the war effort, failed to receive adequate legislative support during the prosperous years of the 1920's. However, it gained momentum during the New Deal years.

A similar movement directed toward regulated use of natural resources emerged in Texas in the early 1900's. Obviously, this Texas conservation movement took some leads from the national movement, and the patterns of momentum were somewhat the same. However, the specific factors that stimulated this concern with management of resources in Texas were the steady increase in the population in the semiarid western two-thirds of the state and the concern with water resource problems that was a result of both this growth and the phenomenal development of oil production. The tremendous floods of 1913 and of several prior years, the severe droughts of 1917 and 1918 that racked West Texas, the momentous oil discovery at Spindletop in 1901, the Ranger and Burkburnett oil booms of the late teens, the discovery of the prolific Yates field in the Permian Basin in 1926, and, finally, the opening of the tremendous East Texas oil field in the spring of 1931 were factors which led to the passage of legislation to help conserve and utilize the water resources to the fullest and to prevent waste in oil production and curtail pollution therefrom.

The first general irrigation and water law in Texas was passed on March 17, 1889 (246). This was the legislative response to the crippling drought of 1886-1888. The first State Irrigation Regulation Act, passed in 1913, established the Board of Water Engineers as the regulatory body (247). This act was a response to the public concern with the floods that had been ravaging the state for 4 years. It declared all surface waters to be State property and authorized the Board of Water Engineers to undertake work under the Federal Reclamation Act of 1902.

Perhaps spurred by the severe droughts of 1917 and 1918, the Legislature passed an act prescribing more fully the powers of the Board of Water Engineers. The most significant piece of conservation legislation passed by this 35th Legislature was the Conservation Amendment (ratified in 1917), which was a response not only to the challenge of drought but also to the growing awareness of the waste accompanying poor oil-field development and oil production. The Conservation Amendment demanded State regulation of all natural resources in the name of conservation and the prevention of waste (248):

> The conservation and development of all of the natural resources of this State, including the control, storing, preservation and distribution of its storm and flood waters, the waters of its rivers and streams for irrigation, power, and all other useful purposes, the reclamation and irrigation of its arid, semi-arid and other lands needing drainage, the conservation and development of its forests, water and hydroelectric power, the navigation of its inland and coastal waters, and the preservation and conservation of all such natural resources of the State are each and all hereby declared public rights and duties; and the Legislature shall pass all such laws as may be appropriate thereto.

The following year (1918) Governor W. P. Hobby called for the passage of legislation to make this amendment effective so as "to meet the unusual conditions that have arisen on account of the severe drought recently prevailing in this State" (249). The enabling legislation passed in March 1918 provided specifically for the establishment of conservation and reclamation districts for the management of water resources, all under the supervision of the Board of Water Engineers and the State Reclamation Engineer (250).

In March 1919, as a solution to the problems of oil wasting that became evident during the rapid and uncontrolled development of the Burkburnett, Ranger, and Desdemona fields, the 36th Legislature passed an act granting authority to the State Railroad Commission to formulate and enforce rules for the conservation of oil and gas (251). The main concern of this act was the waste of natural gas produced along with the oil.

This Oil and Gas Division of the Railroad Commission was established on July 8, 1919, under Major George C. Butte (252). Butte took a year's leave of absence from his law firm to organize this Division and to draft the conservation regulations governing it. However, prior to 1930, these regulations were not rigidly enforced because of the increasing demand for petroleum products and prosperity. By the early part of 1930, the reality of oversupply was brought to the fore by the dropping price of oil, and the realization was becoming generally accepted that open-flow production was depleting reservoir pressures so fast that much oil was being left in the ground. It was plain that production limitations must be enforced, and the waste-prevention clause of the Conservation Amendment and subsequent statutes seemed to offer the legal basis for control and limitation of production.

On the basis of the powers of conservation granted to the Railroad Commission, the first statewide proration order was issued on August 7, 1930 (253). However, before this order could receive a trial and possible acceptance, the discovery well in the great East Texas oil field came in during October 1930, and by the summer of 1931 it was evident that if this field were produced openly, it alone could provide the oil requirements for the entire United States. There were a multitude of producers in this huge field, from major oil companies to greenhorn independents, and any type of voluntary limitation of production was out of the question. The Railroad Commission took the matter in hand by issuing a proration order for the entire field on May 1, 1931. The issuance of this order was followed by a series of injunctions and restraining orders against the Commission and, finally, by the invalidation of the proration order by the Federal District Court in July 1931. This decision was based on the premise that the proration order had been issued to fix prices and not merely to prevent physical waste.

Governor Ross Sterling thereupon called a special session of the Legislature to consider the matter. In his opening address to this Legislature he echoed the words of President F. D. Roosevelt in calling for the greatest good for the greatest number. He emphasized the need for more regulation in every walk of life: "My idea is that we should have a commission that would control or regulate or conserve the natural resources of the State; the soils, the waters, the minerals, and all the natural resources of the State" (254). In August 1931 this special session passed an act empowering the Railroad Commission to limit production only in an effort to prevent physical waste through depletion of reservoir pressures. The Civil Appeals Court ruled in March of 1932 that production in excess of market demand also resulted in physical waste, both above and below the ground. The State Supreme Court upheld this conclusion, stating that the limitation of production to avoid waste would only incidentally affect the price of oil. Finally, the State Legislature passed the "Market Demand Act" in November 1932, which added "to the authority of the Railroad Commission that of inquiring into the market demand for crude petroleum, oil and natural gas to determine whether waste exists or is imminent" (255).

The conservation efforts of the Texas Legislatures that have been described above were limited to regulation of irrigation and reclamation and control of oil production. However, Harbert Davenport postulated that the true purpose of the "Water Conservation" Act of 1917 was to destroy the rights of riparian owners (256). James Hart found that there could be little question but that the primary desire of advocates of State restriction on production was to achieve price stabilization (257). The fact that restriction of production also in fact prevents physical waste was a fortunate coincidence for the proponents of restriction. Although some of the specific enactments described above were indeed special interest legislation, the fact remains that they were passed only because of association with the conservation movement that was a strong force in the state during the first three decades of the 20th century (258).

From the turn of the century up to 1932, many precedents in the fields of conservation and resource management were established by the U.S. Geological Survey, as described earlier in this chapter. As shown above, Texas legislatures also demonstrated a very real concern with the efficiency of natural resource utilization and development during these years. However, the Bureau of Economic Geology failed to assume an active role in the conservation movement, and other State agencies took control of resource management programs. Only recently has the Bureau stepped into the vanguard of state resource planning efforts.

Following the lead established by Major John Wesley Powell as Director of the U. S. Geological Survey, Director E. T. Dumble and W. H. von Streeruwitz of the Texas Geological Survey (1888-1896) campaigned hard for the controlled distribution of the State lands of West Texas on the basis of prior classification of these lands as to irrigation and mineral potential (259). They were interested in helping the state, the public schools, and the University of Texas realize the full market value of their lands and the minerals thereon. An equally strong democratic concern was that these public lands, most of which were located in the semiarid Trans-Pecos country, become a region of small one-family ranches, with enough irrigable acreage to be almost self-sufficient. Like Major Powell, Dumble and von Streeruwitz viewed the small ranch, like the small farm, as the backbone of the American democratic system (260).

Unfortunately, the depression of the 1890's brought to an end the Dumble Survey and its land management schemes and plans for the storage of storm waters for irrigation use (261). All interest in the careful management of public lands seemed to wane after this time, and eventually large ranchers acquired title to these lands or acquired cheap grazing leases on them (262). Interest was rekindled by the development of irrigation on the High Plains and by the discovery of oil in the Permian Basin. The only early Bureau report to deal comprehensively with the question of land planning was C. L. Baker's study of *The Geology and Underground Waters of the Northern Llano Estacado* published in 1915 (263). Baker observed, in tones echoed by Walter Webb in his classic *Great Plains,* that there was not sufficient water to irrigate all of the land underlain by the shallow water without depleting that source (264). He viewed the best development plan as irrigating 80 acres out of each section and utilizing the remaining acreage for producing dry-farming forage crops and grazing (265).

Both Baker and Udden expressed concern over the depletion of ground waters, but neither suggested that the Bureau make water resource studies a vital part of the Bureau's work or called for State control over the use of these resources (266). All of the county reports published by the Bureau did include a section on water resources, but apparently the Board of Water Engineers or the farmers or the legislators did not take note of these, for there were no calls for increasing the number and scope of the Bureau ground-water studies. In 1927 and again in 1929, Sellards did ask University President H. Y. Benedict for additional funds for ground-water studies, declaring that "no resource in Texas has greater permanent value than underground water supply" (267). The request was denied, and in the latter year the Legislature voted an appropriation of $50,000 to enable the Board of Water Engineers to undertake ground-water studies, which it did under a cooperative scheme with the U. S. Geological Survey (268). Perhaps the appropriation would have gone to the Bureau if the Regents had recommended Sellards' proposal to the Legislature. However, the Board of Water Engineers had the most effective lobby.

So, through a series of acts in the first three decades of the 20th century, the State had assumed regulatory control over both surface and underground waters and had established the Board of Water Engineers to take charge of the administration of the regulatory laws as well as the investigations of water resources. With the exception of the work of Baker in 1915 and the water resource studies in the county reports, the only work done by the Bureau in connection with the management of water resources prior to the takeover of such studies by the U.S. Geological Survey in 1929 consisted of some studies of proposed reservoir and dam sites for the Board of Water Engineers (269). This Board had contracted the Bureau to undertake these studies to determine the feasibility of geological conditions for the construction of storage dams on various rivers throughout the state. Actually, this is the kind of advisory role that the Bureau could have performed for the Railroad Commission and the Board for Lease of University Lands (270). This would have precluded the establishment of separate geological departments in these agencies.

The correspondence files of the Bureau are rather devoid of any reference to the Railroad Commission's proration work (271). However, one letter does bring out a crucial point that Peter Flawn emphasized in his *Geology in the State Government of Texas* (272). This letter, written by R. T. Hill to Wallace Pratt in April 1920, stated (273):

> I am not aware of the exact nature of the geologic work which the Railroad Commission proposed to do, but if it has any geologic work and money to do it with, it had best be done under the direction of Dr. Udden, to be handled in a co-operative manner by the two organizations. I have no suggestions as to any other possible means of handling the situation in question. I do not believe in scattering the geological work among various bureaus and it will detract from the building up of what the State needs most — a strongly centralized and single Geological Bureau for the benefit of all. I think gathering opinions of this character from the petroleum geologist could be one of the most useful and beneficial functions of the association.

The fact that the Bureau was part of the University of Texas may explain the reluctance of the Legislature and the Railroad Commission to ask the Bureau to take a major role in the statewide programs and field development programs of the Commission at this time (274). However, the fact that the Bureau was not even given responsibility for the management of oil and gas production on University lands can perhaps only be attributed to lack of interest on the part of the Bureau.

Dr. Udden had been the first to suggest the possibility of a favorable oil structure on University lands northeast of Marathon, and his words led to the drilling of the Santa Rita

No. 1, the first oil well on University lands. Udden had been very conscious of the impact of this fact whenever appropriation time rolled around (275). Instead of taking command of overseeing the development of the oil potential of the University lands, Udden and Sellards allowed control to pass to another body. As the oil production of University lands increased, the Board of Regents felt a real need for organized supervision of mineral leasing on University lands. In 1929, they asked Sellards to present an organizational plan, which he sent to President Benedict on May 3, 1929 (276). In the latter part of that same year, H. P. Bybee, Professor of Geology, was appointed to the newly created position of Geologist of University Lands. This position had been established for the purpose of advising the Board for Lease of University Lands as to the mineral potential of these lands (277).

Of course the magnitude of the geological energies involved in evaluating the mineral potential of University lands is nowhere near that involved in underground water studies or in statewide oil proration work, but in each case it was the failure of the Bureau to assume an active role in providing the necessary geological services for these resource management programs that necessitated the establishment of separate geological departments for each program (278). The attentions and energies of the Bureau were directed elsewhere, toward geological research in support of mineral resource inventory. The geologists of the Bureau were wrapped up in the special geological problems associated with the exploration for oil.

Furthermore, the staff and the budget of the Bureau were too limited to permit any additional work of such scope as underground-water studies. That this budget remained low can be attributed to the fact that neither the Legislature nor the Bureau directors ever envisioned the Bureau as a broad-based, central State geological agency that would concern itself with public service and resource management programs, as well as with geological research in support of the extractive mineral industries (279). As the various State agencies concerned with resource management—the Board of Water Engineers, the Oil and Gas Division of the Railroad Commission, the Board for Lease of University Lands—established their own geological staffs to meet their particular needs, the Bureau became even more cut off from any chance for growth in the area of public services. It might have become the central geological agency for the state as R. T. Hill had hoped, but Bureau directors failed to grasp the significance of precedents such as the U.S. Geological Survey's expansion into these areas of public service and thus did not follow the example this Federal survey set in moving into the fields of land classification, underground-water studies, quality of water studies, and even the *ad hoc* Super Power Project (280). One reason was lack of funds, but perhaps the basic factor was lack of initiative in these areas.

One final point should be made with regard to the opportunity for expansion missed by the Bureau. In 1917 at a crucial point in the development of the conservation movement in Texas, when the main resource regulatory bodies, the Board of Water Engineers, and the Oil and Gas Division of the Railroad Commission, were being given important regulatory powers, the University of Texas was embroiled in a struggle with Governor James E. Ferguson (281). It would have been politically unwise to support increased appropriations for the University's Bureau of Economic Geology. In addition, as noted above, the Bureau's University status might have precluded the Legislature from viewing the Bureau as a central State geological agency capable of handling the geological work demanded by various State agencies.

Thus, of the many factors that worked together to prevent the Bureau from living up to the intent of its parent division—the Division of Conservation and Development of the Natural Resources—perhaps the most significant were lack of initiative for expansion on the part of the directors and the Bureau's affiliation with the University (282). It is ironic that University affiliation would be a drawback, as one of the main reasons for this affiliation was to avoid the direct impact of the legislative appropriation axe. As far as the directors' lack of initiative is concerned, it must be noted that in the pre-New Deal years directors of all other state surveys apparently considered resource management policies to be beyond the scope of a state geological survey.

SEVEN

Emergency Years: Depression and War

During the depression years of the 1930's and the war years of the 1940's, the geologists of the Bureau continued to focus their main research energies on stratigraphic investigations in support of oil exploration. Their outstanding studies of the Ellenburger Group of Central Texas, the Pennsylvanian and Permian strata of West and North-Central Texas, and the Tertiary formations of the Gulf Coast will be discussed later. They also participated in mineral resource surveys funded by the Work Projects Administration (W.P.A.), which provided work for unemployed geologists and laborers. In the later thirties, Dr. Sellards steered these surveys more into the area of vertebrate paleontological investigations.

Sellards put more and more of his research energies into vertebrate and associated archaeological examinations and became rather detached from the fine basic geological work of his three very capable staff geologists — F. B. Plummer, V. E. Barnes, and H. B. Stenzel (283). However, he proved to be a very energetic administrator of these W.P.A. surveys and participated in important state programs for the planned use of the natural resources of Texas. The official correspondence of the Bureau contains great amounts of information on the crucial issues of government subsidization of mineral resource exploration, planned development of these resources, and the exhaustibility of mineral reserves. Consequently, the story of the Bureau during the years 1933-1945 provides perspective for our contemporary concern with these same issues.

Both Federal and State planning organizations administered the distribution and use of the vast Federal funds made available during the Depression and World War II, and the adaptive Sellards adjusted quickly to this new emergency research environment. The many programs for emergency allocation of government funds set up by the Franklin Roosevelt administration were coordinated by the National Resources Board (formerly the National Planning Board). In June 1934, the President appointed Secretary of the Interior Harold L. Ickes as chairman of this Board, the purpose of which was "to collect, prepare and make available to the President with recommendations, such plans, data, and information as may be helpful to a planned development and use of land, water, and other natural resources."

This national planning scheme called for the establishment of regional and state planning boards, and in May 1934 Governor Miriam A. Ferguson appointed a Texas State Planning Board of 14 members, including Dr. Sellards (284). This was the first body ever charged with the specific task of studying the interrelationship of the heterogeneous interests of the state to develop an overall policy of economic development on the basis of the primary natural resources of the state. The Board established five committees—Land Use, Industrial, Water Conservation, Social Welfare, and Mineral Resources. As chairman of the Mineral Resources Committee, Sellards cited as the major goal the reduction of waste of "non-restorable" natural resources. However, before this group could get into operation, Governor Ferguson was defeated in her bid for reelection.

In December 1934, Governor-elect James V. Allred asked Sellards to serve on a proposed committee to supervise public works developments in the state (285). Sellards eagerly accepted and thereupon wrote Everett L. Looney, Chairman of the National Public Works Committee for Texas, urging the creation of a permanent state planning board along the lines of the one appointed by Governor Ferguson (286). The following spring, the 44th Legislature, at the request of Governor Allred, created the Texas Planning Board (T.P.B.) with the specific goal of formulating a comprehensive state program for "the constructive expenditure of Federal funds for the alleviation of suffering and want and for the rehabilitation of worthy indigent families." The T.P.B. was further charged with the duty of establishing a statewide program for "the proper and conservation use of all of the various resources of the State, natural and otherwise" (287).

This very year marked the 25th anniversary of the Bureau of Economic Geology. The *Quarter-Centennial Memorial Volume of the Division of Natural Resources* included a speech that Morris M. Leighton, Illinois State Geologist, delivered at a session of the Quarter Centennial celebration. His talk seemed to embody the essence of the New Deal approach to the planned development of resources (288):

> Wholesale extraction and inefficient and wasteful methods of utilization have marked this [early period of resource exploitation]. Our soils have been depleted, our forests have been largely destroyed, billions of tons of our best coal have been wastefully mined and used, our oil and gas reservoirs have been unscientifically tapped and uneconomically developed, reservoir pressures have been dissipated, our underground water supplies have not been properly husbanded, our fish and game have been massacred, our streams have been polluted, and many other resources have been mistreated or destroyed.
>
> All of this is characteristic of the exploitative stage of development. The time has come, however, when we are entering the next stage, the stage of putting our resources to their highest uses in harmony with economic principles. Those states that take cognizance of this will become the more progressive and the welfare of their people will be more secure.
>
> An integrated and practical program, based on the appreciation of powers of science and technology and on sound principles of economy, should be formulated and set in motion in every state, for the purpose of making as complete an inventory of its resources as is possible, of making studies that will point the way to the correction of wasteful practices of recovery, of finding more efficient methods and forms of utilization, and of determining wise and effective means of conservation and development.

Unfortunately, the pressing issues of unemployment and economic recovery and the war partially overshadowed the intent of these praiseworthy words. Accelerated development of natural resources became the primary goal of geological research, especially during the war. Efficiency of production, the essence of the conservation movement during the New Deal years, was a guideline in the most important Texas mineral industry — the oil industry. Conservation of water resources was a major concern of Texas legislators during the 1930's. In this context, conservation meant the control of flood waters through the construction of impoundment dams and the production of hydroelectric power from these impoundments. Outside of the utilization of these most important natural resources of oil and water, controlled development of the state's resources by the Texas State Planning Board simply meant more governmental support of mineral discovery and exploitation.

Not until the post-World War II boom had slackened would Texas begin to apply to all resources the concept of controlled development that had first characterized the oil industry during the 1930's. (It should be noted that the oil industry itself was responsible for instituting the principle of prorated oil production.) The concept of waste resulting from pollution, in addition to waste due to inefficient recovery methods, would not begin to become a significant public issue until the late 1960's. From 1932 to 1945, the director and staff of the Bureau of Economic Geology failed to take up the cause of conservation — prevention of waste of natural resources — and focused only on mineral inventory in addition to their basic geological studies (289). They were not unusual in this respect, as the U.S. Geological Survey also seemed to have reverted to the exploitative stage of resource development and put aside its program of pollution studies.

Dr. Sellards proved to be a most effective administrator of the emergency mineral inventory programs funded by the Federal government during this period. In 1933 he had actively campaigned through the Association of American State Geologists for allocations of Federal public work funds for undertaking inventories of the mineral resources of the states. His main objective was the useful employment of geologists who had lost their jobs (290). The 1933 and 1934 meetings and bulletins of this association were devoted to this concern. After a specific proposal made through the State Planning Board for Federal funds for Texas was turned down in December 1933, Sellards stepped up his lobbying efforts by corresponding with the local geological societies about resource development projects and the numbers of unemployed geologists (291).

His efforts were rewarded in October 1935 when the W.P.A. director in San Antonio advised the Texas Planning Board that $386,000 had been allocated for a statewide mineral survey to be supervised by the Bureau of Economic Geology (292). Sellards presented his plan for this survey to University President

Benedict in November (293). All geological assistants and laborers were to come from the ranks of the unemployed. The project was well under way by the winter of 1935.

Realizing that this allotment was much greater than the mineral inventory would require, in the latter part of 1937 Dr. Sellards successfully petitioned Acting University President J. W. Calhoun for authorization to change the scope of the W.P.A. project to include the collection of vertebrate fossils and the excavation of a meteor crater in Ector County (294). Sellards had picked the opportune moment for asserting his interest in vertebrate collecting in the state and made possible the beginnings of the fine vertebrate collections of the University and the Texas Memorial Museum. The results of this mineralogical-paleontological project are discussed later.

As German victories on the Continent mounted, the demands on the Federally sponsored mineral surveys began to shift to concern with strategic minerals. An editorial in the *Austin Statesman* on September 17, 1939, announced the start of a sustained exploration of the Central Mineral Region to be led by Dr. Barnes. The objective of this project was to explore the nation's largest supply of granite and marble and "to investigate reported traces of the two war minerals, chromium and nickel" (295). In 1941 the Bureau received a new W.P.A. appropriation of $250,000 for geological investigation of "strategic, critical and essential minerals" (296). By the fall of 1941, Dr. Sellards wrote University President Homer P. Rainey that the Bureau was devoting its full energies to studies that affected the national defense program (297).

As noted above, oil proration and flood control and hydroelectric projects were the outstanding examples of controlled utilization of resources by the state (298). The statewide mineral resource inventory was heartily endorsed by the Texas Planning Board (T.P.B.) as a necessary prerequisite to any program for controlled utilization of the state's resources. Unfortunately, not all of the other recommendations of this Board concerning "conservation" projects reached fruition; possible exceptions were flood control and hydroelectric projects.

E. A. Wood, Director of the T.P.B., sent a report on "Pollution of Texas Waters" to Dr. Sellards in 1935 (299). Wood's report recommended legislation establishing pollution control agencies. He called upon the Federal government to provide the staff and funds for the preparation of a comprehensive plan for the elimination of various kinds of pollution in Texas waters. This plan, to be used in drawing up the State pollution codes, would be prepared in cooperation with the State Board of Health, the Game, Fish and Oyster Commission, the State Board of Water Engineers, the Bureau of Engineering Research, and the Texas Engineering Experiment Station at Texas A&M College. Sellards replied that he had already suggested that a similar study be undertaken with W.P.A. funds (300). However, no pollution studies were undertaken in Texas, nor did the U. S. Geological Survey have such a program (301).

In 1938 Governor James Allred called a meeting to discuss the need for conservation and administrative control of ground waters (302). C. S. Clark, Chairman of the Subcommittee on Ground Water of the Texas Planning Board, subsequently drafted a recommendation that the State assume the regulation of underground waters (303, 304). Clark's committee even drew up a regulatory act, but the State did not initiate the regulation of ground waters until 1949.

Sellards' official correspondence does not include any references to pollution or any "conservation" concerns. The one exception is a letter from Phil Goodrum, Director of Wildlife Restoration of the Game, Fish and Oyster Commission, in which he reported on the progress being made in the establishment of a unified program of conservation education (305). His program emphasized the need to regulate the use of all renewable and nonrenewable resources. Goodrum asked Sellards to join a panel of experts for consultations with the State Department of Education on this conservation program. Sellards gladly participated in this program, along with representatives of the State Forestry Department, the Game, Fish and Oyster Commission, the Department of Agronomy at Texas A&M College, and the coordinator of the Soil Conservation Service (306), but this cooperative effort was short lived.

According to his correspondence, in addition to his interest in early man and vertebrate paleontology and meteorites, Dr. Sellards' main energies were directed toward resource development rather than conservation. He considered Texas to be "the one important section capable of expanding the industrial frontier of the Nation" during its times of trial (307). As a member and 1943 Chairman of the Texas Industrial and Commercial Research Council, Sellards proved to be a strong advocate of Texas industrialization. The Board of Regents created this council in 1938 in response to a request by A. B. Cox, Director of the Bureau of Business Research (308). In the council, the Regents joined together the already existing research facilities of the University — the Bureau of Economic Geology, the Bureau of Engineering Research, and the Bureau of Industrial Chemistry — to facilitate "investigations leading toward the industrial and commercial development of the State" and to increase employment and income (309). In many areas, the council replaced the Texas Planning Board, which expired by statutory limitation in March of 1939. The council's records show the scope of the wartime planning activities of the state (310).

In a letter to Professor D. B. Casteel, head of the University's committee studying research grants, Dr. Sellards declared that the Bureau of Economic Geology was prepared to do regional studies of underground waters, especially a study of the groundwater supplies of the Edwards Plateau (311). University of Texas Publication 4301 contains a paper by F. B. Plummer in which he surveyed the principal ground-water-producing formations within the state (312). This general study is the closest to a regional study made and published by the Bureau. The ground-water investigations of the U.S. Geological Survey continued to dominate this research area and are discussed below. In this same letter to Professor Casteel, Dr. Sellards also mentioned an aspect of water resource development in which the Bureau did play an important role — investigations of dam sites and "that part of the valleys of the Colorado, Red, Brazos and Pease, and other rivers that is to be flooded and permanently concealed by the building of dams during the next few years."

The rapidly expanding program of dam construction for control of flood waters and the production of hydroelectric power created a new demand for the services of the Bureau. The Federal government had established the precedents for reclamation projects, such as the Tennessee Valley Authority, and Texas soon followed. The Brazos River Construction and Reclamation District was the first attempt by Texas to establish flood control and hydroelectric projects for a major drainage system. This district, established in 1929, was permanently organized in 1935 (313). This same year Sellards assigned C. L. Baker to an investigation of the geology of dam sites on the Brazos River in Johnson, Hood, Bosque, and Hill Counties (314). The following year, the district asked the Bureau to give an opinion on the value of the oil and gas reserves underlying the land to be flooded by the construction of the Possum Kingdom Dam. Mr. Plummer supervised this study (315).

The most significant reclamation project in Texas remains the series of dams along the Colorado River. This project was launched in July 1934 at a barbecue held at the site of the future Buchanan Dam. The assembled group was the organizational committee of the Colorado Valley Authority. The committee elected an executive committee that included John A. Norris of the Board of Water Engineers as President, and Dr. Sellards (316). This executive committee drafted a bill and lobbied quite effectively using such men as U.S. Representative James Buchanan and Henry Hunt, Chief Counsel of the W.P.A. In November 1934, the 43d Legislature voted to establish the Lower Colorado River Authority and gave it the "power to control, store, preserve, use, distribute, and sell the waters of the Colorado River and its tributaries; to develop, generate and sell water power and electric energy" (317). By means of loans and grants, the W.P.A. aided the L.C.R.A. in the construction of the series of dams. Buchanan and Inks Dams were completed in 1938, Mansfield in 1940, and Marble Falls and Granite Shoals in 1949. The Authority rebuilt Tom Miller Dam for the City of Austin in 1940.

The drought of 1934 marked the beginning of a devastating succession of droughts and floods that continued until the 1940's. These natural disasters stimulated the demand for ground-water resource studies and flood control projects such as those discussed above (318). The U.S. Geological Survey, with the cooperation of the State Board of Water Engineers and the Bureau of Industrial Chemistry, undertook a study of the ground-water resources of 25 counties in the High Plains (319). From 1936 to 1939, Survey hydrologists made inventories of all wells in the area under an allocation of funds by the W.P.A. The final report discussed the recharge of the Ogallala aquifer and the general and persistent decline of the water table in the irrigation pumping districts during the previous few years. The authors did not suggest any regulatory controls but made the following comment: "With their water supply, the farmers of the High Plains are in a position similar to that of a man who has a large capital but who uses a part of his capital each year" (320).

In addition to the ground-water studies that it was conducting in Texas and in the semiarid Western States, the U.S. Geological

Survey, like the Bureau, was kept busy examining reservoir damsites and surveying the mineral resources of the Boulder Dam region. This work was done for the U.S. Bureau of Reclamation.

One final note of interest with regard to the public resource development programs during the depression years concerns the awakening to the enormous recreational opportunities that developed with the establishment of the large man-made lakes. Again, the T.V.A. took the lead in promoting public recreational facilities by donating shorelines to State and local governmental units for parks (321). In 1936, the National Park Service (N.P.S.) issued a report on "The Recreational Use of Land in the United States," and began to emphasize the recreational aspects of park lands. The N.P.S. cooperated with the Forest Service in 1938 in constructing the Appalachian hiking trail, which stretches from Maine to Georgia. The following year, the N.P.S., in cooperation with the Bureau of Reclamation and the L.C.R.A., began a comprehensive study of the recreational possibilities of the impounded waters of the Colorado River of Texas. Financing was by the W.P.A. (322).

In this connection, the growth of Texas State parks during the 1930's was noteworthy (323, 324). In 1931 there were six State parks receiving a total appropriation of $11,715. By 1939, there were 26 parks with an appropriation of $68,000. A 1936 centennial year project was the creation of 500 roadside parks (325). The N.P.S. assisted the states in the development of the State parks program. Charles N. Gould, State Geologist of Oklahoma, was the regional geologist for the N.P.S. for the State Parks Division that included Texas.

EIGHT

Dr. Sellards and the Search for Early Man

Although Dr. Sellards' organization and administration of the many facets of the W.P.A.-funded programs were indeed praiseworthy, his interest in these projects, other than the vertebrate collecting, and his concern with the geological research of the staff geologists were mainly administrative. Sellards himself became more and more interested in the search for knowledge of early man in Texas. His involvements with vertebrate collecting and the Texas Memorial Museum were offshoots of his vital concern with early man.

Something of the nature of Dr. Sellards' scientific curiosity is revealed in a birthday letter he wrote to the noted paleobotanist Dr. E. W. Berry on February 8, 1945 (326): "I wish you were out here in Texas on this day, we would certainly go some place and see something of the perpetually interesting things of this world." The following excerpt from a memorial to Dr. Sellards by his good friend Glen L. Evans illustrates his tremendous capacity for work (327):

> Dr. Sellards' devotion to work and his ability to carry an extraordinary load of work were recognized by all that knew him. . . . He served capably (from 1938 to 1945) as director of both the Bureau of Economic Geology and the Texas Memorial Museum, and at this time both institutions were especially active.

During a substantial part of that period he also directed a state-wide paleontological and mineralogical project and made many extended field trips that were required of him in that capacity. In addition to his regular duties he also served during 1938 as president of the Society of Economic Paleontologists and Mineralogists; from 1938 to 1940 as councilor of the Geological Society of America; during 1942 as president of the Paleontological Society; and during 1943 as vice-president of the Geological Society of America.

Dr. Sellards had a dignified, rather formal presence, characteristic of his generation of scientists, but beneath it lay tremendous scientific curiosity and zeal.

Dr. Sellards was born in Kentucky in 1875, and his family moved to Kansas shortly thereafter. After receiving his Bachelor's and Master's degrees from the University of Kansas, he attended Yale University, where he received his Ph.D. in 1903. The title of his dissertation was "The Fossil Insects and Plants of the Late Paleozoic of Kansas." Then he taught at both Rutgers University and Florida State University before becoming the first director of the Florida Geological Survey in 1907, a post he held until he joined the Bureau in 1918.

Dr. Sellards' interest in early man began with the discovery near Vero Beach, Florida, of artifacts in association with extinct mammals. This find indicated that the antiquity of man in America was much greater than was then generally supposed. His published accounts of these discoveries attracted such wide attention and controversy that the Journal of Geology published a symposium of papers on the topic in 1917. J. W. Gidley of the National Museum subsequently made similar finds to corroborate those of Sellards. The result was a revival of scientific interest in the search for evidence of ancient human cultures in all of North America (328). This interest remained keen in Sellards. At the time of his death in 1961 he was engaged in an extensive revision of his book *Early Man in America — A Study of Prehistory* (329).

In 1934 J. E. Pearce, Chairman of the Anthropology Department at the University of Texas, discovered artifacts in stream terrace deposits along Brushy Creek near Round Rock just north of Austin. The discovery launched Sellards into his search for the remains of early man in Texas. After studying the physiographical history of the terraces that contained Pearce's finds, Sellards reported that this important archeological discovery placed the time of the appearance of man in North America 10,000 to 20,000 years ago rather than the 5,000 to 10,000 years that had been accepted up to that time.

In 1937, Sellards delivered a paper on "The Significance of the Vero (Florida) Finds" at the International Symposium on Early Man on the occasion of the 125th anniversary of the Academy of Natural Sciences of Philadelphia. That same year a team of workers began excavating "a fossil watering hole" in Roberts County, Texas, where in December 1934 Judge J. A. Meade had discovered a spear point in association with an elephant skeleton (330). Glen Evans supervised this work, which was part of the W.P.A. mineralogical-paleontological project under the administration of the Bureau. The artifacts proved to be similar to those of the "Folsom culture," which were discovered on a tributary of the Cimarron River in Union County, New Mexico, in 1926. The "Folsom man" was considered the oldest human inhabitant of North America.

Sellards was also alerted to the discovery in 1936 of artifacts associated with vertebrate fossils along Blanco Creek east of Beeville on the Buckner Ranch. In 1938, Glen L. Evans and his coworkers began the excavation of this terrace site as part of the W.P.A. project. In 1944 Evans and Grayson Meade, vertebrate paleontologist with Texas Technological College, in the course of studies of the Pleistocene geology of the High Plains for the Bureau discovered a vertebrate locality in a gravel pit near Plainview that contained a new type of spear point associated with the bones of more than 100 bison (331). These are the three "early man discoveries" of importance that were so closely studied by Dr. Sellards (332).

Efforts to date these finds led Sellards into detailed studies of the physiographic history of alluvial terraces (333). Many archaeologists criticized the Pleistocene dating of these artifacts on the basis of association with extinct mammalian remains. They maintained that this coexistence only proved that these mammals lived into recent times. Therefore, Sellards felt that the only definite clues to the relative age of the artifacts would come from studies of the development of the terraces (334).

In 1936 Dr. Sellards corresponded with the Chairman of the Division of Geology and Geography of the National Research Council about the need for studies of the development of stream terrace deposits in the Southwest (335). He pointed out that the main objective of such studies would be the correlation of the stream terraces with terraces along the major rivers reaching into glaciated areas. This correlation would enable geologists to tie the late Cenozoic history of the streams of the Southwest into the glacial history of the North. Sellards supported W. Armstrong Price's unsuccessful application for Geological Society of America funds for the undertaking of a study of the stream and marine terraces of South Texas in an effort to correlate them with glacial periods (336).

In his 1938 G.S.A. paper on the Roberts County site, Sellards, in his reserved way, declared that the age of the artifacts had not been determined, but he could say that important faunal and topographic changes had occurred since the artifacts and the elephant remains were included in the terrace deposits (337). Sellards and Glen Evans coauthored a 1940 G.S.A. paper on the Buckner site, in which Evans wrote "Notes on the Terrace Deposits" (338). They concluded that the terraces in question were the upstream equivalents of the Pleistocene Beaumont Formation. The following year Sellards reviewed developments in terrace studies and was confident that the Coastal Plain terraces would be soon correlated with the glacial periods and would thereby afford a means of establishing the age of man's appearance in America (339). Thus Sellards' interest in early man provided a great stimulus to the important, and heretofore rather neglected, field of terrace geology (340).

Although the discovery of artifacts of early man was of vital concern to Sellards, the main focus of the work of the W.P.A. mineralogical-paleontological project from 1936 to 1939 was the collection of vertebrate fossils. As in the case of the terrace studies, Sellards' interest in early man provided a great impetus to vertebrate paleontology in Texas (341). As supervisor of the W.P.A. project, Glen Evans became the most active vertebrate paleontologist in Texas. He traversed the state checking reported vertebrate localities and supervising excavations such as those in Bee and Roberts Counties.

Grayson Meade, then a doctoral candidate at the University of Texas, was in charge of Bureau investigations in Howard County from 1939 to 1941 and made vital additions to the understanding of the Triassic and Pliocene stratigraphy of the Llano Estacado. In 1941 Dr. L. T. Patton, Chairman of the Department of Geology at Texas Technological College, upon the recommendation of Sellards, added Meade to his department's staff (342). Texas A&M College, like Texas Technological College, had also entered the field of vertebrate paleontology in 1938 by appointing Dr. Curtis J. Hesse Assistant Curator of the Museum in Charge of Vertebrates (343). In September 1940 Horace John Sawin, who had just received his doctorate from Harvard, joined the Bureau in a part-time capacity to aid in collecting vertebrate fossils in Howard County. He became a full-time Bureau geologist in 1943, joining many others in one of the prime vertebrate collecting areas in America (344).

The name E. H. Sellards looms large in the story of the creation of the Texas Memorial Museum. In 1928, in a Bureau newsletter, he spoke of the critical need for a natural resource museum and made an appeal for private endowment funds (345). Following the establishment of the McDonald Observatory agreement between the University of Texas and the University of Chicago in 1934, Dr. Sellards sent a request to the Carnegie Foundation for the Advancement of Teaching for funds to support "A Proposed Institute of Geologic Research at the University of Texas" (346). He also made the same request of the University administration. In this proposal, he noted that specimens of the rich vertebrate faunas of the state were found in leading museums all over the country, and that vertebrate fossils were still being taken from the state partly because Texas lacked any facilities for restoring and exhibiting them. Sellards envisioned that this geological foundation would channel its energies into basic research, such as vertebrate paleontology, and would thereby supplement the more economically oriented work of the Bureau. It would provide services for the geological sciences comparable to the services that the McDonald Observatory provided for the astronomical sciences (347).

It is not clear when Dr. Sellards abandoned his concept of an Institute of Geologic Research and threw his support to the drive for the creation of the Texas Memorial Museum, which could serve some of the same purposes. In 1926 Dr. Sellards had been one of the five founders of the Texas Museum Association, along with University professors J. E. Pearce, Eugene Barker, W. J. Battle, and W. N. W. Splawn (348). This group proved to be very effective lobbyists, and finally, in 1935, their campaign reached fruition. That year the Federal government, at the urging of Senator Thomas Connally, appropriated $300,000 of the Texas Centennial fund for the construction of the Texas Memorial Museum (349). Dr. Pearce, the first director of the Museum, died in October 1938, 10 days before the planned opening, and the post of director was filled by Dr. Sellards.

As director of both the Bureau and the Museum from 1938 to 1945, at which time he became Director Emeritus of the Bureau,

Sellards received his salary from the two on a 3-to-2 ratio. Carl Chelf was the first Curator of Geology at the Museum; W. D. Mateer and A. H. Witte were early collectors for the Museum and often accompanied the W.P.A. collecting parties. In 1941, at Sellards' request, Glen Evans, supervisor of this statewide project, resigned as geologist with the Bureau to become Assistant Director of the Museum. As noted, in 1945 Sellards attained emeritus status with the Bureau and devoted full time to the Museum (350).

Dr. Sellards also had a keen interest in the study of meteorites, as did V. E. Barnes, whose significant pioneer work on tektites will be discussed later (351). Of particular interest in the meteorite studies is the work done by Dr. Sellards and Glen Evans in investigating a meteor crater 9 miles southwest of Odessa, in Ector County. In the late 1920's, Sellards and others established the fact that what local ranchers thought was a "blow out" was in reality a meteor crater 600 feet in diameter. Numerous meteorites weighing up to 100 pounds were also found in and around the crater.

Sellards was successful in obtaining W.P.A. funds for making extensive studies of this crater, which not only would provide vital scientific information but also could be a "public educational exhibit" (352). Excavations were begun in September 1939 under the supervision of Glen Evans. The crater was the subject of a field excursion conducted by the West Texas Geological Society in connection with the annual meeting of the G.S.A. in Austin in the winter of 1940. The following excerpt from the mimeographed guide by Drs. Sellards and Barnes describes the stage of the investigations (353):

> Much of the crater rim has been uncovered exposing steeply tilted beds, some of which are overturned. Soil auger tests and wells show 90 feet of unconsolidated silt at the center of the crater. This silt is resting on about 75 feet of coarser material which is presumed to be the detritus that fell back into the crater shortly after the impact. A magnetometer survey, contributed by the Humble Oil and Refining Company, indicated a magnetic high slightly north of the center of the crater. Six test wells, contributed by Ector County, were drilled on the location indicated by the magnetometer survey. Two of these, nos. 1 and 6, 10 feet apart, encountered at depth 164 feet a mass which was essentially unaffected by the pounding of the 1500-pound drilling bit. This is believed to be the meteorite. Two wells encountered "iron shale" and the other two did not encounter meteoritic material. The actual size of the meteorite is undetermined beyond the fact that it exceeds 10 feet in one dimension. The observed anomaly was 15 gamma. Calculations made under direction of M. M. Slotnick indicate that a sphere of cast

iron, 18 feet in diameter, under the same conditions would produce this anomaly. It is expected that the actual size will be determined by additional drilling.

Sellards had high hopes that a shaft would encounter the main meteorite mass at a depth of 164 feet. In fact, the shaft disclosed that the wells had hit hard quartzitic concretions, and so work on laterals from the shaft was discontinued in September 1942. The Barringer family of Philadelphia, owners of the meteor crater, encouraged and financed a new program of drilling in 1958 and 1960. The results of all the explorations led to the conclusion that the crater-forming mass must have disintegrated upon impact (354).

NINE

Multifaceted Geological Research in Texas

When Dr. E. H. Sellards assumed the post of Director of the Bureau of Economic Geology in 1932, the staff included geologists F. B. Plummer, C. L. Baker, and W. S. Adkins. Plummer and Adkins, along with Sellards, were the authors of the classic stratigraphic study, University of Texas Bulletin 3232, a fitting monument to their geological acumen and range. The importance of this first volume, *The Geology of Texas: Stratigraphy*, cannot be overstressed. The following review by Edgar Owen is a good one (355):

> To co-ordinate the geological data of an area exceeding a quarter million square miles is a formidable undertaking under any circumstances. When it is considered that Texas comprises several diverse geologic provinces and probably includes more unsolved problems than any other part of the United States, the magnitude of the task increases. Added to the enormous scope of the surface evidence is the information obtained from many thousand oil wells which have been drilled in the state.
>
> One of the most serious handicaps with which the Texas geologist has to contend is the multiplicity of formation names which have attained more or less general usage, involving the literature in a network of nomenclature that almost precludes intelligent usage. In attempting to bring order out of this chaos the present volume has had to make use of 165 formation names, after disposing of a much greater number of obsolete terms. The immensity of this effort can be

better understood when it is stated that for the Upper Pennsylvanian of north-central Texas alone, 120 member, formation, and group names occurring in the literature have had to be relegated to their proper position in the geologic column. The authors have concentrated on the presentation of an orderly and comprehensive geologic history, leaving much of the detailed lithologic description to other publications.

The Bureau staff was soon altered. In 1934 Adkins left to join the Shell Petroleum Company in Houston (356). The premier Cretaceous paleontologist returned to Austin after retiring from Shell in 1950. He worked on a manual of Cretaceous guide fossils, completed his revision of Mesozoic stratigraphy, and compiled a comprehensive work on the paleontology of the Texas Mesozoic. The Bureau provided secretarial assistance for Adkins' formidable projects, and he was officially a consultant on the Bureau staff from 1952 until his death in 1956.

C. L. Baker, like his equally brilliant former associate on the Bureau staff, J. W. Beede, found himself and his family in dire financial straits with the onset of the Depression. He petitioned University President Benedict for an increase in salary, and when that was not forthcoming, he accepted an offer to fill the vacancy created by the resignation of Dr. John T. Lonsdale (357) as Chairman of the Department of Geology at Texas A&M College. Lonsdale returned to his home state to become head of the geology department at Iowa State College. Baker remained at A&M until 1943, when he parted ways, a victim of "puerile struggles for power" within the administration (358). He then joined the South Dakota Geological Survey.

Dr. Virgil E. Barnes became Research Fellow for the final year of American Petroleum Institute Project 25B on earth temperatures and oil field waters in Texas after receiving his doctorate from the University of Wisconsin. Following this, he joined the U.S. Geological Survey and in 1935 replaced Baker as resident authority on petrographic and mineralogic investigations for the Bureau. In 1936 Barnes initiated his investigations of tektites. The supervisor of a W.P.A. mineral survey in Grimes County had obtained the first samples of curiously shaped black glass, which Dr. H. B. Stenzel of the Bureau identified as tektites (359). Intrigued by tektites, Dr. Barnes began the studies that would culminate in his excellent and definitive 1973 publication, *Tektites* (360). In his now classic 1940 paper, "North American Tektites," Dr. Barnes first described the occurrence of tektites in North America (361).

In 1938 the building of Marshall Ford and Buchanan Dams necessitated a thorough exploration of the Central Mineral Region, for it was believed that with the cheap hydroelectric power, marginal mineral deposits might become economically feasible. Dr. Barnes was assigned to investigate the mineral resources of each county. Other studies going on in this same area included Dr. Stenzel's work on the Precambrian rocks of the area, the zoning of lower Paleozoic rocks by Dr. Josiah Bridge of the U.S. Geological Survey, and the mapping of the Carboniferous formations fringing the Central Mineral Region by F. B. Plummer.

In the December 1943 Bureau Newsletter, Dr. Sellards reported that he and the Bureau staff had met with H. D. Miser, Chief of the Fuels Section of the U. S. Geological Survey, and other Survey representatives to discuss an enlarged program of petroleum studies being undertaken by the Survey. These discussions led to plans for enlarging the Ellenburger studies that Dr. Bridge, Dr. Barnes, and Leo Hendricks of the Bureau had already started (362). Barnes' studies of the Cambro-Ordovician strata of this area and Hendricks' work on insoluble residue would be enlarged by the Federal survey's assigning a paleontologist, a geologist, and a laboratory assistant to this project.

Sellards had originally suggested this cooperative project to U.S.G.S. Director W. E. Wrather in a letter of September 10, 1943, in which the former noted that the Ellenburger was becoming increasingly important as a petroleum reservoir (363). Miser assigned Preston E. Cloud to undertake surface work in coordination with Barnes; A. J. Crowley joined Hendricks in studying insoluble residue of both surface and subsurface samples of the Ellenburger (364). In addition to his collaboration in the cooperative project, Dr. Barnes worked on maps and reports of Gillespie and Blanco Counties and conducted a study of the iron ores of the Llano region for the Bureau (365).

Dr. H. B. Stenzel, who received his doctorate from Schlesische Friedrich Wilhelms University in Breslau, Germany, in 1922 and was a member of the Geology Department at Texas A&M College from 1925 to 1934, joined the Bureau in 1934 as a replacement for Adkins (366). Stenzel, who had a strong reputation as a Tertiary paleontologist and stratigrapher, first completed the map and report on Leon County that he had begun at Texas A&M College (367). He then undertook similar studies in Houston County, using aerial photographs with a scale of 8 inches to the mile — the first time such detailed mapping had been done in the Texas Coastal Plain. This area is structurally important because it lies in the transition zone between the Coastal Plain and the Tyler Basin (368). Stenzel also joined Julia Gardner, who was working on the revised geologic map of Texas for the U.S.G.S., in making field investigations in the Coastal Plain, and he undertook stratigraphic studies of the low-grade iron ores of East Texas. Throughout, he continued to publish studies on Tertiary paleontology (369).

Dr. Stenzel, along with F. E. Turner and C. J. Hesse of Texas A&M College, conducted some important investigations of brackish and nonmarine Miocene beds near Burkeville, Newton County, Texas. Here they correlated subsurface and surface strata, marine and nonmarine facies, and vertebrate and invertebrate faunas. Like the Ellenburger studies cited above, "the results obtained through the geologic investigation of the Burkeville locality are of importance to the stratigraphic interpretation of well samples" (370). It is a fascinating study.

From 1933 to 1936, Professor Plummer was a full-time geologist with the Bureau during the summer only. During the rest of the year most of his time was devoted to the affairs of the Department of Petroleum Engineering. However, in 1936 when Harry Power and George Fancher were added as professors in the department, he was able to devote his full energies to the Bureau. His first task in his full-time capacity was the completion of a paper on "The Paleozoic Ammonites of Texas," coauthored with Gayle Scott of Texas Christian University (Part I of Volume III of *The Geology of Texas* series) (371).

Mr. Plummer continued to serve as the Bureau's authority on the Carboniferous rocks of North-Central Texas (372). He extended his studies down into the Central Mineral Region, where he undertook the mapping of the Carboniferous formations that fringe this region on the north and west in Burnet, Brown, and San Saba Counties. This work led to the publishing of several papers, culminating in the posthumous publication in 1950 of University of Texas Publication 4329, *The Carboniferous Rocks of the Llano Region of Central Texas*, which included a fine geologic map (373).

Volume II of *The Geology of Texas* series planned by Dr. Sellards, *Structural and Economic Geology (Exclusive of Petro-*

leum) (374), appeared in December 1935. It was accompanied by the first statewide structural map, which was printed in early 1936 (375). The discussions of structure were necessarily confined to major features, trends, and regional structural conditions. Dr. Sellards sent out five circulars to selected geologists to provide a forum for discussion of certain outstanding structural features. Although the excellent map went through two editions, the text material was eventually outdated, and the volume was never reprinted. However, like Volume I (University of Texas Bulletin 3232), this work maintained the excellent research status of the Bureau.

Some of the sections on economic geology were written by U.S.G.S. geologists, the rest by Sellards and Baker (376). Some of the outstanding economic discussions include "The Iron Ores of East Texas," by Edwin B. Eckel and Paul E. M. Purcell, and the "Texas-New Mexico Potash Deposits," by G. R. Mansfield and W. B. Lang, all of the U.S.G.S. (377). University of Texas Publication 4301, *Texas Mineral Resources*, which Sellards also planned, was not issued in its entirety until November 1946. In part it replaced some of the material in University of Texas Bulletin 3401. Some of the notable sections in this bulletin are E. H. Sellards' and Leo Hendricks' "Occurrence of Oil and Gas in Texas" and Dr. Barnes' "Earth Temperatures and Oil Field Waters of North-Central Texas," which was a final report on an A.P.I. project begun before World War II (378). Another important economic report of great merit is University of Texas Publication 4246 (379), *Building Stones of Central Texas*. C. L. Baker and George A. Parkinson, of the Bureau of Engineering Research, had begun the collecting and testing of specimens in 1933 and 1934. Parkinson continued his work on this cooperative project in 1935 and 1936. In 1936, Dr. Barnes determined the geologic setting of each sample and made extensive petrographic studies.

The wealth of oil in the Pennsylvanian and Permian formations of Texas was the principal driving force for the numerous studies of the stratigraphy of the Permo-Carboniferous strata of the state. In University of Texas Bulletin 3501, F. M. Bullard and Robert Cuyler, of the University's Department of Geology, described the stratigraphy of the Upper Pennsylvanian and Lower Permian rocks exposed along the Colorado River in McCulloch County (380). They found the section to be quite similar to the one Plummer and Moore had described in Brazos

River exposures, although there were many lithologic gradations and thickenings. That same year the Bureau published *The Geology of Palo Pinto County*, by F. B. Plummer and Joseph Hornberger, Jr. (381). It is an excellent stratigraphic study of the Pennsylvanian beds in this area and is based on extensive fossil collections. The Cooperative Mapping Committee of the American Association of Petroleum Geologists and the Bureau of Economic Geology joined in the compilation of the fine geologic map accompanying the report. Plummer had first studied this area in 1918-1920; he and Hornberger made their detailed investigations in 1932.

In the summer of 1934, the U.S.G.S. also conducted investigations of the Pennsylvanian and Permian strata in North-Central Texas using funds allocated to the Survey by the W.P.A. (382). The Federal geologists consulted freely with the staff of the Bureau and the officers of the North Texas and Fort Worth Geological Societies. Wallace Lee, senior member of the Survey team, concentrated his work on the beds of the Cisco Group along the Brazos River, and C. O. Nickell investigated equivalent strata exposed in the Colorado River Valley (383). James Steele Williams attempted to correlate the two sections on the basis of invertebrate fossils. Lee felt that these faunal correlations were "perfectly sustained by the stratigraphy." The complexity of the stratigraphy of these beds and the pitfalls of attempting fossil correlations are best described in Nickell's discussion of the Colorado Valley equivalent to the Palo Pinto Formation of the Brazos Valley (384):

> [The] exact equivalent in the Colorado River Basin has not yet been determined with complete satisfaction. For purposes of discussion, Cheney's provisional identification is herein accepted with the realization that further revision will probably be necessary.

One final work on the Carboniferous stratigraphy of the Osage Plains is *The Carboniferous Rocks of the Llano Region of Central Texas* by F. B. Plummer, which was published posthumously in 1950 (385). Plummer had studied these rocks intermittently from 1918 to 1937 and also had done detailed mapping on the western side of the uplift during the summers of 1937 to 1941. This report contains studies on the stratigraphy and paleontology of the Carboniferous rocks. The underlying beds are discussed only briefly, but the overlying Cretaceous rocks are studied more closely.

The Bureau published several outstanding monographs on the paleontology of the Pennsylvanian and Permian formations within the state. As already noted, Plummer and Gayle Scott coauthored a paper on "Upper Paleozoic Ammonites," which appeared in 1937 as Part I of Volume III of *The Geology of Texas* series (386). This paper included data on an almost complete sequence of Texas ammonites from the Upper Mississippian to the top of the American Permian. The sequence enabled the authors to propose worldwide correlations of the Permo-Carboniferous. They also included an excellent survey of previous work on these fossils.

Carl O. Dunbar of Yale and J. W. Skinner of Humble Oil & Refining Company coauthored Part II of Volume III of *The Geology of Texas* series, University of Texas Bulletin 3701, "Permian Fusulinidae of Texas" (387). They presented a checklist of some 540 described species and varieties collected chiefly in Trans-Pecos Texas, in which is found the most complete section of North American Permian. They observed that the structural complexity and rapid evolution of these large foraminifers give them peculiar possibilities as zone fossils, as do their wide distribution and numerical abundance (388).

It should be emphasized again that the amount of research funds and energies devoted to Permo-Carboniferous studies can be attributed to the fact that these beds in Texas are major petroleum reservoirs. Both the National Research Council and the American Association of Petroleum Geologists sponsored studies on the correlation of the Pennsylvanian and Permian rocks of North America. A major step toward the establishment of equivalent time-stratigraphic divisions of these systems in various geological provinces was the 1944 "Report on the Correlation of Pennsylvanian Formations of North America," compiled by the N.R.C. subcommittee headed by R. C. Moore, and "Correlation of the Permian Formation of North America," compiled by the subcommittee headed by C. O. Dunbar (389).

Dr. H. B. Stenzel was in charge of the Coastal Plain investigations for the Bureau; he made outstanding contributions to the knowledge of the geology of this area. He was a premier descriptive geologist, but at times he got so caught up in detail that he carried his passion for description to extremes. His outstanding publications during this period were *The Geology of Leon County* (390) and "The Yegua Problem." The former, which appeared in 1939, is a beautifully detailed stratigraphic study of the Claiborne and Jackson beds. Of special interest is his discussion of the depositional history of the Woodbine Sandstone, in which he contrasts the northern section in the East Texas field with the more southerly Leon County section (391). Stenzel began work on this report while at Texas A&M College and spent about 9 years on it.

In the Leon County report, Stenzel included a correlation chart on the base of the Yegua, as placed by several authors. He discussed this stratigraphic problem in detail in a paper published the following year—"The Yegua Problem" (392). He had been investigating the nature of the upper and lower boundaries of this formation over the years and finally prepared a paper on it for presentation to the A.A.P.G. meeting in March 1938. He noted that a sharp erosional break, quite important in the depositional history of the Coastal Plain, marked the top of the Yegua and concluded that the lower boundary should be placed at the top of the uniform marine fauna of the underlying Claiborne beds. The discussions are extremely detailed.

In 1937 Stenzel also began work on a geologic report on Houston County, which lies just to the east of Leon County, across the Trinity River. There was a good deal of interest in the mineral resources in these counties. Lignite deposits occur in the Wilcox beds in both Leon and Houston Counties, and the Weches beds in both counties contain extensive low-grade iron ores (393). Domes and zones of faulting also made both of these counties interesting to the oil prospector. The only publication coming out of Stenzel's years of work on Houston County, other than quadrangle maps, was a short paper accompanied by a map on "Faulting in Northwestern Houston County" (394). This neat little study was based on field studies done in 1941 and 1942 and was probably prompted by the Elkhart field just to the north.

In a 1945 report prepared for Acting University President T. W. Painter, Dr. Sellards discussed the principal results that were expected to be accomplished in the Coastal Plain region during the next 4 years (395):

1. Report on the geology and mineral resources of Houston County accompanied by a map of the county published in quadrangle form including 13 quadrangles. Most of the field work on this county is finished but there remains a large drafting job and some additional checking in the field.

2. East Texas iron ores and stratigraphy of the ore-bearing formation. This study is under way.

3. A dam site on Neches River. The necessary field work for this report has been largely completed.
4. Continuation of "Card Catalogue of North American Early Tertiary Fossils of the Gulf and Atlantic Coastal Plain." Four parts of this project have been issued.

All of these were Stenzel's projects. Unfortunately, his concern with detail and thoroughness precluded the publication of much of his very significant work in the Coastal Plain.

Other publications on the geology of the Coastal Plain that appeared in reports of the Bureau were *The Van Oil Field*, a notable report by R. A. Liddle, former geologist with the Bureau and Chief Geologist of the Pure Oil Company, Texas Division, and *The Jackson Group and the Catahoula and Oakville Formations in a Part of the Texas Gulf Coastal Plain*, by B. Coleman Renick, District Geologist for Magnolia Petroleum Company in San Antonio (396). Liddle's study of the discovery and development of this Woodbine sand reservoir overlying a deeply buried dome in the East Texas Embayment is a notable contribution to the literature of oil geology. He also included a section on the precedent-setting unitization of this field. Renick's report is a very significant regional stratigraphy study of the geology of the San Marcos Arch region.

A most important addition to the geological knowledge of the state with which the Bureau was not involved should be mentioned at this point. In March 1933, Donald C. Barton, geophysicist with Humble Oil & Refining Company, read a paper before the annual meeting of the A.A.P.G. in Houston in which he and C. H. Ritz and Maude Hickey first announced their hypothesis of the existence of a Gulf Coast geosyncline, which was based on geophysical evidence (397).

In December of 1930, Barton had presented before the annual meeting of the Geological Society of America his observation that the Coastal Plain of southeast Texas was of deltaic origin, composed of the coalescent deltas of the Pleistocene Trinity and Brazos rivers (398). He also noted that subsidence along with deposition was one of the significant features of sedimentation in this region, as demonstrated by the marginal nature of most of the Pleistocene sediments. In his 1933 paper, Barton suggested that loading by sediments might be an important cause of subsidence in the geosyncline.

Shortly thereafter, the geologists of the Louisiana Geological Survey and the School of Geology at Louisiana State University began their significant studies of the Mississippi River delta deposits. Henry V. W. Howe, Director of the School of Geology and Director of the Research Division of the L.G.S. (Cyrus K. Moresi was State Geologist), and his colleague at L.S.U., Richard Joel Russell, contributed several outstanding papers on recent sediments that pointed out the evidence for the existence of the Gulf Coast geosyncline (399). Russell was a strong proponent of the view that subsidence within the geosyncline was caused by the loading of sediments stripped from the earth's second largest degradation area (400). Once more, these tremendous contributions to the understanding of the geology of the Gulf Coast were outgrowths of investigations of the "petroliferous" sediments and structures of the Coastal Plain.

The cooperative studies of the Ellenburger strata of Texas by the U.S.G.S. and the Bureau of Economic Geology occupied a considerable amount of the energies of the Bureau geologists during the Sellards years. In 1912 Sidney Paige of the U.S.G.S. had named and described the Ellenburger Group in his investigations of the Llano and Burnet Quadrangles in the central part of the Llano region, or the Central Mineral Region (401). In 1930 C. L. Dake of the Federal survey spent 6 weeks in the field studying the Ellenburger. Josiah Bridge joined him in 1932 in making the first zonations and faunal correlations of the Ellenburger and other older Paleozoic beds with equivalents in other states (402). Their work formed the cornerstone for all subsequent Ellenburger studies.

Aided by a grant from the Penrose Bequest of the G.S.A. in 1937, Bridge expanded his studies of the lower Paleozoic on the western side of the Llano region, and subsequently Dr. Barnes joined him in these investigations (403). In 1943, therefore, when the geologists of the Bureau and of the U.S.G.S. began their collaboration on the Ellenburger, they found that their job was to introduce refinements to a problem already generally understood. Preston E. Cloud and Josiah Bridge of the Federal survey and V. E. Barnes of the Bureau published a progress report on the "Stratigraphy of the Ellenburger in Central Texas" in the winter of 1946, and Cloud and Barnes authored a final report, which appeared in December 1948 (404). It was "a basic study of Ellenburger rocks with emphasis on features of possible significance to the search for new sources of oil." In short, it is a brilliant stratigraphic study that includes regional correlations and "lithologic and paleoecologic speculations." Leo Hendricks of the Bureau and A. J. Crowley of the U.S.G.S. supplemented the surface work of Barnes and Cloud with insoluble residue studies of surface and subsurface Ellenburger samples in the Llano area and downdip to the north. In 1940 Hendricks had divided the Ellenburger found in wells of North-Central Texas into zones on the basis of the character of the chert found in the samples. This work appeared in University of Texas Publication 3945 (405). Later, in 1945, he and Crowley coauthored "Lower Ordovician and Upper Cambrian Subsurface Subdivision in North-Central Texas," which subdivision they based on insoluble residues of typical Ellenburger beds from the Llano region to the Red River (406). In 1946, Crowley published the results of a study of a more densely drilled area, the K.M.A. field in Wichita County, which in fact was a practical application of the 1945 results (407).

The vertebrate fossil collecting that was a major part of the mineralogical-paleontological project funded by the W.P.A. stimulated interest in the Triassic and Cenozoic strata of the High Plains in the late 1930's. These stratigraphic studies were launched when Grayson Meade, supervisor of the W.P.A. project in Howard County, found a vertebrate locale southeast of Big Spring in which he discovered numerous remains of a Triassic reptile—*Trilophosaurus*—in the Dockum beds. Dr. Joseph T. Gregory, of the W.P.A. vertebrate laboratory in San Antonio, later described this fossil in a paper published in University of Texas Publication 4401, and Dr. H. J. Sawin described the amphibians from the same locality (408).

From 1940 to 1944, Glen Evans and Grayson Meade conducted field studies of the Quaternary deposits and physiography of the High Plains for the Bureau. They presented their results in "Quaternary of the High Plains," which was included in University of Texas Publication 4401 (409). After making detailed studies of the Blanco beds of Crosby County, they found conclusive evidence that these deposits were laid down in a lake basin rather than in aggrading stream valleys, as postulated by C. L. Baker and W. D. Matthew (410):

> It became apparent that basins similar to those occupied by modern playas have been forming and filling on the Texas plains throughout most if not all of Quaternary time, and that the lake deposits if properly interpreted and related to the main stages of canyon cutting, valley filling, and deposition of the eolian sand sheets might establish a Quaternary chronology unexcelled outside of the glaciated regions.

Baker had also advanced the hypothesis that the large volumes of material represented by the recent basins of the High Plains had collapsed into subterranean caverns within the salt and other soluble rocks of the underlying Permian formations. However, Evans and Meade suggested that deflation during relatively arid stages had been the primary force in the origin and modification of most of the basins (411):

> Deepening by wind action was made possible by a lowering water table. Lateral expansion of the playas proceeds during relatively humid substages after deepening has ceased at the water table. Lateral enlargement of the playas is now taking place and is effected by a combination of wave action and deflation.... As the waters lower, the freshly eroded materials become exposed and are attacked by the wind and the finer particles are removed.

They postulated that the 60 to 70 feet of the older Blanco beds must have been deposited during a relatively humid cycle of considerable duration, perhaps a glacial stage. Meade's studies of the Blancan fauna disclosed that though the fauna contained Pliocene forms, its Pleistocene genera constituted a fossil assemblage that indicated an earliest Pleistocene age for these beds. Accordingly, they considered the Blanco beds to be of Nebraskan age (412).

Bureau attention to Cretaceous studies waned, and the only articles dealing with the Cretaceous that the Bureau published after the departure of Adkins were paleontological studies. The most significant of these was Dr. L. W. Stephenson's *The Larger Invertebrate Fossils of the Navarro Group of Texas,* which appeared in 1941, some 13 years after Dr. Sellards had first suggested that the author undertake this study (413). Dr. Stephenson noted that the Cretaceous-Tertiary unconformity represented an important time interval in the Gulf region during which revolutionary changes of the marine life of the Gulf region occurred. During this break, the strandline retreated to the relatively steep outer slope of the continental shelf, throughout much of the length of the Atlantic and Gulf Coastal Plain (414).

In resurveying the tremendous additions to the knowledge of the geology of Texas that came through the pens of Bureau geologists during the years 1933-1945, it becomes quite clear that most of the research efforts were focused on the primary oil-producing horizons: the Pennsylvanian and Permian beds of North-Central Texas and the Permian Basin, the Ellenburger strata exposed in the Llano region and producing in the subsurface of the Osage Plains, and the Tertiary beds of the Coastal Plain. One of the primary functions of a state geological organization is to assist private enterprise in the development of the natural resources of that state. Increasing the knowledge of the geology of a state through basic geological research is a necessary prerequisite to fulfilling this function. As shown in this chapter, the Bureau of Economic Geology concentrated its research energies in those regions of special interest to oil geologists. Some critics suggest that the Bureau should have spread its research energies evenly throughout the various geological provinces of the state, showing no favoritism to any one mineral resource. However, if a lot of research had been economically unproductive, appropriations would have been jeopardized. Also, more energies could have been pumped into public service programs. These shortcomings were erased in the 1970's under Director Peter T. Flawn.

During the frantic years of the Depression and World War II, the cooperation between the U.S.G.S. and the Bureau improved tremendously. The joint Ellenburger study discussed above is solid evidence of this. In part, this new thrust of cooperation was the result of meeting national needs shoulder to shoulder, but the vital new spirit of the Federal survey was also an outgrowth of changes in administration—Hugh Miser's takeover of the Section of Geology of Fuels and W. E. Wrather's appointment as director to succeed W. H. Mendenhall in April 1943. By 1938, correspondence between the U. S. Geological Survey and the Bureau demonstrated a solid spirit of communication and cooperation (415).

The Federal survey's primary thrust during the years 1933-1945 was to aid with the national problems of economic recovery and World War II. In the 1930's, the Government allocated funds to be used by the U.S.G.S. for mineral resource and land classification studies that would help stimulate economic recovery, and during the war the development of strategic and critical mineral resources received top priority.

Congress made general economy cuts that reduced the Geological Survey appropriations for 1933-1934, but the W.P.A. allocated funds for programs that completely absorbed the energies of the Survey geologists (416). Early in 1934 the W.P.A. earmarked $276,000 for mineral resource and land classification studies in 19 states east of the Rockies, with a stipulation that

$34,500 be spent in Texas. The U.S.G.S. projects set up under this grant included studies of the Terlingua and Shafter mining districts, investigations of the "brown" iron ores of northeast Texas (Cass, Marion, and Morris Counties), analyses of the bleaching clays south of San Antonio, and research on the oil, gas, and coal reserves of the Osage Plains.

Clyde P. Ross, Professor at the College of Mines and Metallurgy at El Paso, Texas, took charge of the studies of the mining districts. He began work in February of 1934, with the assistance of W. E. Cartwright and C. H. Caldwell. Their preliminary report, prepared at the close of the field season, appeared in University of Texas Bulletin 3401, and Ross' final report on *The Geology and Ore Deposits of the Shafter Mining District, Presidio County, Texas*, appeared in 1943 as U.S.G.S. Bulletin 928-B (417). It included a reconnaissance map of the region around Shafter and a detailed geologic and topographic map of the eastern part of this mining district.

Edwin B. Eckel of the U.S.G.S. conducted extensive field work from April to November 1934 exploring the iron ores of Cass, Marion, and Morris Counties. His report, *The Brown Iron Ores of Eastern Texas*, published in 1938, was very comprehensive and included maps of parts of the counties explored showing the areal geology and distribution of iron ore deposits (418). He reported that ores of commercial grade were confined almost entirely to the outcrops of the Weches Formation and were derived from glauconite by ordinary weathering processes.

While working on the clay deposits of the San Antonio area in 1934, Philip King continued his studies of the geology of the Trans-Pecos area (419). He spent 8 months in the Guadalupe Mountains in 1934 and 1935, during which his field expenses were paid by a grant from the Penrose Bequest of the G.S.A. Publication of the results of his work was delayed by the outbreak of the war; *Geology of the Southern Guadalupe Mountains, Texas* did not appear until 1948 (420). King published many other fine reports on the Trans-Pecos during the years 1933-1945.

The cooperative program between the U.S.G.S. and the Bureau on studies of the Ellenburger has been mentioned, as has the W.P.A.-funded work by the geologists of the Federal survey on the Pennsylvanian and Permian beds of the Osage Plains.

The year 1937 marked the publication of the new geologic map of Texas compiled by N. H. Darton, Julia Gardner, and L. W.

Stephenson of the U.S.G.S. (421). Miss Gardner prepared the eastern part of the map; Dr. Stephenson, the Upper Cretaceous of the east-central area; Dr. Darton, the information on the outcrops of the Upper Cretaceous and older formations (422). Darton exhibited this map at the annual meeting of the G.S.A. in Cambridge in December 1932 for criticism by local geologists. He also circulated the map among geologists in or near Texas "who might be able to supply corrections or betterments" and also invited them to Austin for a critique of the preliminary edition.

In an all-out effort to discover and develop new sources of minerals vital to the war effort, the U.S. Bureau of Mines launched a program of exploratory testing and establishment of pilot production plants. In 1943 Congress appropriated funds for these programs with $65,000 earmarked for work in Texas. Also in 1943, the Bureau of Mines began exploratory work on fluorspar deposits, which had been investigated in depth in the Eagle Mountains of Hudspeth County by Glen Evans of the Bureau of Economic Geology (423). Since the results of trenching projects were good, the Bureau of Mines core-drilled the area. Both Evans and A. B. Needham, District Engineer for the Bureau of Mines, cooperated closely in studying these fluorspar deposits, which could provide another source of this mineral so needed in the production of steel and aluminum (424). Unfortunately, no commercial fluorspar production of note resulted from these cooperative efforts between the Bureau of Economic Geology, the Bureau of Mines, and the U.S.G.S.

A similar cooperative undertaking was the exploration of the magnetic iron ores of the Llano region conducted from 1942 to 1944 (425). Needham again headed the Bureau of Mines contingent, and V. E. Barnes represented the Bureau. Three prospects were selected by the Bureau of Mines for exploration, but none proved commercial. Barnes, Samuel S. Goldich of the U. S. Geological Survey, and Frederick Romberg of Lacoste and Romberg, Inc., wrote a definitive report on these iron ore deposits in 1949, which the Bureau published as Report of Investigations No. 5 (426).

Eckel's report on the iron ores of eastern Texas was followed by Bureau of Mines experiments in 1943 at a semicommercial plant at Longview that prompted the Madras Steel Corporation to build a pilot plant to reduce iron ore to sponge iron using natural gas rather than coke and limestones. The sponge iron was converted into steel ingots by the construction of a commercial plant in 1949, as large-scale production of East Texas ores began after the war. In 1947 ores from Morris County were being processed in the Lone Star Steel plant at Daingerfield, built in 1943 with Federal money (427).

In the fall of 1944, Dr. Sellards corresponded with then Congressman Lyndon B. Johnson about the possibilities of Texas' receiving a congressional appropriation for "the construction and operation of demonstration plants to produce liquid fuels from coal, oil shale, agricultural and forestry products . . . in order to aid the prosecution of the war and conserve and increase the oil reserves of the nation" (428). The initial $5 million grant for this 5-year Federal synthetic fuel program was to be used to construct and operate laboratories and demonstration plants under the auspices of the U.S. Bureau of Mines. Sellards suggested that the government plant northwest of Austin, in which Ellenburger dolomites were converted to magnesium, and which was scheduled for shutdown October 15th, could easily be converted into an ideal pilot plant for processing oil shale (429). Sellards felt that the facilities could also serve for testing ores and research in alloys and mineral products (430).

Finally, the Bureau played a role in the program established by the Committee for Cooperation with American Republics in 1941, which outlined plans for the U.S.G.S. to undertake examinations of deposits of strategic and critical minerals in Brazil, Mexico, Cuba, and other Latin American countries. This program, sponsored by the Interdepartmental Committee for the Scientific and Cultural Cooperation of the U. S. Department of State, also stipulated that the U.S.G.S. and Bureau of Mines give counsel on the development of any important mineral deposits found (431). Dr. Stenzel participated in this program in the capacity of geologic adviser.

The most important development in the area of nonpublic research during the 1933-1945 period was the establishment of the Penrose Bequest for the G.S.A. in 1931. When R.A.F. Penrose, Jr., died in July 1931, at the close of a very successful career, he left several million dollars to the American Philosophical Society and to the G.S.A., the income from which was to be used at the discretion of these organizations (432). The G.S.A. used its portion of the Penrose Bequest to support geological research and to strengthen the publication outlets of the society.

The official biography of Penrose, written in 1951 for the G.S.A., showed that through that year, the society had issued 649 grants-in-aid for a total of $989,488 (433). Thus, the Penrose Bequest provided vital funds for basic geological research during the years when public geological research energies were being devoted mainly to stimulating economic recovery, easing unemployment, or aiding the war effort. Some of the significant works made possible by the funds of the bequest include F. P. Shepards' work in "Physiography, Current Dynamics, and Sedimentation in Selected Areas of the Sea Floor of the California Coast," a pioneer oceanographical study ($10,300), and studies of the eruption of the volcano Paricutin in Mexico under N.R.C. supervision ($15,000) (434).

Before and during the period of his directorship of the Bureau of Economic Geology, Dr. Sellards was an extremely active member of the Geological Society of America. He was a Councilor, a member of the Membership and Budget Committees, a representative of the society on the Division of Geology and Geography of the National Research Council, and in 1942 was Third Vice-President of the Society. In December 1940, the Bureau hosted the 53d Annual Meeting of the G.S.A. in Austin. This was an extremely exciting event for the Bureau staff.

In his report on the work of the Division of Geology and Geography of the N.R.C., which appeared in the Proceedings of the G.S.A. for 1934, Dr. Sellards commented that the main work of the technical committees in its best form was coordinating research (435). He found that their reports actually provided only a useful digest of work done during the year in a particular field. There were some outstanding reports, such as the Report of the Committee on Sedimentation (Parker Trask, Chairman) and the Report of the Committee on Paleoecology (W. H. Twenhofel, Chairman), which appeared in 1936 (436). For the most part, however, the main successes of the research efforts of the N.R.C. were accomplished in conjunction with the research branches of the G.S.A. and the A.A.P.G., as will be shown below.

In fact, the effectiveness of the N.R.C. as a stimulator of research was sharply curtailed when the Rockefeller Foundation terminated grants of research monies. From 1929 to the end of 1933, the Foundation made grants totaling $351,198 in all of the sciences represented in the N.R.C. The figure for 1934 was only $48,400, which included an allotment of $7,595 for the

Division of Geology and Geography. Finally, the Foundation gave $80,000 for the years 1935-1937, which was to be regarded as a terminal appropriation (437). From that time on, the only research monies available from the N.R.C. came in the form of postdoctoral fellowships.

F. B. Plummer of the Bureau was active in the affairs of the A.A.P.G., serving on the Research Committee during 1933 to 1938 and 1940 to 1942 (438). This committee produced positive results. One outstanding research project sponsored by the Association was the special volume on *The Permian System of the United States West of the Mississippi River*. This extensive project had been suggested at the El Paso meeting of the A.A.P.G. in 1939. Part I of this volume was published in the Bulletin in 1940 and Part II in 1942 (439); both were edited by R. K. DeFord of Argo Oil Company. In 1941 the Research Committee extended financial aid in support of a study of the relations of crude oil to stratigraphy ($1,500) and for the completion by the G.S.A. of the revision of the Geologic Map of North America ($1,000). In 1942 aid was given for the completion and publication of the Tectonic Map of the United States, which was a project of the Division of Geology and Geography of the N.R.C. ($1,000) (440).

The Research Committee of the A.A.P.G. also tried to encourage research studies in which the Association served only in the role of coordinator (441). In 1935, D. C. Barton, Chairman of the Research Committee, wrote Joseph M. Dawson, President of the San Antonio Geological Society, asking that San Antonio geologists participate in a cooperative study of the occurrence of oil and gas in relation to variation of formational facies in the Tertiary of the Gulf Coast (442). Dawson replied that his society's geologists were too busy with their own affairs and suggested to Barton that the Bureau of Economic Geology was the logical organization to conduct this research (443).

The project was shelved until 1943, at which time John R. Sandidge of the South Texas Geological Society wrote to University President Homer P. Rainey, urging that the Bureau assume the responsibility of coordinating and disseminating the data gathered by the local geological societies and the A.A.P.G. for the benefit of all Texas geologists (444). Sandidge referred to the Tertiary study suggested by Barton and noted that Dr. Stenzel was the best qualified man to supervise these studies. Subsequently, Sellards turned to the U.S.G.S., which was conducting its own research on petroleum geology, and asked Director Wrather to consider certain projects on which the Bureau and the Survey could cooperate, with a preponderance of Federal funds (445). One project that he submitted for consideration was a detailed surface-subsurface stratigraphic study of the Frio, Catahoula, and marine Oligocene strata, based on 200 to 500 core holes. Dr. Sellards also requested Federal funds for Stenzel's work in Houston County. The Bureau director was also the first to suggest the comprehensive cooperative Ellenburger studies mentioned earlier. Wrather heartily endorsed the Ellenburger project, but felt that there was not enough priority for the Frio or the Houston County studies (446).

The many research projects on the Permo-Carboniferous beds of the Osage Plains and the Trans-Pecos illustrate the good cooperation between the various private geological research organizations. In 1933 Dr. Sellards petitioned the G.S.A. requesting a grant-in-aid from the Penrose Bequest to sponsor detailed studies of these strata (447). In 1934 the society did grant $3,000 for "Group Study to Establish More Satisfactory Stratigraphic Correlation of the Mississippian, Pennsylvanian, and Permian Sediments in the Texas Region" (448). F. B. Plummer and Robert King began studies of the macrofaunas of these strata and Helen Jeanne Plummer studied the microfossils. Dr. Sellards also sought the cooperation of all of the geological societies in these West Texas areas in drawing cross sections across the Permian Basin. P. B. King of the U.S.G.S. and C. L. Baker and Dr. Sellards of the Bureau examined the Marathon exposures to search for clues to the placing of the Pennsylvanian-Permian boundary (449). At the same time, King had a $1,400 grant from the G.S.A. for undertaking a detailed study of the southwest Guadalupe Mountains.

Sellards corresponded with Carl O. Dunbar, Chairman both of the Committee on Stratigraphy of the N.R.C. and of the Subcommittee on the Permian, and with R. C. Moore, Chairman of the Subcommittee on the Pennsylvanian, on the progress of the studies undertaken by the Bureau under the G.S.A. grant (450). In 1932 W. H. Twenhofel, Chairman of the Division of Geology and Geography of the N.R.C., had inaugurated the Committee on Stratigraphy. Subcommittees were established for each geologic system and assigned the task of summarizing the knowledge of the stratigraphy of that system in North America (451). Concurrent with these studies, A. I. Levorsen, Chairman of the

Research Committee of the A.A.P.G., appointed a subcommittee on the Permian "to prepare a digest and recommendations for the classification of the Permian west of the Mississippi." As noted previously, R. K. DeFord was the editor of this proposed Permian volume.

One of the outgrowths of these many interrelated studies was the publication in 1960 by the Permian Subcommittee of the N.R.C. of the "Correlation of the Permian Formations of North America" (452). The preliminary draft of this paper had been made in 1940, "but because of many unsettled problems its publication was long delayed." This paper is a tribute to the splendid, unselfish cooperation among all geological research organizations, both public and private, that characterized the cooperative spirit of the Depression and war years.

TEN

The Bureau and Resource Management

Lt. Col. John T. Lonsdale officially accepted the post of Director of the Bureau of Economic Geology on June 7, 1945 (453), and served in that capacity until his sudden death in October 1960. Dr. Lonsdale achieved the rank of Lieutenant during World War I, became a Lieutenant Colonel during World War II, and was an active reservist. He was a staunch patriot, and in recognition of his services to his country, his ashes were interred at Arlington National Cemetery.

He was a "hard rock" geologist who found his real challenge in exploring the complex igneous geology of the Big Bend area. He began these studies while still Chairman of the Geology Department at Texas A&M College, which he helped establish. In addition to this work, the University of Texas faculty committee that wrote his memorial for the American Association of Petroleum Geologists Bulletin cited that one of his greatest achievements as Bureau director was the initiation of a program of surveying the state's mineral resources, concentrating on minerals other than oil and gas.

Though the Bureau did produce outstanding pieces of research work during his tenure, the accomplishments could have been even greater had it not been for Lonsdale's enigmatic protection of his projects. He was a gentleman of the first order, and a very respected scientist, but he tended to run the Bureau as if it were his own.

As director, Lonsdale was involved with assisting in the industrial development of the state, and the Bureau's numerous studies of the mineral resources of particular river drainage basins were intended to provide a basis for predicting future industrial growth and resultant demands on water resources. For example, Lonsdale was a member of the Natural Resources Division of the Houston Chamber of Commerce, whose goal was to increase the area's industrial growth. The middle name of the Bureau is *Economic*, and this category has always been one of the central concerns of the Bureau directors.

In 1945 the U.S. Army Engineers requested that the U.S. Geological Survey report on the mineral resources within the drainage basin of the Trinity River. Such a report would provide a basis for appraising the potential tonnages of mineral raw materials and products that might be handled as a result of the construction of a navigable 9-foot canal along the upper Trinity from Liberty to Fort Worth (Congress had already authorized the construction of a canal from Liberty to Houston). The Bureau agreed to cooperate with the U.S.G.S. in this program, and Dr. Lonsdale assigned H. B. Stenzel and H. C. Fountain to the task of writing the descriptions of certain of the important mineral resources of the area. A minimum of field work was done specifically for the 1948 report, which was published by the Bureau — *Geological Resources of the Trinity River Tributary Area in Oklahoma and Texas.* A. E. Weissenborn directed the compilation by U.S.G.S. geologists of mineral resource data in the Oklahoma area (454).

Shortly thereafter, the Bureau undertook another study of industrial minerals in the eastern half of Texas to provide figures on present and future industrialization for use in long-range water-use planning. This work was done in connection with surveys of the mineral resources of the Canadian and Red River basins. In 1953 President Eisenhower directed the establishment of the Arkansas, White, and Red River Basins Interagency Committee to coordinate the work of Federal and State agencies in making a comprehensive survey of the water and related land resources of this large midcontinent basin area. Dr. Lonsdale was a member of the Geology and Mineral Committee, whose report outlining the resource development potentialities for the next 25 years appeared in June 1955 (455).

Joseph M. Perkins was recruited by Dr. Lonsdale in 1954 under a temporary appointment as principal geologist in charge of a survey of the mineral resources of the Texas coastal area. The goal of this project, which was directed by Dr. Lonsdale under a $10,000 contract for the U. S. Bureau of Reclamation, was to obtain accurate figures on important mineral commodities in their relation to present and future industrialization for use in long-range water-use planning. The Bureau of Reclamation was undertaking a series of studies of water conservation in Texas that included present and future requirements for municipal, irrigation, and industrial development. Lonsdale and Perkins coauthored a preliminary report published by the Bureau in 1955 as Mineral Resource Circular No. 38 — *Mineral Resources of the Texas Coastal Plain* (456).

In 1957 the Bureau accepted a grant of $1,650 from the Colorado River Industrial Development Association to conduct a mineral resource survey of a 10-county area from San Saba County to Matagorda County. John W. Dietrich headed this survey. His report — *Mineral Resources of the Colorado River Industrial Development Association Area* (457) — was published by the Bureau in 1958. In 1960, the Bureau received a grant-in-aid of $5,775 from the Corpus Christi Chamber of Commerce to conduct a survey of the mineral resources of the Corpus Christi trade area. Ross A. Maxwell supervised this project. His report — *Mineral Resources of South Texas: Region Served Through the Port of Corpus Christi* — appeared in 1962 (458).

These five mineral resource surveys conducted by the Bureau were parts of Federal and private projects aimed at anticipating future industrial growth and resultant future water supply demands. They were resource management programs, the first attempts at planning or at least predicting resource or industrial development. With the exception of the W.P.A. mineral resource studies during the 1930's and 1940's, the aforementioned projects marked the entry of the Bureau into resource planning programs. Today, these programs receive a significant part of the Bureau's energies.

The State's entry into the area of planned resource development was precipitated by the crippling drought that gripped over half of Texas from 1949 to the mid-1950's. Such organizations as the Texas Water Conservation Association, "dedicated to the task of developing, conserving, controlling, protecting, and utilizing the water resources of Texas," were active before the drought (459). The T.W.C.A. sent out monthly mimeographed bulletins that included coverage of legislative activities pertaining to water. Dr. Lonsdale was a member of the Mapping Committee of this organization. However, no lobbying organization could match the powerful impact that the drought had on alerting the lawmakers to the need for controlled resource development. Such works as Walter Webb's *More Water for Texas: The Problem and the Plan* were indicative of the growing strength of the drive for State control of water resources (460).

The first positive step toward State control of water resources came in 1949 with the passage of the Texas Underground Water Conservation Act, discussed earlier (461). In 1952 and 1954, the School of Law at the University of Texas sponsored conferences on water law. Experts from the faculty of the Law School were joined by experts from the U.S.G.S. and the Texas Board of Water Engineers, and by leaders of the water conservation movement, such as Guy Jackson, Jr., President of the Texas Water Conservation Association, and Senator John Bell.

Then the final catalytic agent was introduced — the worsening of the drought in 1956. The Legislature was prodded into action and passed the Water Planning Act of 1957, which created the Texas Water Resources Planning Division within the Board of Water Engineers (462). The legislators called for an inventory of all surface and underground water resources and authorized the Water Resources Planning Division to enter into contracts with any public agency to carry out a joint program of topographic and geologic mapping of the watersheds of the state and to compile all of the geologic and hydrologic information on underground water reserves. The absence of Bureau involvement is conspicuous, as this act stipulated that three geologists be on the staff of this Planning Division. Dr. Peter T. Flawn, the fifth director of the Bureau, observed in his paper *Geology in the State Government of Texas* that as of 1965 the Texas Water Development Board (the new name of the Board of Water Engineers) employed about 40 geologists (463). This trend toward building separate geological staffs within the various State agencies is discussed later in this chapter.

It is interesting to note that the only energies that the U.S.G.S. funneled into programs of planned resource use and development in the post-war decade were water resource studies. In his 1946 annual report, Geological Survey Director Wrather declared that the Water Resources Branch had been expanding studies of the availability and chemical quality of water resources as a necessary prerequisite to the successful planning and construction of projects related to irrigation, power production, industrial installations, municipal supplies, and flood control. In 1948 Congress enacted a 5-year program of grants to states for pollution studies, and thereafter the pollution work of the Water Resources Branch picked up sharply (464).

As the drought continued to grip many of the southwestern states, the need for a coordinated water-use policy became more and more obvious, and a major contribution toward the meeting of this need was the report of the President's Advisory Committee on Water Resources Policy. This report, prepared for President Eisenhower, was a major contribution toward the development of a national water conservation policy. In 1965 Congress passed the Water Resources Planning Act, which created the Water Resources Council at Cabinet level (465). For the first time, the Federal government had an instrument for coordinating a comprehensive planning policy for use of water resources (466). It is noteworthy that this action was followed by 8 years of similar action by the State government of Texas. In spite of the fact that the U.S.G.S. worked closely with the Texas Board of Water Engineers in water resource studies, neither the Federal survey nor the Bureau of Economic Geology dealt with land-use studies prior to the 1960's, although it is obvious that the topographic and geologic maps the two produced would be major tools in any resource planning program.

Periodically, paleontologists, stratigraphers, and economic geologists of the U.S.G.S. were at work in Texas on such diverse research projects as the conodonts of the Llano region, quicksilver mining in Texas, and the lower Paleozoic stratigraphy of West Texas. However, in the postwar decade-and-a-half, two Texas projects received special attention by the Federal survey. Both were concerned with energy sources — oil and uranium, specifically.

In light of their irreplaceable nature and their limited volume, minerals and fossil fuels are fundamental to our mechanical and industrial civilization, and should be used with zealous care. These resources received a tremendous and unexpected drain in World War II. With the onset of the Cold War and the Korean

War in particular, the annual reports of the Secretary of the Interior and the Director of the Geological Survey became increasingly filled with statements such as the following taken from the 1951 report of Secretary of the Interior J. A. Krug (467):

> This nation is engaged today in an entirely new kind of defense program. . . . Our survival as a people may well depend on our ability to turn our great potential strength into military striking power on short notice.

The Secretary went on to state that an adequate supply of electrical power was at the very heart of the defense program, and that all fuels were especially vital. He noted that the U. S. Bureau of Mines and the Geological Survey were hard at work at the problem of augmenting this fuel supply (468):

> In Western Texas, where large oil and gas reserves have recently been discovered, investigation of the occurrence and distribution of reefs in the oil bearing rocks has been started under the sponsorship of the Petroleum Administration for Defense.

The Secretary was speaking of studies of oil-bearing reefs in the Pennsylvanian rocks of the Scurry County area of West Texas. The need for surface and subsurface studies of these strata in an effort to explain the causes of reefing had been discussed several years earlier in a letter from Dr. Lonsdale to Dr. Harold Wanless of the University of Illinois, who was the recipient of a Geological Society of America grant for the study of regional variations in Pennsylvanian lithology (469). In 1949, there was a symposium on reefs at the annual meeting of the A.A.P.G. in St. Louis. Obviously, Lonsdale and U.S.G.S. representatives discussed matters because Watson Monroe, the Survey's Fuels Branch representative in Texas, had corresponded with Lonsdale about a cooperative study of Carboniferous stratigraphy and reefing in Texas between the Federal and State organizations (470).

D. Hoye Eargle and W. A. Heck of the Geological Survey began work on this project in the summer of 1949. In 1952, after coauthoring the first published report of this project—*The Boundary of the Pennsylvanian and the Permian in the Subsurface Scurry Reef, Scurry County, Texas* (471)—Heck resigned to establish a service company in Midland. The major published report of this Geological Survey reef study project (the Bureau did make a small financial contribution in addition to extensive consulting assistance)—*The Geology of the Late Paleozoic Horseshoe Atoll in West Texas,* written by Donald Myers, Philip

Stafford, and Robert Burnside—appeared in 1956 as University of Texas Publication 5607 (472). About this time a shift in the emphasis of this project was noted in a letter to Lonsdale from John A. Reinemund, of the Mid-Continent office of the Fuels Branch (473):

> The program of investigations of Pennsylvanian and Lower Permian rocks in Texas, into which the Geological Survey has now poured more than a quarter of a million dollars during the last five years, and out of which the principal product has been the comprehensive bulletin on the horseshoe atoll . . . is now shifting from a dominantly subsurface study of the atoll to a dominantly surface mapping and stratigraphic study that will provide correlations and data on ancient environments for use in future subsurface studies.

Survey geologists Myers, Stafford, and Eargle were involved in these surface studies in 1960 (474).

In this same letter, Reinemund declared that this detailed mapping of the Pennsylvanian and Permian rocks in North-Central Texas could be directly incorporated into a new geologic map of Texas. He added that he thought this should be the next major objective of a cooperative program between the Survey and the Bureau. Reinemund also noted that the explorations of the uranium ore potential in Texas supported by the Atomic Energy Commission were providing additional stratigraphic information that would also be useful in putting together a new state map. The nucleus of the staff for this project would be the geologists of the Fuels Branch who had gained considerable competence in Texas geology.

Lonsdale had made his thoughts on this subject clear in a letter to Chief Geologist W. H. Bradley of the Geological Survey on May 6, 1954 — "obviously the preparation of a new geologic map is beyond the resources of the Bureau and logically the Geological Survey should again handle this project [as in the case of the 1933 state map]" (475). Furthermore, in February 1956 the Geological Survey transferred Reinemund and his group (with the exception of Eargle) to Denver and thereby ended talk of any cooperative mapping project (476). However, Lonsdale seems to have had a change of heart, for Dr. Peter Flawn, in a discussion of the Texas Geologic Atlas Project in the Bureau's annual report for 1961, noted that Dr. Lonsdale had initially conceived this statewide geologic mapping program. This Geologic Atlas Project was launched in September 1961 (477) with V. E. Barnes as supervisor; this extensive project was supported by monies from oil companies and professional geological groups. Many oil companies have also contributed geologic data of enormous dollar value. These maps attest to the high quality of the geologic and cartographic personnel of the Bureau of Economic Geology.

The U.S.G.S. investigations of uranium ores in South Texas for the Atomic Energy Commission (A.E.C.) also received high priority because of concern with the national defense effort during the Cold War (478). The Geological Survey had maintained a laboratory at the University of Texas Balcones Research Center from the inception of the reef study in 1949, and in February 1953 this laboratory also became the base of operations for field studies of radioactive rocks and uranium-ore deposits in Texas. D. Hoye Eargle assumed supervision of investigations of uranium-bearing rocks for the Geological Survey (479). Eargle investigated the radioactive Eagle Ford clays of West Texas and then started to study the stratigraphy of Karnes and adjacent counties where uranium ores were discovered (480). Eargle's detailed stratigraphic studies were directed toward discovering (as an aid to prospecting) the physical and chemical phenomenon that controlled the deposition of uranium minerals.

The first published results of these studies appeared as the Bureau's Report of Investigations No. 17, *Preliminary Report on the Stratigraphy of the Uranium-Bearing Rocks of the Karnes County Area, South Central Texas* (with coauthor J. L. Snider, 1957) (481). Eargle found the ores occurring as disseminations in the sands and tuffaceous sands of the Eocene Jackson Formation, the Miocene Catahoula tuff, and the Pliocene Goliad sand. The authors postulated that "the source of the uranium was in the tuffaceous sediments and the uranium, leached from the tuffaceous sediments, was concentrated by evaporation or by interaction with clays and carbonaceous or other materials where ground waters were concentrated in stratigraphic or structural (fault) traps or channels" (482).

In May 1955, the Division of Raw Materials of the A.E.C. established an Austin sub-office for the evaluation of the uranium production potential of Texas (483). The men of this office made studies of the economic and engineering problems associated with the mining of Texas uranium ores. On the basis of their studies, in April 1958 the A.E.C. announced that it would enter into contract with a qualified mining or milling company for the construction of a uranium-ore processing plant in Karnes County. In July 1960 the A.E.C. and Susquehanna-Western, Inc., reached agreement on the construction of a mill capable of processing approximately 200 tons of ore per day (484).

In conjunction with the mapping studies supervised by Eargle in 1957, the Geophysical Branch of the U.S.G.S. initiated the "Texas Coastal Plain Project" with the goal of correlating airborne radioactivity and magnetic data with ground studies of radioactivity and all other stratigraphic and structural data (485). The aerial survey was of a 20,000-square-mile area in south-central Texas. Alice Weeks of the Geological Survey authored several papers based on these geophysical studies, and Eargle and R. M. Moxham compiled the *Airborne Radioactivity and Geologic Map of the Coastal Plain Area, Southeast Texas* (486). A comprehensive report on the uranium-bearing rocks of this area was prepared and submitted to the Geological Survey for review in 1960 as noted by Eargle in the annual report of the Balcones Research Center (487).

During this uranium excitement and exploratory work by the Geological Survey, the Bureau of Economic Geology provided informal cooperation in the form of office space, use of library, and consultation. The Bureau also published Eargle's preliminary report as well as a paper on *Prospecting for Uranium in Texas*, prepared by P. T. Flawn and G. H. Anderson (488).

The B.E.G. and U.S.G.S. did keep each other advised of their geological activities in Texas, thus avoiding overlapping efforts. However, there was only a minimum of cooperation between the two (489).

ELEVEN

From the Ouachita Foldbelt to *Eohippus*

While the staff geologists of the Bureau were involved in some of the previously mentioned programs, most of their energies were expended in basic geological research projects. It is clear that studies of this type are fundamental to any resource planning program. The Bureau has always maintained the respect of fellow scientists because of the high quality of the research work of its geologists. In spite of low budgets and the resultant minimum geological staff, the research output has been tremendous.

As of January 1945, Glen Evans was in the middle of an extensive program of investigations of mineral deposits in Trans-Pecos Texas. In a report to Director Sellards, Evans proposed to expand his studies to (1) the Van Horn-Allamoore district where silver-bearing copper ores appeared in several localities, (2) the lead and zinc ores in the Quitman Mountains of Hudspeth County, and (3) the silver-bearing lead ores in the Altuda Mountains of Brewster County and the Chinati Mountains of Presidio County.

However, in that same year Evans transferred from the Bureau to become the full-time Assistant Director of the Texas Memorial Museum. He was replaced by Dr. Samuel S. Goldich, a graduate of the University of Minnesota and a member of the faculty at Texas A&M College before World War II. Goldich immediately undertook the supervision of an extensive program of detailed geological mapping work in the Trans-Pecos. The Geological Society of America made an initial contribution of $1,300 from the Penrose Fund to cover some of the field expenses of this "West Texas Project" — studies of the Tertiary igneous geology of the Trans-Pecos. Among the geologists involved in these quadrangle studies were some of the first doctoral candidates at the University of Texas to receive some financial support from the Bureau (490). Today, this arrangement continues to be one of great benefit to both the Bureau and the University's graduate program (491).

Ross A. Maxwell, who resigned as first Superintendent of Big Bend National Park to join the Bureau in 1952, and Dr. Lonsdale continued their in-depth studies of the geology of the Big Bend area as part of this West Texas project (492). Lonsdale had first worked in this area in 1928, and Maxwell joined him in 1936 under a joint grant. Roy T. Hazzard, retired Gulf Oil Company geologist, later assisted them in stratigraphic studies (493).

In the summer of 1952, in a letter to the Secretary of the G.S.A. concerning the progress of the G.S.A.-funded project, Dr. Lonsdale reported that "the latest development in the work which has called for additional work was the discovery of what appears to be very important Tertiary vertebrate fossils" (494). Lonsdale felt that this major discovery necessitated the undertaking of a collecting expedition under John A. Wilson, of the University's Department of Geology. The Bureau's Report of Investigations No. 14—*New Paleocene and Lower Eocene Vertebrate Localities, Big Bend National Park, Texas*—by coauthors Wilson, Maxwell, Lonsdale, and James H. Quinn, is the official announcement of the discovery and identification of these important Tertiary vertebrates (495).

Ronald K. DeFord, who left the oil industry to join the Department of Geology faculty at the University of Texas in 1948, was responsible for the expenditure of more research energies in the Trans-Pecos area. As head of the Department's graduate program and consultant to the Bureau during the summers, DeFord launched a geologic mapping program in the Rim Rock Country between the Sierra Vieja and the Rio Grande in Jeff Davis and Presidio Counties. Graduate students under DeFord's supervision mapped this area between 1954 and 1957 and received some financial support from the Bureau (496). In looking back, one can say that during the Lonsdale years, the Trans-Pecos — a vast, remote, and complex geologic region — received such a concentration of geologic research energies that by 1960 the Trans-Pecos geologic frontier had been conquered (497).

When Goldich left the Bureau in 1948 to assume a faculty position at his alma mater, the University of Minnesota, Peter T. Flawn, a doctoral candidate at Yale, replaced him to begin an illustrious career with the Bureau that would culminate in his directorship in 1960 (498). His first official project was a study of the Precambrian rocks in the Van Horn area, which became his doctoral project. The results of Flawn's work on these crystalline Precambrian rocks and his studies of the copper and silver mines in the area, along with Philip King's work on the younger Precambrian sedimentary rocks of the Sierra Diablo area, were published by the Bureau as University of Texas Publication 5301 (499). Working between 1931 and 1939, King had mapped the northern part of the Van Horn area, and his map was drafted along with Flawn's. This excellent cooperative effort between the Bureau and the U.S. Geological Survey presents a comprehensive record of the Precambrian geology of the Van Horn area (500).

Upon completion of this study, Lonsdale directed Flawn to begin a study of the subsurface Precambrian beds in West Texas and southeastern New Mexico (501). Oil companies freely contributed samples and other well information to this project, which Flawn began in the spring of 1951. His work in the Franklin Mountains was temporarily postponed because of military training there, but the final report appeared in 1956 (502).

In the final basement rock paper, Flawn devoted three pages to a discussion of the Ouachita System, especially the formulation of new ideas concerning the origin of this geologic feature (503). He was very interested in Virgil Barnes' research on the nature and behavior of geosynclines; this research involved a study of well cores and cuttings from the Ouachita Foldbelt in Central Texas. These studies enabled Barnes to outline belts of metamorphism that evidenced progressive increase in the metamorphism of these buried Paleozoic rocks from northwest to southeast, and he suggested that the "schists" encountered under the Cretaceous cover in the Luling field in Caldwell County were not Precambrian, and thus that the southeastern boundary of the geosyncline was completely unknown.

In his short discussion of the Ouachita Foldbelt, Flawn observed that the old concept of Llanoria as an ancient crystalline landmass separated from the North American continent foreland to the north by the Ouachita geosyncline had also been challenged. In 1952, H. J. Morgan, Jr., had contended that Llanoria was "a bundle of mountain ranges possessing a long linear but limited lateral extent . . . [which] served as a source for the clastic sediments in the geosyncline which they partly occupied" (504). This concept is in line with recent theories on the role of island arcs in geosynclinal development.

Flawn closed his discussion of the Ouachita System with the comment that the fundamental nature of it was still in question and needed further study on the basis of new data and a review of all old data. Subsequently, a Bureau news release dated June 17, 1954, announced that Flawn was in charge of a major research project, a petrological approach to the tectonic interpretation of the origin of the Ouachita Foldbelt (505).

Since the study depended largely on information from wells, Dr. Lonsdale contacted the major oil companies in search of support, and they all proved very cooperative. J. Ben Carsey, Chief Geologist of Humble Oil & Refining Company, informed Lonsdale that Peter Masson of that company had made a very detailed study of thin sections of cuttings and cores of all of the wells that penetrated the Ouachita System, and that he and Raymond Wood would probably publish the results (506). Carsey was afraid of overlap and duplication of effort. Lonsdale replied that Flawn had arranged a meeting with these geologists and August Goldstein, Jr., of Stanolind Oil & Gas Company, to discuss the coordination of work and the possibilities of joint authorship (507). Lonsdale added that this comprehensive study was a logical one for the Bureau to undertake. Once begun, there was open interchange of information between company geologists and Flawn, and throughout the course of the project preliminary results were made available to all. This was another very positive display of successful voluntary cooperation between industry and the Bureau (508).

University of Texas Publication 6120, *The Ouachita System*, stands as one of the outstanding Bureau research publications

(509). The complex geologic history of this foldbelt is well presented, and the map is excellent. Flawn was unable to define the southeastern limit of the foldbelt—he ran out of well control—but he was able to define two tectonic provinces within the Ouachita System, a frontal zone bordering the craton where deformation is characterized by flexure and an interior zone where deformation is mainly by shear.

In the summer of 1951, the director of the Exploration Department of Humble Oil & Refining Company informed Dr. Lonsdale that Humble would support an extensive survey of the stratigraphy of the subsurface pre-Simpson rocks of Texas. He suggested also that other oil companies be asked for financial support. Eventually Atlantic Refining Company, Gulf Oil Corporation, Shell Petroleum Company, Sun Oil Company, and the Texas Company joined Humble in contributing a total of $58,522 in support of this project (510). The thorough and excellent researcher, Dr. Virgil Barnes, supervised this project, which was to some extent a continuation of his work with Preston Cloud on the Ellenburger Group in Central Texas (511). The research target included all of the sedimentary rocks in Texas and southeast New Mexico from the top of the Precambrian to the base of the Middle Ordovician Simpson Group.

In a letter describing the project to Acting University President T. S. Painter, Dr. Lonsdale described the Ellenburger as a prolific source of oil in West Texas and noted that this cooperative project evidenced the growing importance that the oil industries were attaching to fundamental research, and the willingness of individual companies to support projects that would benefit the whole industry (512). Elsewhere, Lonsdale reported that Humble Company officials were emphatic in stating that the delineation of local structures or commercial application of the criteria would not be a part of the project: "They looked to the project as possibly promising tools, techniques or information which the industry, if smart enough, might use in trying to find more Ellenburger oil" (513).

The project extended from May 1952 to March 1957 and involved the use of many special fields of investigation in an effort to establish criteria for subdividing these rocks. Barnes and his assistants were unable to come up with any shortcut method by which subdivisions of the pre-Simpson beds could be recognized in the subsurface (514). Some helpful criteria were color, insoluble residues, and limestone and dolomite grain sizes, but the most dependable information was obtained through thin section studies.

During his systematic mapping projects in Central Texas, Dr. Barnes also conducted many surveys of the economic geology of the Llano Uplift area. Some of the published results include "Iron Ore in the Llano Region," "High Purity Marble Falls Limestone, Burnet County, Texas," "Phosphorite in Eastern Llano Uplift of Central Texas," "Lead Deposits in the Upper Cambrian of Central Texas," and, with S. E. Clabaugh, "Vermiculite in Central Texas" (515). The study of the iron ores was an outgrowth of studies of the gravity anomalies in the basement rocks of Central Texas by Barnes and Frederick Romberg, a geophysical consultant. These studies were funded by the Office of Research and Investigation of the Navy Department (516). The aim of this research was to discover methods to predict accurately the shape of basement rock masses exposed in Central Texas and to characterize the type of gravity anomalies such masses produce (517).

In 1960, Dr. Barnes received a National Science Foundation grant for extensive studies on the composition and origin of tektites (518). The results of Barnes' longtime studies of tektites, which began in the late 1930's, are encompassed in his *Tektites*, published in 1973 (519). Dr. Barnes' name is synonymous with the quality of the Bureau's research (520).

In 1950 and 1951, several oil companies contributed funds for a study of the Edwards Limestone in fault-line oil fields along the Balcones flexure. Dan E. Feray, who was appointed to the Bureau staff in 1948, initiated this project, which he continued after he became associated with Mobil Research Laboratory (521). He solicited $500 donations from Humble Oil & Refining Company, Shell Petroleum Company, Magnolia Petroleum Company, Atlantic Oil & Refining Company, Gulf Oil Corporation, Stanolind Oil Company, and Standard Oil Company of Texas. Henry F. Nelson, a University of Wisconsin graduate, worked on this project during his short stay with the Bureau. (Nelson resigned to join Mobil Research Laboratory, as did Feray after a short stay at Tulsa University.)

As late as February 1951, Feray was active on this project, but his interest later waned (522). However, Nelson continued these Edwards studies after joining Mobil and submitted a paper entitled "Description and Alteration of the Edwards Limestone" for the Bureau's 1959 Symposium on the *Edwards Limestone in Central Texas* (523).

Publication of this symposium was made possible through the establishment in 1951 of the East Texas Geological Fund for the Bureau. The nucleus for this fund came from the sales of University of Texas Publication 5116, *The Occurrence of Oil and Gas in Northeast Texas* (524). The first mention of a comprehensive survey of oil fields in Texas is in a letter from Lonsdale to Earl Noble, President of the American Association of Petroleum Geologists and of the Union Oil Company of California (525), requesting a copy of a comprehensive report on the oil fields of California published by the California State Division of Mines. In January 1948, Lonsdale secured Frank A. Herald, a retired geologist, as editor-in-chief of the Petroleum Resources of Texas Project. Lonsdale clearly viewed the cooperation of oil companies and independent geologists as essential, and asked them to supply information and assume authorship of articles on individual fields. The industry gladly extended their services through the East Texas Geological Society. Sixty-nine geologists contributed 135 papers in this report, *The Occurrence of Oil and Gas in Northeast Texas*. By 1951, the West Texas Geological Society was working on a similar study of oil fields in West Texas, which appeared in 1957 (526). Both publications illustrate a unique public service made possible through the coordinating efforts of the Bureau.

In the 1950's, the Pleistocene deposits of the Texas High Plains became the object of an academic controversy between Dr. John C. Frye and Dr. A. Byron Leonard, of the Illinois State Geological Survey and the University of Kansas, respectively, and Dr. James H. Quinn, of the University of Arkansas. (Quinn had received his doctorate in vertebrate paleontology [527] at the University of Texas in 1954 under the supervision of Dr. John A. Wilson.) The studies at issue were both sponsored in part and published by the Bureau, which again served a meaningful role as coordinator of research in Texas geology.

In a Bureau Report of Investigations published in November 1957, *Studies of Cenozoic Geology along the Eastern Margin of the Texas High Plains, Armstrong to Howard Counties,* Frye and Leonard used fossil molluscan faunas and fossil seed floras along with physical stratigraphic evidence to develop correlations of the Pliocene and Pleistocene (late Cenozoic) strata of the eastern margin of the Texas High Plains with beds in the northern Great Plains and with the Midwestern glacial section (528). They concluded that the important High Plains Blancan strata were laid down by streams of very low gradient flowing across a semiarid terrain. They challenged the opinion in 1945 of Evans and Meade that the Blancan beds were deposited in a large lake.

In the same year, Quinn, in his excellent paleontological study, *Pleistocene Equidae of Texas*, reported that the teeth and bones of fossil horses "substantiate the concept of distinct, large-mammal assemblages belonging to different stages of the Pleistocene" (529). He postulated that inclement conditions during glacial periods decimated large mammals over nearly the whole of continental North America, and he was able to relate several horse faunal assemblages to Quaternary glacial chronology.

Quinn took time from his paleontological work to challenge the stratigraphic work of Frye and Leonard. He wrote to Lonsdale, bemoaning the fact that this pair failed to see that during glacial times of excess humidity in nonglaciated regions, there would be no deposition except in basins such as the Blancan lake basin postulated by Evans and Meade (530). Quinn also concluded that the Blancan beds were deposited during the Kansas glacial period and not Nebraskan in age as postulated by Frye and Leonard (531).

Glen Evans wrote a critique of the manuscript by Frye and Leonard, which he found to have a great deal of merit, especially the fossil flora and molluscan work (532). He added that he did not feel that they had established the fluvial origin of the Blancan beds to his satisfaction. In a negative reply to Dr. Lonsdale's suggestion of a conference on the Pleistocene of the High Plains, Frye stated that there was no way in which he and Leonard could agree with Quinn (533). In a subsequent letter to Frye, Lonsdale stated that he would publish Quinn's work on the merit of the great volume of paleontological work and also publish the report by Frye and Leonard with the statement that "they are contributions toward a solution of problems concerning which there are still great differences of opinion" (534).

The vertebrate work of Quinn pointed up a new development in vertebrate paleontology in Texas. Just after World War II, Dr. Lonsdale and A. H. Deen, Chairman of the Geology Department of the University of Texas, had corresponded about the need for a vertebrate paleontologist whose time could be divided between the Bureau and the Department (535). Shortly afterwards, Dr. John A. Wilson, from the University of Michigan, joined the geology faculty and immediately began to stimulate a resurgence in vertebrate work.

During World War II, a plant had been constructed just north of Austin for the purpose of extracting magnesium from dolomite rocks quarried near Burnet. It was to be an emergency source of magnesium. However, only a small amount of magnesium was actually refined from the dolomite, and after the war the plant was closed by the Government (536). This plant and 393 acres were acquired by the University of Texas under a contract dated December 10, 1949, which stipulated that "the Research Center must be utilized to extend the research function of the University's education program and that this utilization must result in benefits to the people of the United States." The premises had been under lease to the University from September 21, 1946, to December 10, 1949, and were designated Off-Campus Research Center (537).

Dr. Sellards had continued to be a staunch supporter of vertebrate studies in Texas. He was head of the Laboratory for Research in Vertebrate Paleontology, located at the Off-Campus Research Center (renamed Balcones Research Center in 1952), which in 1948 was under the joint supervision of the Bureau, the Department of Geology, and the Texas Memorial Museum. Lonsdale, Sellards, and Deen had successfully petitioned Professor J. Neils Thompson, Director of the Off-Campus Research Center, for the housing of this research and teaching laboratory, which was endowed with large and valuable Work Projects Administration collections made from 1934 to 1941 at a cost of over $300,000 (538).

As the first director of this Vertebrate Paleontology Laboratory, Dr. Sellards supervised two principal exploring and excavating projects during 1950 — a continuation of the study of the geology and prehistory of the Southern High Plains near Lubbock, and the excavation and study of the site of the French settlement made by La Salle in 1685 on the Texas Gulf Coast (539). The veterans of the W.P.A. work, Evans and Meade, were in charge of these studies. However, their active participation in the projects of this laboratory decreased, and Dr. John A. Wilson, who had been teaching classes in advanced vertebrate paleontology in 1948, 1949, and 1950, was appointed Paleontologist-in-

Charge of the laboratory in 1951. (Incidentally, Wilson assumed the Presidency of the Society of Vertebrate Paleontologists in February 1952.)

Wilson's first studies were of the Miocene Fleming faunas of the Coastal Plain; in the summer of 1952, as noted earlier, he began his study of the sizable collection of Paleocene and Eocene vertebrates from the Big Bend region (540). This Bureau-sponsored collecting work led to the discovery of the first five-toed horse, *Eohippus,* to be found in Texas.

In 1957, Dr. Sellards retired as nominal director of the laboratory and was replaced by Dr. Wilson (541). Dr. Ernest Lundelius, Associate Professor of Geology, was added to the permanent staff at this time.

The Bureau also maintained a Mineral Technology Laboratory at the Balcones Research Center for the identification and classification of minerals and rocks being studied by the Bureau. R. M. Wheeler was in charge of this laboratory; he was succeeded by Dan Schofield from the Bureau of Chemical Engineering. The Well Sample and Core Library of the Bureau is also housed at Balcones Research Center. This library dates back to Dr. Udden's collections of cuttings from the earliest oil drilling tests; however, it was not until 1937 that the State Legislature, under the prodding of Dr. Sellards and Hal P. Bybee, granted appropriations for the establishment of a Well Sample Library under Bureau supervision. Dr. Leo Hendricks was the first supervisor of this library, which was moved in 1942 from the Little Campus at 18th and Red River Streets (the Bureau's home for 40 years), to Sutton Hall and the Old Library Building on the Main Campus; a final move was made in 1949 to the Balcones Research Center (542).

Hendricks left in 1946 to join the faculty at Texas Christian University and was succeeded as supervisor by J. L. Starnes, who was in turn succeeded by Dan E. Feray in 1948. Feray wrote an article on the Well Sample and Core Library for the *Oil and Gas Journal* in October 1948 and in 1950 wrote with Starnes the commendable *Index to Well Samples* (543). When Feray left to join the faculty at the University of Tulsa, Director Lonsdale secured Philip S. Morey, a consultant and operator of a commercial laboratory, to supervise the library (544).

Several other members of the Bureau's staff involved in important geological research during the Lonsdale years were destined to become eminent in Bureau activities and publications. In the spring of 1957, Dr. L. Frank Brown, Jr., resigned from the Standard Oil Company of Texas to join the Bureau and began work on a detailed stratigraphic study of the Upper Pennsylvanian Cisco Group of the Brazos Valley. No one had done extensive surface study of these rocks since M. G. Cheney's work in 1940, and since the death of F. B. Plummer, none of the Bureau's geologists had concentrated on Pennsylvanian geology (545, 546). Dr. Brown, a highly capable stratigrapher, made detailed studies of restricted intervals across the Brazos Valley and found thin but persistent mappable units in northern Stephens County. The published results of his work, *Stratigraphy of the Blach Ranch — Crystal Falls Section, Northern Stephens County, Texas,* which appeared in April 1960 was another of the excellent, classical stratigraphic studies produced by the Bureau (547). Brown joined the faculty at Baylor University in the fall of 1960 but in 1966 rejoined the Bureau and is now Associate Director of Research.

Dr. H. B. Stenzel continued to produce outstanding work on the paleontology and stratigraphy of the Gulf Coastal Tertiary formations of Texas until he resigned in the fall of 1954 to take the chairmanship of the Geology Department at the University of Houston. In January 1946, Stenzel accompanied Dr. Lonsdale on a tour of the Troup district in northeast Cherokee County, where ball clay of economic importance had just been uncovered. Lonsdale directed Stenzel to check out the stratigraphy of the area and map the ball clay deposits. Typically, Dr. Stenzel became so wrapped up in this project that he spent several years on it. The published results of his efforts, *Ball Clay of the Troup District, Texas* (1950) and *The Geology of Henrys Chapel Quadrangle, Northeastern Cherokee County, Texas* (1953), demonstrate the depth and quality of his stratigraphic investigations (548). Stenzel's final work for the Bureau included studies of the Eocene Claiborne strata of East Texas. Pelecypoda *from the Type Locality of the Stone City Beds of Texas* (1957) is a fitting tribute to the paleontological and paleoecological skills of Dr. Stenzel (549).

When Dr. Brown resigned to go to Baylor, William L. Fisher, who later received his doctorate from the University of Kansas, replaced him (550). Dr. Fisher undertook regional studies of the Tertiary beds of the Texas Coastal Plain. They were the basis for his (1963) economic study *Lignites of the Texas Coastal Plain* (551). Dr. Fisher succeeded Dr. Flawn in 1966 to become the sixth director of the Bureau of Economic Geology.

After Professor W. S. Adkins left the Bureau in the 1930's, Bureau geologists gave only minor attention to research on Cretaceous stratigraphy. The Bureau did publish many fine studies by various university faculty members, such as *Cretaceous of the Llano Estacado of Texas* (1953) and *Comanchean Stratigraphy of Kent Quadrangle, Trans-Pecos Texas* (with R. K. DeFord, 1958), by Dr. John P. Brand of Texas Technological College; *Handbook of Cretaceous Foraminifera of Texas* (1954), by Don L. Frizzell of the Missouri School of Mines and Metallurgy; *Geology of Parker County, Texas* (1957), by Leo Hendricks of Texas Christian University; and especially the work on Cretaceous ammonoids by Dr. Keith Young of the University of Texas, including *Upper Albian (Cretaceous) Ammonoidea from Texas* (1957) (552).

Following his retirement from Shell Petroleum Company, W. S. Adkins returned to Austin and became a consultant on the Bureau staff. He was the clearinghouse for information about inquiries on Texas Cretaceous paleontology and stratigraphy (553).

Lonsdale had searched for another Mesozoic paleontologist; in January 1947, Dr. Bernard Kummel of Columbia University was engaged in that capacity. Kummel began some studies of the Cretaceous strata of Central Texas for the Bureau, but before publishing any results, he resigned to join the U. S. Geological Survey. In April 1958, Dr. Peter U. Rodda of the University of California at Los Angeles accepted an appointment to the Bureau as a Cretaceous and Tertiary paleontologist and stratigrapher. Rodda became one of the mainstays of the Bureau during the 1960's. He and Dr. Fisher collaborated on several outstanding research projects.

Other research contributions of interest during the Lonsdale years include Roselle Girard's excellent *Bibliography and Index of Texas Geology, 1933-1950*. Subsequently, in 1971, Elizabeth T. Moore (with Margaret D. Brown) continued the bibliographic compilation through 1960, and in 1976, Mrs. Moore continued this work through 1974. (A bibliography of papers published up to 1933 was published as part of University of Texas Bulletin 3232 [554].)

In 1959, Dr. William H. Matthews, of the Geology Department at Lamar State College in Beaumont, initiated work on a series of popular publications that represented a new public service offered by the Bureau. His *Texas Fossils: An Amateur Collector's Handbook* (1960) has gone through eight reprintings (555). Matthews also authored *The Geologic Story of Longhorn Cavern* (1963) and *The Geologic Story of Palo Duro Canyon* (1969); Roselle Girard wrote *Texas Rocks and Minerals: An Amateur's Guide* (1964, with three reprintings) (556). Ross A. Maxwell compiled the excellent guidebook on *The Big Bend of the Rio Grande* (1968, with three reprintings) (557). This guidebook was an offshoot of the outstanding study, *Geology of Big Bend National Park, Brewster County, Texas* (1967), by Maxwell, John T. Lonsdale, Roy T. Hazzard, and John A. Wilson. Dr. Lonsdale had worked with Dr. Maxwell from the inception of this study in 1946 until his death in 1960. Senior editor Maxwell worked continuously on the project from 1952 until 1963. He also coauthored, with L. F. Brown, Jr., G. K. Eifler, Jr., and L. E. Garner, the *Geologic and Historic Guide to the State Parks of Texas* (1970).

Finally, another outstanding step forward was taken by the Bureau during the Lonsdale years when James W. Macon was added to the staff in 1950. In charge of the drafting division, he now fills the budget position of cartographic engineer (558). Macon's reputation, together with that of his chief assistant, Dan F. Scranton, is outstanding among his colleagues. Macon's cartographic group has added new dimensions and professionalism to Bureau publications and mapping projects, such as the Geologic Atlas of Texas Project, initiated in 1961, and the Environmental Geologic Atlas Project, initiated in 1969.

TWELVE

The Maturation of Oil Company Research Attitudes

The discovery of most of the structurally controlled oil fields through surface mapping and later through geophysical techniques (which were first suggested by Dr. Udden and then developed by the oil industry researchers) necessitated a more sophisticated type of oil exploration based on subsurface stratigraphic studies. This search for stratigraphic oil traps encouraged the development of company research in sedimentation in an effort to find methods of predicting the environment of deposition of ancient sediments. This transformation in type of exploration was paralleled by the establishment of research and development departments in the exploration divisions of the major oil companies following World War II (559). Late in the summer of 1945, Shell Petroleum Company established its Exploration and Production Research Center in Houston, and Humble Oil & Refining Company expanded its exploration research at this time. (L. T. Barrow, former head of the Humble Company's Exploration Department, noted that there was greater opportunity to employ mature trained men in these postwar years [560].)

The large expense of monies and manpower involved in basic sedimentation studies precluded the Bureau's entry into this type of research. Flawn's study of the Ouachita System was similar in scope, and logistically it required the cooperation and assistance of the major companies. For the most part, the scientific manpower of the companies alone was equal to this new research task. (The main exception was research funded by the American Petroleum Institute and the Geological Society of America, described below.) The Bulletin of the American Association of Petroleum Geologists began to include more and more subsurface stratigraphic and sedimentation studies by company geologists. Typical was the article by August Goldstein, Jr., and Duane Reno of Stanolind Oil and Gas Company, "Petrography and Metamorphism of the Sediments of the Ouachita Facies" (561). Goldstein was later a coauthor of Flawn's *Ouachita System.* As noted previously, Humble geologists were also doing research on the Ouachita geosyncline.

Two interesting research projects initiated by the active Research Committee of the A.A.P.G. further reflect the oil industry's growing awareness of the need for basic research in sedimentation. In the late 1940's, Sheppard W. Lowman, Chairman of the A.A.P.G. Research Committee, sponsored a research proposal for the study of recent and near recent sediments (562). Eventually, in 1951, the Advisory Committee on Fundamental Research of the A.P.I. supported this proposal and agreed to fund Project 51 through the auspices of the Scripps Institute of Oceanography of the University of California. The goal of the project was to establish criteria for the study of recent shallow-water marine and estuarine sediments that could be used as aids in identifying the environment of deposition of ancient sediments.

The first research funded under this project was by F. P. Sheppard and D. G. Moore of the Scripps Institute. Their very extensive and thorough report, "Central Texas Coast Sediments; Characteristics of Sedimentary Environment, Recent History, and Diagenesis," was published in the A.A.P.G. Bulletin (563). It is noteworthy that this report appeared in 1955, the same year that Dr. Robert L. Folk of the University of Texas contributed his first publication on Texas geology (564). Folk, a nationally renowned sedimentation geologist, has established an excellent sequence of courses in sedimentation and sedimentary petrography at the University.

In August 1952, Marcus Hanna, Chairman of the Research Committee of the A.A.P.G., wrote to Dr. Lonsdale, a member of that committee, about a symposium on the geology of the Gulf of Mexico that the committee had planned for the 1953 Houston meeting of the association (565). Hanna wrote that at the annual meeting of the G.S.A. in Washington in 1950, there had been a symposium on sedimentary volumes of the Gulf Coastal Plain of the United States and Mexico. The various papers were published as "Sedimentary Volumes in the Gulf Coastal Plain of United States and Mexico" in the December 1952 Bulletin of the G.S.A. (566). Dr. Grover Murray of Louisiana State University was the editor of this volume. Since this symposium was concerned mainly with tectonic activity outside the coastal area, Hanna felt that the study of the sedimentary framework and the history of sedimentation of the "Gulf of Mexico" geosyncline, which would be the direction of the A.A.P.G. symposium, would complement that of the society. One of the papers later presented at Houston was by Norman Greenman and Rufus LeBlanc, of the research division of Shell Petroleum Company, who had worked with researchers of the Woods Hole Oceanographic Institute. Their report, "Recent Marine Sediments and Environments of the Northwest Gulf of Mexico," appeared in the May 1956 Bulletin of the A.A.P.G. (567). This is another example of the co-operative spirit of geologic researchers.

As noted before, in the postwar decade-and-a-half, oil companies were quite willing to contribute money to research projects conducted by the Bureau of Economic Geology, such as the pre-Simpson study and the Edwards Limestone study, and to contribute their geologists' time and information to such projects as Flawn's Ouachita study and the oil and gas field reports. Company geologists are involved in research projects under the auspices of the A.A.P.G. and the G.S.A., and company exploration research divisions are conducting their own basic geological research projects. Geological research supported by private funds has assumed gigantic proportions. The Bureau has always encouraged and counseled private geological research and in return has often received funds from private sources (568).

One note of discord between private geologists and governmental geologists was sounded in a letter from Frank Conselman, independent geologist, to Dr. Lonsdale concerning the U.S. Geological Survey's study of the Pennsylvanian reefs. Conselman was disgruntled because private geologists who had the practical experience in exploring for these reefs had not been consulted during the planning of the Federal survey. Perhaps he voiced the attitudes of many independents: "I hope that in the not too distant future that the present policy in planning federal and state economic projects will be revised to include the viewpoint of the practicing geologist for whom the work is intended" (569). Conselman told Lonsdale that it was common knowledge that the Federal survey had lagged painfully behind the practicing geologist in its approach to economic problems.

Another outstanding manifestation of the oil industry's sophisticated support for basic geological research was the establishment of the Geology Foundation for the Department of Geology at the University of Texas, which was announced in a departmental newsletter of March 1954. This was in fact the realization of a dream of the late Hal P. Bybee, one-time departmental chairman. Keith Young noted that Bybee's dream had been conceived in 1939 when he was corresponding with E. L. DeGolyer and P. T. Moore about methods of endowing the Department (570). As Young further observed, the travel and research and scholarship monies made available through the Foundation have increased research output and have bettered teaching by making the Department more attractive to prospective faculty members.

All of the major companies have contributed generously to the Geology Foundation, and the Humble Oil & Refining Company especially has been a long-time benefactor of the Department. In 1947, in a letter to Acting University President T. S. Painter, L. T. Barrow declared that "in recognition of the splendid training in chemical engineering, chemistry and geology offered by the University of Texas and the relatively large number of Texas graduates that have been employed by the Humble Oil and Refining Company, the Company would like to establish graduate fellowships in the departments of chemistry, chemical engineering and geology" (571).

Another major note of distinction for the Department was sounded in 1961 by the grants of funds by Mrs. Lillie Rice Farish of Houston for the establishment of a Professorship in Geology at the University in memory of her late husband, William Stamps Farish. Farish had been one of the organizers of the Humble Company and had gone on to become Chairman of the Board of the Standard Oil Company of New Jersey. Dr. J. Hoover Mackin was the first geologist selected to fill this chair in 1962.

The Geology Foundation was established at a time when the Bureau and the Department were finally developing a close relationship, which was symbolized by their joint occupancy of the new Geology Building in 1967. This air of mutual respect and coordination of efforts that came to full flowering during the directorship of Peter Flawn was slow in developing. Misunderstanding and jealousy between the members of the two organizations were solidly established early in the life of the Bureau during Udden's directorship. Udden had a total lack of respect for the credentials of the geology faculty; in addition, a kind of intellectual jealousy characterized the scientific climate of Udden's day. He persistently opposed the housing of the Bureau in the same building with the Department, and thus they were not located in the first Geology Building completed in 1932 (572). Udden's successor, Dr. Sellards, adhered to the premise that no geologist could be both an effective researcher and a teacher, an attitude that further hindered the establishment of a working relationship between the two organizations.

The first thawing of this scientific cold war came with the appointment of a secret *ad hoc* committee by University President Homer P. Rainey to investigate the relations between the Bureau and the Department and the relation between research and teaching. The report of this committee has not been located, but it was alluded to in the following words in the departmental minutes of October 1942 (573):

> The other members of the budget council concur with Dr. Sellards . . . in his opinion that research workers in the Bureau cannot function satisfactorily as regular instructors in the Department, and likewise, that teachers in the Department cannot function as regular researchers in the Bureau.

These words do not indicate a thaw, but the minutes went on to declare that one area of cooperation that could be developed was the contracting of graduate students and faculty members to do field work projects for the Bureau. Furthermore, the budget council asserted that the physical separation of the two organizations had seriously handicapped any cooperation and should be ended.

This nascent spirit of cooperation received a great boost during Dr. Lonsdale's directorship. In 1947 he wrote Acting University President T. S. Painter that "the Bureau should be regarded as an important element in the instructional function of the University as well as an active research and public service unit" (574). He was fighting not only for better housing and salaries for the Bureau, but was also calling for increased cooperation between the Bureau and the Department. He was a member of the Budget Council of the Department, as Sellards had been, and he helped in the building up of the Department's reputation and faculty. He became especially active in fostering the graduate program.

Following World War II, there was an increased demand for trained geologists by the booming oil companies. This made possible a tightening up of the Department of Geology's undergraduate curriculum, and the tremendous growth of the graduate program was to a large degree the result of the efforts of Ronald K. DeFord, who left private industry to join the Department of Geology in 1948 and who became Graduate Adviser in 1951. However, Lonsdale and the Bureau also deserve a great deal of credit for the successful growth of graduate work in the Department. In a 1953 letter to University President Logan Wilson, Lonsdale reported that the Bureau had been very helpful in the development of the graduate program of the Department during the difficult post-war years (575). The first three doctorates conferred in these years — those of Richard R. Bloomer, Roy Graves, and Ralph Erickson—depended largely on funds provided by the Bureau in support of field work (576). In a letter to Graduate Dean Albert P. Brogan 3 years later, Lonsdale reported that during the postwar years, the research of eight full-time faculty members had been supported by the Bureau in the amount of $20,000, and the research problems of 7 of the 14 successful candidates for the doctorate were supported by the Bureau in the amount of $27,000 (for salaries, field expenses, and vehicle use) (577). This approach and cooperative attitude became much more pronounced under Lonsdale's successor, Dr. Peter T. Flawn.

EPILOGUE

It is difficult to evaluate and characterize the spirit and accomplishments of the Bureau of Economic Geology during the directorships of Drs. Phillips, Udden, Sellards, and Lonsdale. There was a great deal of continuity, and projects initiated under one director often reached fruition under another. In an effort to understand where the Bureau is today and how it got there, the historian tends to emphasize changes and new developments rather than continuities. Only Josephine Casey, who served the Bureau for more than 45 years under four directors, can attest to the subtleties in the transformation of the Bureau from a small group of geologists working in old "B Hall" to the sophisticated research organization it is today.

The years of Udden's leadership witnessed the emergence of the oil industry. This feature, along with the pioneer nature of much of the geological research done by the Bureau's geologists, made this period a lively and productive one. The special demands on the Bureau resulting from the Depression and World War II and the W.P.A. grants for research on mineral resources and vertebrate paleontology gave the years of Dr. Sellards' directorship their own special flavor.

Historians have labeled the years following the war as the "Crucial Decade" in American history, crucial because the American nation was undergoing the throes of diplomatic, political, and economic changes under the duress of the threat of "Communistic imperialism." The American society's concern with economic growth and with national defense—the Cold War and the Korean War—had definite influences on the research work done by Bureau geologists during the 15 years of Dr. Lonsdale's leadership. The direction and goal of the Bureau generally mirrored the concerns of American society as a whole, as described earlier. However, some of the activities of the Bureau gave hints of new direction. Such things as the mineral resource evaluations done by the Bureau for the Trinity River, the Red River, and the Colorado River drainage basins were intended to provide bases for predictions of future industrial growth and the resultant increased demands on water resources. They were programs in aid of industrial growth, but they also represented early attempts at planned usage of resources.

The decade of the 1960's was dominated by the frustrations of the Vietnam War, and the questioning of American foreign policy soon was transformed into the questioning of the direction of American society and economy. The wastefulness and unhealthfulness of pollution—air, land, and water pollution—became vital concerns. Under the direction of Dr. Flawn, the Bureau also began to take a new direction in the 1960's. Land use planning and resource management became important goals of geological research.

Actually, this profound change in the mission of geological research was heralded during the Lonsdale years by the creation of geological staffs in the various State agencies concerned with special resource problems, such as oil and water, and with the application of geology to engineering problems (578). As early as 1929, the Board for Lease of University Lands had established a geological advisory group, and in 1935 the newly activated Oil and Gas Division of the State Railroad Commission had hired its first geologist. The Texas Highway Department, the Board of Water Engineers (later Texas Water Development Board), and the General Land Office followed this trend in the decade following World War II.

In some states, the geological surveys expanded to meet the new problems of resource management and performed the work through increased appropriations or contracts from the appropriate agency. The directors of the Bureau prior to Lonsdale limited Bureau participation in special resource projects. Perhaps this was intentional or necessitated by the limited funds of the Bureau, but the Bureau has been more of a research agency for inventorying Texas mineral resources and less of a public service organization than other state surveys. However, on several occasions Bureau geologists have served as consultants to geological staffs in other State agencies (579). In 1965

Director Flawn had the following comments to make on the situation as it developed in Texas (580):

> Perhaps the most serious fault of the concept of building separate geological staffs within the several agencies is that these staffs are specialized and have a narrow view. They are not likely to be aware of the breadth, interrelationships, and complexities of conservation and resource problems throughout the State. This can be overcome by the creation of a joint council or some other system to improve the liaison between the separate agencies.

Dr. Flawn himself began to remedy this situation by making the Bureau the source of the basic data for all land use and resource planning projects (581). In the Bureau's annual report for 1961, Flawn discussed the vital need for detailed systematic mapping such as that of the Bureau's Geologic Atlas Project (582):

> Texas' rapid advance toward a technical and industrial, metropolitan economy requires that the State make such preparation as is needed for the development forecast for the coming decades.

In the 1964 Bureau's annual report, Flawn, quoting from an address to the Governor's Conference on Natural Resource Management and Development, spoke of the conflicts between conservation and resource management groups and of the need for long-term planning to reconcile these conflicts (583).

His most direct response to this need was the project "Environmental Geologic Atlas of the Texas Coastal Zone," which was initiated in 1969. In the 1970 annual report, this project is described as follows (584):

> The project is designed to provide environmental geologic maps of the coastal area in order to document the distribution of modern environments, landforms, sedimentary facies and older Pleistocene facies, [and] the maps are the basis for many special derivative maps of environmental importance in the delicately balanced coastal zone where development of industry and population are increasing spectacularly.

The role that the Bureau assumed under Flawn's directorship, that of a central data bank for all resource management projects, is clearly described in the following comment on the "Environmental Geologic Atlas" in the same annual report (585):

> [It] is intended to provide fundamental environmental information on which planning, legislative guidance, and more specialized scientific studies can be based during the next decade. The environmental geologic map and its special derivatives provide a common language and vehicle that can bring together

many divergent specialists and governmental groups and allow their collective contribution to be simultaneously focused on vital coastal environmental problems.

As the Bureau historically provided the basic tools for mineral exploration and production in the form of geologic maps (along with the U. S. Geological Survey) and stratigraphic studies, today the Bureau provides another vital service in the form of these environmental geologic maps. The Bureau remains basically a research organization, but it has broadened the scope of its research projects to include ecological considerations.

Actually, this environmental approach was initiated in 1964 in a project on the geology of Austin and vicinity under the general direction of Keith Young of the Department of Geology and Peter Rodda of the Bureau staff. This project was focused on engineering and environmental geology and was intended to serve as a basic reference for engineers and planners. A discussion of this project was included in the Conference on Application of Geology and Soil Science to Urban Environmental Problems held in Austin in September 1965 (586). L. E. Garner and Keith Young coauthored the final report, *Environmental Geology of the Austin Area: An Aid to Urban Planning* (587), which was published in 1976. In 1974, the Land Resources Laboratory was created to establish a continuing program of land resources and environmental research at the Bureau.

Typical projects underway in 1977 at the Bureau include:

Ecological Studies—

Environmental Geologic Atlas of Texas Coastal Zone under the direction of L. F. Brown, Jr.

Shoreline Changes on Central Padre Island by R. A. Morton and M. J. Pieper (8th study in series)

Land Resources of Fayette County, Texas by Charles M. Woodruff and Thomas C. Gustavson (Acting Coordinator of the Land Resources Laboratory, which was established in 1974 for a continuing program of land resources and environmental research)

Energy Studies—

Mineral and Energy Resources of Texas Atlas by Thomas J. Evans, L. E. Garner, and Ann E. St. Clair

Evaluation of Geothermal Resources, Texas Gulf Coast by Don G. Bebout, Robert G. Loucks, and A. R. Gregory

Basic Research—

Pennsylvanian/Permian Facies, Eastern Shelf, North-Central Texas by L. F. Brown, Jr.

Lower Cretaceous Carbonate Facies and Depositional Environments—South Texas by Don G. Bebout

Geologic Mapping—

Geologic Atlas of Texas directed by Virgil E. Barnes (nearing completion)

Geologic Quadrangle Mapping in Central Texas directed by Virgil E. Barnes

Guidebook—

Guide to Points of Geologic Interest in Austin by A. R. Trippet and L. E. Garner

Another aspect of the accelerated public service role of the Bureau was the initiation in 1969 of a long-term project to survey and evaluate the mineral resources of all State-owned lands (588). The first phase, under the direction of W. R. Stearns, is entitled "Estimate of Future Oil and Gas Production from Established Fields of Texas Gulf Coast, Offshore, Bays, and Estuaries." The report on "Mineral Resources of University Lands (Exclusive of Oil and Gas)," compiled under the supervision of C. G. Groat and W. L. Fisher, was submitted to the University's Office of Investments, Trusts, and Lands in the fall of 1970.

Dr. W. L. Fisher became Director of the Bureau of Economic Geology on August 1, 1970, succeeding Dr. Flawn, who had been appointed Vice-President for Academic Affairs at The University of Texas at Austin. Dr. Fisher, who joined the Bureau in 1960 under Lonsdale, had done research primarily on the stratigraphy, sedimentology, and mineral resources of the Texas Gulf Coast Basin. *Rock and Mineral Resources of East Texas* by Dr. Fisher typifies his thorough and practical research (589). His studies of the Wilcox beds and other Gulf Basin units indicate the significance of deltaic and related deposition to the filling of the Gulf of Mexico Basin and the importance of these systems in oil and gas exploration. Fisher, along with Frank Brown, Charles Groat, and J. H. McGowen, initiated the work on the Environmental Geologic Atlas Project. Under his leadership, the Bureau continues to provide basic data for resource planning programs in Texas. At this time, the Energy Resources Investigations, under the directorship of Don G. Bebout, received high priority because of the energy crisis that gripped the nation. Finally, the basic geological research carried out by the geologists of the Bureau still represents a large percentage of the Bureau's research energies. Dr. Fisher's participation in the initiation of the Environmental Atlas Program, his fine study on *Lignites of the Texas Gulf Coastal Plain* (590), and his important basic research on ancient Gulf Coast clastic sediments mark his outstanding qualifications for leading the Bureau of Economic Geology in the three major research areas—ecology, energy, and basic research—in the fast-moving and challenging 1970's.

Under Dr. Fisher the Bureau today is a "beehive" of activity, as evidenced in the preceding short summary. This activity was probably equaled only by that of a much smaller Bureau during the emergence of the oil industry under Dr. Udden. The size and housing of the Bureau have changed tremendously from Udden's day to the present. However, throughout its life, the Bureau has always upheld the highest standards of basic geologic research and has always risen to play a major part in meeting whatever the resource demands of the day might be. The Bureau led the way in micropaleontology for the oil explorers in the 1920's and now serves in the forefront of the programs for meeting the environmental and energy crises of the 1970's.

Former Director Peter Flawn's words tell the story of the versatile, resilient Bureau of Economic Geology:

> The Bureau survived over the years because its leadership was able to anticipate change, adapt to it and provide new directions when they were called for. In my view, a close association with the Department of Geology is essential as long as the Bureau remains a part of The University of Texas. I think it is also worth making the point that all the directors of the Bureau have been scientists first and have done scientific work that received national recognition. For an organization as small as the Bureau, and with resources as limited as the Bureau's, to make the kind of contribution it has made over the years required maximum utilization of its people and its resources and cooperation with other state and federal agencies. If the directors had a common talent, it was making much out of little (591).

William Battle Phillips
Bureau director, 1909-1915

Johan August Udden
Bureau director, 1915-1932

Elias Howard Sellards
Bureau director, 1932-1945

Photo courtesy of AAPG

John Tipton Lonsdale
Bureau director, 1945-1960

Photo courtesy of AAPG

Peter T. Flawn
Bureau director, 1960-1970

William L. Fisher
Bureau director, 1970-present

J. A. Udden (left), **E. T. Dumble**, and **W. F. Cummings** in East Texas, 1916 (photograph by Wallace Bostick).

Charles Laurence Baker at Iguassú Falls on the Río Iguassú, a tributary of the Río Paraná, Argentina, 1922.

Emil Böse

Offices of the Bureau of Economic Geology (1928-June 1967), Little Campus, University of Texas, Austin, c. 1928.

J. A. Udden (left) and **R. T. Hill**, 1929.

W. S. Adkins (left) and **E. H. Sellards**, 1931.

Photo courtesy of AAPG

Joshua William Beede (1871-1940).

Virgil E. Barnes and pet, c. 1940.

Elias H. Sellards, 1956.

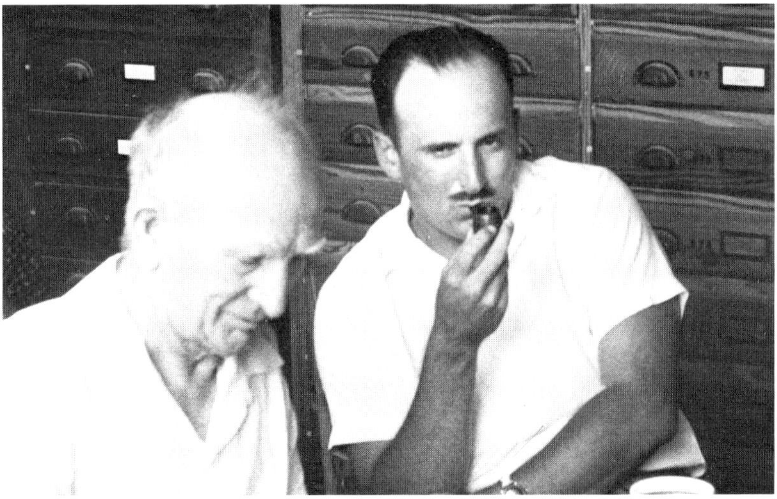

Frank A. Herald (left) and **Peter T. Flawn**, c. 1957 (photograph by Roselle Girard).

John T. Lonsdale, 1959.

Bureau of Economic Geology staff, December 1960: first row (left to right) — Dan Schofield, Ross Maxwell, W. L. Fisher, Philip Morey, Barbara Hartmann, Nancy Pierce, James Macon; second row — Peter T. Flawn, Josephine Casey, Mary Elizabeth Wheeler, Charles Hoker, John Robinson, Eloise Hill, Nancy Keeton; third row — Marce Morrow, Grant Milligan, Cader Shelby, James Pegg, Bill Harris, John Dietrich, Tom Doebbler, Roselle Girard.

Photo courtesy of AAPG

Esther Richards Applin
1895-1972

Virgil E. Barnes
1959

L. F. Brown, Jr.
c. 1957

Photo courtesy of AAPG

Halbert Pleasant Bybee
1888-1957

Mary Josephine Casey
1957

Photo courtesy of AAPG

Alexander Deussen
1882-1959

Photo courtesy of AAPG

Alva Christine Ellisor
1892-1964

W. L. Fisher
c. 1962

Photo courtesy of AAPG

Julia Anna Gardner
1882-1960

Photo courtesy of GSA

Ralph Alexander Liddle
1896-1963

Ross A. Maxwell
1952

Photo courtesy of AAPG

Leroy Thompson Patton
1880-1957

Photo courtesy of AAPG

Frederick Byron Plummer
1886-1947

Photo courtesy of AAPG

Helen Jeanne Plummer
1891-1951

H. B. Stenzel
1899-1980

NOTES

1. For further information on the attacks on the University by Governor Ferguson, *see* **Ralph W. Steen,** *The Political Career of James E. Ferguson:* Univ. Texas, Master's thesis, 1929.

2. A concise biography of W. B. Phillips is included in Univ. Texas Bull. 3501 (Quarter-Centennial Memorial Volume), p. 10. E. H. Sellards probably wrote it, since he collected a folder on Phillips (found in Box 44-92, Bureau of Economic Geology, University of Texas Archives). This folder includes an informative letter from University President **W. J. Battle,** a relative of Phillips, to **E. H. Sellards** (May 6, 1935, Bureau of Economic Geology).

For the best account of Udden's life prior to joining the staff of the Bureau, *see* **Monica Heiman,** *Johan August Udden, A Biography* (privately printed, Kerrville, Tex., 1963).

See also **Wallace Pratt** to **W. K. Ferguson,** Aug. 11, 1973:

> To me Dr. Udden was something like a hero. To me he was always a modest, thoroly [sic] gentle man. No official could have been more approachable. When I went to him in 1916, a lowly assistant geologist with Producer's Oil Company, on my problems with the just discovered deep production in the Bend in Stephens County (a full year earlier, note please, than the spectacular Ranger Field discovery in 1917), Dr. Udden received me cordially, on terms of full equality. He was patient, candid and vastly informative. He knew the Bend and the Bend Arch well. Few others did at that time. A few years later, in 1919 I went to him to ask about a small oil well at the base of the 1000-foot section of beds which fill the Pecos River trough at Pecos, Texas. I described the rocks to him and asked him to identify them. He immediately confessed that he didn't know. He was interested and curious but the rocks were new to him. They were clearly not Permian but what their age was he did not know and frankly said so. We learned subsequently, of course, that the Pecos River fill accumulated as the Permian salt was removed by solution during the late Tertiary.
>
> When I saw the Bureau's Annual Report for 1972 I thought of Dr. Udden. How gratified he would be to know of all its activities and its ample staff in contrast with its slender resources during his administration. Then a single undertaking, *The Deep Boring at Spur,* a landmark in the development of technology in the oil industry in Texas, occupied the entire staff for months.

3. For further information on the University of Texas Mineral Survey, *see* **W. Keene Ferguson,** *Geology and Politics in Frontier Texas, 1845-1909:* Univ. Texas Press, Austin, Tex., p. 114-131, 1969.

4. Professor Frederick W. Simonds knew both the Battle family and Phillips during his tenure in the Geology Department at the University of North Carolina. He and Professor William J. Battle probably worked hard to get Phillips to join the University of Texas (Keith Young, personal comment).

5. **W. B. Phillips** to **Louis J. Wortham,** House of Representatives, Jan. 21, 1911. Correspondence of the Bureau of Economic Geology, University of Texas Archives.

6. **W. B. Phillips** to **E. W. Parker,** June 29, 1910, Bureau of Economic Geology.

7. Between his years with the Mineral Survey and the Bureau, W. B. Phillips was a consultant for the Chisos Mining Company in the Trans-Pecos. *See* **Kenneth B. Ragsdale,** *Quicksilver: Terlingua and the Chisos Mining Company:* Texas A&M Press, College Station, Tex., 1976.

8. For an introduction to the history of laws regarding mineral development on State lands, *see* **W. K. Ferguson,** *Geology and Politics in Frontier Texas, 1845-1909, op. cit.* (3), p. 110-113, 122-123.

9. **W. B. Phillips** to **Richard Burgess,** Jan. 18, 1913, Bureau of Economic Geology. Phillips noted that the annual amount received from sale of mineral lands during the previous 29 years had been $1400, which was insignificant to the school fund but sufficient for the examination of unsold mineral lands. Phillips also presented a plan to University President S. E. Mezes outlining a survey of the University lands in West Texas to be undertaken by Udden "to classify the lands with reference to water, and perhaps make a study of the structural geology" (**Phillips** to **Mezes,** June 8, 1911, Bureau of Economic Geology).

10. **W. B. Phillips** to **W. M. Courtis,** Sept. 29, 1914, Bureau of Economic Geology.

11. *See* Appendix IV for resume of early fuel investigations by Bureau geologists.

12. **E. H. Sellards,** *William Battle Phillips* [biographical sketch]: Univ. Texas Bull. 3501, Quarter-Centennial Memorial Volume of the Division of Natural Resources, p. 10, 12-13, 1935.

13. **J. A. Udden** to **J. L. Newman,** Feb. 14, 1920, Bureau of Economic Geology.

14. There are many letters by Udden that indicate that he was a man with a real sense of humor. Some examples are: "You've heard of the three sweetest words in the English language, I suppose? Enclosed find check!" (**J. A. Udden** to **O. C. Wheeler,** Sept. 1, 1920, Bureau of Economic Geology).

"I never intended (the term Devils River limestone) as a serious geological term. I had no time for making any discriminating study of this series of rocks. My use of the term must be acknowledged to represent a retrograde moment in stratigraphy" (**Udden** to **R. T. Hill,** Dec. 18, 1920, Bureau of Economic Geology).

15. **J. A. Udden,** *Suggestions of A New Method of Making Underground Observations:* Am. Assoc. Petroleum Geologists Bull., v. 4, no. 1, p. 83-85, 1920.

16. **J. A. Udden,** *Oil Bearing Formations in Texas:* Am. Assoc. Petroleum Geologists Bull., v. 3, p. 82-98, 1919.

17. **J. A. Udden,** *The Deep Boring at Spur [Dickens County, Texas]:* Univ. Texas Bull. 363 (Sci. Ser. 28), 101 p., 1918.

18. **E. W. Owen,** *The Trek of the Oil Finder: The History of Petroleum Exploration:* Am. Assoc. Petroleum Geologists, Tulsa, Okla., 1647 p., 1975.

19. **J. A. Udden** to **Homer A. Noble,** Apr. 21, 1921, Bureau of Economic Geology.

Director Phillips had initiated this sample library service, and the Bureau correspondence for 1912 is filled with letters by Udden asking operators for samples and driller's logs. The description of samples remained a vital service offered to oil companies until the companies began their own subsurface laboratories in the early 1920's. As late as 1921, Udden wrote: "We are glad to receive, examine and describe drilling samples from any part of the State; quite a number of companies are availing themselves of our service" (**Udden** to **Edward Sells,** May 21, 1921, Bureau of Economic Geology).

20. **J. A. Udden,** *Aids to Identification of Geological Formations:* Univ. Texas, Bur. Econ. Geology and Technology, Handbook no. 1, 69 p., 1919(?).

Udden described the purpose of this handbook in a letter to University President **W. M. W. Splawn** (May 16, 1925, Bureau of Economic Geology):

> I trust it can be claimed for me that I have been instrumental in bringing about a certain amount of cooperation and a better understanding among the geologists on one side and the drillers and operators on the other, I have aimed to interest the oil trade in taking more accurate observations in making deep explorations. . . .

In this well-done little book, Udden describes the appearance of the most common minerals found in cuttings, and the formations in which they are found.

21. For a history of the development of micropaleontology in the Southwest, *see* the excellent study by **Carey Croneis,** *Micro Paleontology — Past and Future:* Am. Assoc. Petroleum Geologists Bull., v. 25, no. 7, p. 1208-1255, 1941. Croneis commented that most agreed with the words of Hedwig Kniker (p. 1224):

> One fact that has always impressed me is that in those early years Dr. Udden trained several micropaleontologists, one after another, and then sent them out to work for oil companies, thus furnishing trained personnel for the work he considered of prime importance. He realized that his lab could not begin to look at all of the samples that should be worked, so this is a happy solution.

22. *Ibid.*, p. 1221.

23. **J. A. Udden** and **D. M. Phillips,** *A Reconnaissance Report on the Geology of the Oil and Gas Fields of Wichita and Clay Counties:* Univ. Texas Bull. 246 (Sci. Ser. 23), 308 p., 1912.

24. **J. A. Udden,** *The Deep Boring at Spur [Dickens County], op. cit.* (17), p. 75-87.

25. **J. A. Udden** and **H. P. Bybee,** *The Thrall Oil Field:* Univ. Texas Bull. 66, p. 3-78, 1916.

26. Most of this information is taken from **Carey Croneis,** *Micro Paleontology — Past and Future, op. cit.* (21).

E. W. Owen presents a very perceptive analysis of the pioneer work of the Rio Bravo Oil Company under Dumble in *The Trek of the Oil Finder: History of Petroleum Exploration, op. cit.* (18), p. 523-525.

See also **Wallace Pratt** to **W. K. Ferguson,** Aug. 11, 1973:

> His name is J. Wallace Bostick (he is proud of it and he is a cousin of Dr. Dumble, not his "son-in-law"). J. Wallace wrote me recently that "cousin" Theodore Dumble had influenced him to specialize in micropaleontology. He did so, first at the University of California at Berkeley where he took his degree, then at Texas under Dr. Udden. I have reason to believe that J. Wallace, rather than Dr. Udden himself, trained Esther Richards and others of the early micropaleontologists in the oil industry in Texas. J. Wallace is one of the most remarkable men I ever knew. John Suman who knew him well said of him, "He tells the God damdest lies and you know all the time they are lies, but the hell of it is they turn out to be the absolute truth." J. Wallace was a doer as well as a thinker. Someone should write his biography.

27. Edmund Bell Stiles, who had just returned to the Bureau after service in France during World War I, was hired to replace Waite in 1918 (**J. A. Udden** to University President **R. E. Vinson,** Aug. 4, 1918, Bureau of Economic Geology). Udden and Stiles identified the more common foram assemblages. Stiles resigned as Assistant in Charge of the Subsurface Laboratory on Nov. 30, 1921.

Paul Theodore Seashore replaced Stiles until his resignation in January of 1923. Drue Christener assisted Seashore. (*See* the Catalogue of The University of Texas for the years concerned.)

28. **Carey Croneis,** *Micro Paleontology — Past and Future, op. cit.* (21), p. 1223. This paper was later published as **J. A. Udden** and **V. V. Waite,** *Some Microscopic Characteristics of the Bend and Ellenburger Limestones:* Univ. Texas Bull. 2703, 8 p., 1927.

29. **Carey Croneis,** *Micro Paleontology — Past and Future, ibid.*

30. **Hedwig Kniker** and **J. W. Beede,** *Species of the Genus* Schwagerina *and Their Stratigraphic Significance:* Univ. Texas Bull. 2433, 96 p., 1924.

The Bureau also published the following pioneer studies by another leading woman scientist, **Helen Jeanne Plummer:** *Foraminifera of the Midway Formation of Texas* (Univ. Texas Bull. 2644, 206 p., 1926), which includes an excellent discussion of the depositional history of the Midway; *Calcareous Foraminifera in the Brownwood Shale near Bridgeport, Texas* (Univ. Texas Bull. 3019, 90 p., 1930), which includes a discussion of the relationship of the environment to shell composition and alteration of tests; and *Some Cretaceous Foraminifera in Texas,* in *Contributions to Geology, 1931* (Univ. Texas Bull. 3101, p. 109-203, 1931).

31. Dr. F. L. Whitney initiated one of the earliest micropaleontology courses at the University of Texas in 1923. His assistant, Dorothy Ogden Carsey, authored *Foraminifera of the Cretaceous of Central Texas* (Univ. Texas Bull. 2612, 56 p.) in 1926. This is an excellent study of the various foraminifera, their distribution, and their use as key horizon markers. Other notable Gulf Coast micropaleontologists who received degrees in geology at the University include Alva Ellisor, Hedwig Kniker, and Esther Applin (nee Esther Richards) (**Keith Young,** *History of Geology at The University of Texas:* Univ. Texas, Bur. Econ. Geology Geol. Circ. 67-3, p. 14, 1967).

An interesting anecdote on these pioneer women scientists is found in **Alva Ellisor**, *Rockhounds of Houston* (privately printed, Houston, Tex., p. 13, 1947):

In 1920 micropaleontological work was launched in Houston. On Labor Day of that year Esther Richards, a graduate of the University of California, started to work for Dr. E. T. Dumble and on the last day of November, Alva C. Ellisor of the University of Texas joined the Humble geologic department. In 1921, the Texas Company employed Hedwig T. Kniker, also from the University of Texas, as micropaleontologist. Laura Lee Lane was associated with Esther Richards of the Rio Bravo during the summer of 1922, and after graduating from the University of Texas in June, 1923, she became micropaleontologist for Alexander Deussen. Grace Newman of California became Esther Richards' assistant in the fall of 1922 and Emma Jane Coffman of Oklahoma University, now Mrs. John Miller, joined the paleontological laboratory of Humble in June 1923. One more in this group at this time was Elisabeth Stiles, also from the University of Texas. She started with the Gulf Company as geologist and later was one of the pioneers of the geophysical work of that company.

In 1928, Dr. Sellards tried unsuccessfully to induce the father of micropaleontology in the United States, Dr. Joseph A. Cushman, to join the faculty of the University of Texas **(E. H. Sellards** to **J. A. Cushman**, Jan. 5, 1928, and **Cushman** to **Sellards**, Jan. 10, 1928, Bureau of Economic Geology).

32. **J. A. Udden** to University President **R. E. Vinson**, Dec. 11, 1918, Bureau of Economic Geology.

33. **Otis Smith**, Forty-First Annual Report of the U.S. Geological Survey, p. 10. For a thorough discussion of this topic, *see* **W. K. Ferguson**, *Geology and Politics in Frontier Texas, 1845-1909, op. cit.* (3) p. 114-131.

34. Minutes of the Board of Regents, Vol. C (June 7, 1909), p. 463.

35. The ever-present Robert T. Hill also appears to have had some influence upon the creation of the Bureau. In a letter to University of Texas President H. Y. Benedict several decades later, Hill mentions having addressed the Board of Regents late in 1909 on the subject of reestablishing a geological bureau under the control of the University **(Robert T. Hill** to **H. Y. Benedict**, Nov. 30, 1934, Benedict Coll., Univ. Texas Archives, Austin, Texas).

36. Catalogue of the University of Texas, 1909-1910, p. 209.

37. Minutes of the Board of Regents, Vol. C (June 11, 1909), p. 209.

38. *Ibid*, Vol. C (Oct. 21, 1909), p. 476.

39. For a lively and thorough review of the history of the Department of Geology, *see* **Keith Young**, *op. cit.* (31). Young discusses the lack of cooperation between the Bureau and the Department of Geology that undoubtedly hampered the effectiveness of both organizations. This breach lasted until the late 1940's (p. 12-13).

40. **J. A. Udden** to **George E. Bartholomew**, Dec. 31, 1917, Bureau of Economic Geology.

41. *See* **Miss M. E. Stiles**, Annual Report for the Year Ending December 31, 1916: Univ. Texas Bull. 1916, no. 35 (June 20, 1916), p. 6.

42. **J. A. Udden, C. L. Baker,** and **Emil Böse,** *Review of the Geology of Texas:* Univ. Texas Bull. 44, 267 p., 1916, revised 1919.

43. **F. B. Plummer** to **J. A. Udden,** July 15, 1921, Bureau of Economic Geology.

44. **F. B. Plummer** to **J. A. Udden,** Feb. 6, 1922, Bureau of Economic Geology. *See also* **E. W. Owen** to **W. K. Ferguson,** July 16, 1973.

Fred Plummer was important, but his complex personality is hard to portray. I met Fred first when he was in charge of the Shell work in 1918; I had an office next door to him for a few days in 1926, and saw him often during the 1930s. He was a different man each time, but always excited and exciting. On every occasion he gave the impression that I was his best and most intimate friend; this was his peculiar way of being "outgoing." A man of rare talents, he was often miles ahead of the field during his early years. Some of his latest work got somewhat foggy.

See also **Wallace Pratt** to **W. K. Ferguson,** Aug. 11, 1973:

Fred Plummer I knew intimately and vastly admired. To this day I am indebted to him for what my contact with him taught me of the geology of both the Paleozoic rocks of North-Central Texas and the Gulf Coast Tertiary section. Although his complexion was florid he was a frail man of unusually slender stature — an infinitely gentle creature. He fairly radiated intelligence and his knowledge of geology was profound. Helen Jeanne, his wife, was a competent micropaleontologist and an able assistant. But the Plummer team found their paramount interest in "pure" geology rather than the search for oil and Fred's effort to find a position on the Bureau staff, even at reduced compensation, is wholly in character.

45. **F. B. Plummer** and **Raymond C. Moore,** *Stratigraphy of the Pennsylvanian Formations of North-Central Texas:* Univ. Texas Bull. 2132, 228 p., 1921.

46. *See* **Wallace Pratt** to **W. K. Ferguson,** Aug. 11, 1973:

Charles Lawrence [Laurence] Baker worked for Dr. Dumble in East Texas in 1920. With him in the field were John Suman, J. Wallace Bostick, and others. Charles Lawrence [Laurence] dominated the group as a geologist. He was outspoken, satirical and cryptic. Already he had written extensively of the geology of Texas and Mexico and they brought him no little prestige. We looked upon him as an authority. You will want to note the unusual tribute in the form of a special publication, reproducing some of his early papers on Mexico, paid to Baker recently by the Geological Society of America.

47. **C. L. Baker,** *Major Structural Features of Trans-Pecos Texas,* in *The Geology of Texas, Vol. II, Structural and Economic Geology:* Univ. Texas Bull. 3401, Jan. 1, 1934, p. 137-214, Dec. 1935.

Baker produced very significant works on this area over the years.

48. There is considerable correspondence about Böse's plight. Perhaps the most penetrating is a letter from **J. A. Udden** to **T. C. Chamberlain** (Sept. 27, 1917, Bureau of Economic Geology) from which the following excerpts are taken:

Some time after Diaz left Mexico, Böse wrote asking for a position as conditions were such that he could no longer remain in Mexico. . . . Dr. Böse is an officer in the Naval Reserve of Germany and it was impossible for him to re-enter the U.S. after leaving in March. I had some conversation with him a couple of years before he got the position. In our talk I learned that he was intensely partisan for Germany and the German government and that he was also imbued with intense contempt for democratic government. . . .

I have heard from other sources several years ago of the behavior of the Germans in Mexico at the time of the meeting of the International Geologic Congress. I have also sometimes thought that I saw some evidence of the war-like intentions of Germany in the attitude they took with regard to the fixing of the date for the next Congress after the meeting at Toronto. There seemed to be a knowledge of uncertainty as to the possibility of arranging for an international meeting in Belgium at the usual interval of time, but this may be a wrong guess. My opinion is that Dr. Böse is innocent of any spying while here. . . .

49. Something of the self-assured personality of the noted German paleontologist is revealed in a letter he wrote to Dr. Udden thanking him for coordinating the work of W. S. Adkins with that of Böse in North Texas (**E. Böse** to **J. A. Udden,** Dec. 18, 1918, Bureau of Economic Geology):

I am certainly glad that you sent Mr. Adkins and **Mr.** Winton to me. I am sure you will get a first class map of Tarrant County. I see that they know every horizontal layer of every fossil in the Washita and the Upper Fredericksburg as far as it is developed here in the north, which is certainly not much.... They were somewhat astonished to find out that my stratigraphic results gained within the first week were entirely corresponding to those they had gained in the last five years, but of course, I have the advantage of a vaster experience and a greater knowledge of European literature.

One of the unofficial services provided by the Bureau was to function as a clearinghouse for information on the paleontology of the State. In the Adkins Papers in the University of Texas Archives, there is a considerable amount of correspondence between Adkins and Udden, and Adkins and Böse, concerning the paleontological work being done by the two Cretaceous paleontologists in both Texas and Mexico. Udden often served as the middleman in the exchange between the two. Dr. T. N. Stanton of the U.S. Geological Survey was also consulted by Udden as a vital source of definitive paleontological information.

50. **Emil Böse,** *Cretaceous Ammonites from Texas and Northern Mexico,* in Univ. Texas Bull. 2748, 357 p., 1927.

51. Alexander Deussen of the U.S. Geological Survey was also considered as a replacement for Böse (**Deussen** to **J. A. Udden,** Sept. 5, 1927, Bureau of Economic Geology). For Deussen's early background, *see* "Our Honorary Life Members" *in* **Alva Ellisor,** *Rockhounds of Houston, op. cit.* (31), p. 22-24.

52. **E. H. Sellards** to University President **R. E. Vinson** (Aug. 21, 1922, Bureau of Economic Geology).

An interesting anecdote on Beede is found in a letter he wrote to Dr. David White, Chief Geologist of the U.S. Geological Survey (Feb. 19, 1921, Bureau of Economic Geology):

I have been wishing to thank you for your kindness in referring the position in the Government University in Peking and the Geological Survey of China to me. Had I been within reach of the telegraph or rail I would have accepted at once; but I was at that time 1200 miles beyond either rail or telegraph, on the Mackenzie River.... It rarely falls to the lot of a man to be able to start a science like paleontology among a race of people like the Chinese.

53. **J. A. Udden** to **E. H. Sellards** (June 13, 1925, Bureau of Economic Geology). The story about Dr. Barbour's influence on Sellards is in **Alva Ellisor,** *Rockhounds of Houston, op. cit.* (31), p. 26-27.

54. **E. H. Sellards,** *On the Underground Position of the Ellenburger Formation in North-Central Texas, With a Preliminary Contour Map:* Univ. Texas Bull. 1849, 32 p., 1918.

55. **E. H. Sellards, B. C. Tharp,** and **R. T. Hill,** *Investigations on the Red River Made in Connection with the Oklahoma-Texas Boundary Suit:* Univ. Texas Bull. 2327, 174 p., 1923.
See also Appendix III for short overview of this work.

56. Another man who served with the Bureau during the first decade was: **R. A. Liddle,** Assistant Geologist, 1919-1920, who was actually hired as a topographer (**J. A. Udden** to **Liddle,** Dec. 4, 1919). Udden had felt the need for topographic mapping to be so urgent during the war when the energies of the Geological Survey were focused on the war effort that he concluded that the Bureau would have to engage the State Reclamation Department to make such maps, or do its own topographic work. He felt that the best plan would be to contract the "well trained and well organized corps of topographers of the State Reclamation Department" (**Udden** to University President **R. E. Vinson,** Sept. 12, 1916). This contract work was to begin in January 1918 with the mapping of Upton and Crane Counties. However, the resumption of Geological Survey mapping after the war cancelled the need for this work.
Thomas L. Bailey and **Leroy Patton** were hired as Associate Geologists in 1922 and served with the Bureau through 1925. Bailey was hired to replace Adkins. His training at the University of California included special work in Tertiary paleontology.

57. **J. A. Udden** to **R. E. Vinson,** July 18, 1921, Bureau of Economic Geology.
Udden also was rankled that the Bureau staff members did not have the ranking or salary accorded other members of the Faculty:

> I consider it of the greatest importance to the future progress of the Bureau that men of such standing as [Dr. Sellards, Dr. Beede, and Mr. Adkins] have should be accorded equal rank with the members of the teaching staff; the title of geologist should be rated equal to that of full professor, with accompanying salary, and that of associate geologist rated as associate professor with accompanying salary. . . .
>
> (J. A. Udden, no date, in "President" folder for 1921, Bureau of Economic Geology)

58. The controversial Hill, who devoted his life to Texas geology, had served with the U.S. Geological Survey from 1892-1902. Then he drifted from New York to California to Dallas as a consultant. In 1930 there was a move to obtain a position for him with the Bureau (**E. H. Sellards** to **R. T. Hill,** May 20, 1930, Bureau of Economic Geology). Nothing came of it.
In 1931, **David White** of the U.S. Geological Survey wrote a letter to **Sellards** (Feb. 16, 1931) in deserved glowing praise of Hill and asked:

> Is it now possible that if a research fund sufficient to carry Hill through his remaining period with opportunity, free from worry, to give all his time to his manuscripts were raised it might grow into a permanent Robert T. Hill research fellowship in your University when, in later years, some petroleum geologists were able to contribute adequately for its permanent foundation. . . .

Hill finally did receive a grant from the Penrose Fund of the Geological Society of America (cf. 434).

59. **F. W. DeWolf** to **D. C. Barton**, Jan. 13, 1927, Barton Coll., Houston Public Library, Houston, Texas.

60. Anonymous, "Texas Niggardly With Funds for Geological Department," *Oil Weekly*, Apr. 29, 1927, p. 43, 46.

61. For the story of this lobbying campaign, *see* the correspondence in the Donald Clifton Barton Collection, Houston Public Library. Note especially **Barton** to **M. E. Foster**, Editor of the *Houston Press*, Feb. 12, 1927, and form circular letter to the Presidents of the Geological Societies of Texas, Feb. 12, 1927.

62. **E. H. Sellards** to University President **H. Y. Benedict**, Jan. 13, 1928, Bureau of Economic Geology.

63. **E. H. Sellards** to **E. P. Swenson**, July 9, 1930, Bureau of Economic Geology. Sellards asked the Swenson Land and Cattle Company to contribute to this fund as compensation for F. B. Plummer's consultations on the geology of Throckmorton County. *See* Report to the President, Apr. 29, 1927, for the role of the Fort Worth Geological Society. For the subsequent contributions, *see* **R. S. Powell** of the North Texas Geological Society to **Sellards**, Jan. 8, 1932; **Leo C. Haynes**, Secretary of the Board of Regents, to the San Antonio Geological Society, Aug. 6, 1930 (all correspondence in Bureau of Economic Geology files).

Josephine Casey wrote summaries of the beginnings of the several revolving publication funds. These are found in the account books.

64. **J. A. Udden** to **W. S. Sutton**, Acting President, Jan. 8, 1924, Bureau of Economic Geology.

65. **E. H. Sellards** to **Wallace Pratt**, May 9, 1927, Bureau of Economic Geology.

66. **J. V. Howell**, of the Marland Oil Company, to **E. H. Sellards**, Dec. 23, 1927, Bureau of Economic Geology.

67. **Wallace Pratt** to **E. H. Sellards**, June 13, 1927, Bureau of Economic Geology.

Back in 1909, Dr. W. B. Phillips had approached the Weatherford, Mineral Wells, and Northwestern Railway Company whose tracks traversed the coal fields (**Phillips** to **P. E. Bock**, Second Vice-President of the W., M. W., and N. Ry., Nov. 27, 1909, Bureau of Economic Geology). He also "made a pitch" to the Texas and Pacific Railway Company. He got negative replies from both. The reply from L. L. Thorne, Vice-President of the T. and P. Ry., was typical: "At first thought, this hardly seems to us a fair proposition, as the railroads already are carrying a heavy load in the way of taxes, advertising, etc." (**L. L. Thorne** to **Phillips**, Nov. 30, 1909, Bureau of Economic Geology).

68. **E. H. Sellards** to **W. M. W. Splawn**, June 16, 1925, Bureau of Economic Geology. The Division of Conservation and Development of the Natural Resources of Texas became the new name in 1928.

69. **L. S. Brown** to **E. H. Sellards**, June 8, 1932, Bureau of Economic Geology.

70. **E. H. Sellards** to **H. W. Harper**, Dean of the Graduate School of the University of Texas, Jan. 26, 1927, Bureau of Economic Geology. It should be added as a point of interest that the Bureau moved to its location at the "Little Campus" off 19th Street and Red River in the summer of 1927 (**Sellards** to **K. C. Heald**, Aug. 29, 1927, Bureau of Economic Geology).

The Petroleum Production Engineering Department was established in 1927 by Dr. Sellards. The scope of its courses included "prospecting for oil, bringing in and developing new fields, and producing and transporting oil to market" (Catalogue of the University of Texas, 1927-1928, Univ. Texas Bull. 2817, May 1, 1928, p. 271). F. B. Plummer took over three of the courses in 1928, serving as Acting Professor of Petroleum Production Engineering. Sellards continued to teach the stratigraphy course. Plummer became full professor in 1930, and the department offered a Master of Science degree for the first time.

71. *See* Appendix II, Memorial to J. A. Udden by C. L. Baker.

72. Forty-First Annual Report of the U. S. Geological Survey, 1920.

73. **L. S. Brown** to **E. H. Sellards**, June 8, 1932, Bureau of Economic Geology.

74. *See* Appendix VII, **E. H. Sellards,** *Some Events in the Geologic History of Texas* (manuscript, undated).

75. **Alexander Deussen,** *Thirty Five Years of Progress in the Knowledge of the Geology of Texas,* in *Quarter-Centennial Memorial Volume of the Division of Natural Resources:* Univ. Texas Bull. 3501 (Jan. 1, 1935), 298 p.

This includes interesting sketches on the discovery of the geology of the Luling-Mexia fault zone and the Sabine uplift.

76. **P. B. King,** *The Geology of the Glass Mountains, Texas, Part I, Descriptive Geology:* Univ. Texas Bull. 3038, 167 p., 1930; and **R. E. King,** *The Geology of the Glass Mountains, Texas, Part II, Faunal Summary and Correlation of the Permian Formations with Description of Brachiopoda:* Univ. Texas Bull. 3042, 146 p., 1930.

77. **F. B. Plummer** and **R. C. Moore,** *Stratigraphy of the Pennsylvanian Formations of North-Central Texas, op. cit.* (45). For data on the micropaleontological studies, *see* footnotes 21 and 30, above.

78. For an excellent review of the geological investigations in the Gulf Coastal Plain, *see* **F. B. Plummer,** *Cenozoic Systems of Texas,* in *The Geology of Texas, Vol. I, Stratigraphy:* Univ. Texas Bull. 3232 (Aug. 22, 1932), p. 520-524, 1933.

A complete discussion of the Cenozoic geology of Texas is covered on p. 518-817.

79. However, **E. T. Dumble,** *The Geology of East Texas* (Univ. Texas Bull. 1869, 325 p., 1918); **R. A. Liddle,** *Geology and Mineral Resources of Medina County* (Univ. Texas Bull. 1830, 172 p., 1918); and **E. H. Sellards,** *Geology and Mineral Resources of Bexar County* (Univ. Texas Bull. 1932, 202 p., 1919) were significant stratigraphic reports on this area published by the Bureau in the late teens.

80. **T. L. Bailey,** *The Geology and Natural Resources of Colorado County:* Univ. Texas Bull. 2333, 159 p., 1923.

81. **T. L. Bailey,** *The Gueydan, A New Middle Tertiary Formation from the Southwestern Coastal Plain of Texas:* Univ. Texas Bull. 2645, 187 p., 1926.

82. **B. C. Renick** and **H. B. Stenzel,** *The Lower Claiborne on the Brazos River, Texas,* in *Contributions to Geology, 1931:* Univ. Texas Bull. 3101, 129 p., 1931.

83. H. B. Stenzel received his doctorate from Schlesische Friedrich Wilhelms University in Breslau, Germany, in 1922. He was an assistant lecturer at the Geologische Institut at this school until 1925, when he joined the faculty at Texas A&M University.

84. **R. A. Jones,** *The Pratt Well in Webb County,* in *Contributions to Geology, 1929:* Univ. Texas Bull. 2901, 226 p., 1929.

85. Twenty-First Annual Report of the U.S. Geological Survey, Part 7, 1901.

86. For an outstanding description of geological work in the Cretaceous rocks of Texas, *see* **W. S. Adkins,** *The Mesozoic Systems of Texas,* in *The Geology of Texas, Vol. I, Stratigraphy:* Univ. Texas Bull. 3232 (Aug. 22, 1932), p. 159-517, 1933.

87. **E. W. Shuler,** *The Geology of Camp Bowie and Vicinity* (Univ. Texas Bull. 1750, 14 p., 1917); **E. W. Shuler,** *The Geology of Dallas County* (Univ. Texas Bull. 1818, 54 p., 1918); **R. A. Liddle** and **T. M. Prettyman,** *Geology and Mineral Resources of Crockett County, With Notes on the Stratigraphy, Structure, and Oil Prospects of the Central Pecos Valley* (Univ. Texas Bull. 1857, 97 p., 1918); **W. M. Winton** and **W. S. Adkins,** *The Geology of Tarrant County* (Univ. Texas Bull. 1931, 122 p., 1919); **E. H. Sellards,** *The Geology and Mineral Resources of Bexar County, op. cit.* (79); **W. M. Winton** and **Gayle Scott,** *The Geology of Johnson County* (Univ. Texas Bull. 2229, 68 p., 1922); **W. S. Adkins,** *Geology and Mineral Resources of McLennan County* (Univ. Texas Bull. 2340, 86 p., 1925); **F. M. Bullard,** *The Geology of Grayson County, Texas* (Univ. Texas Bull. 3125, 72 p., 1931); **Gayle Scott** and **J. M. Armstrong,** *The Geology of Wise County* (Univ. Texas Bull. 3224, 77 p., 1932).

88. **W. S. Adkins,** *Handbook of Texas Cretaceous Fossils* (Univ. Texas Bull. 2838, 385 p., 1928); **W. S. Adkins,** *The Weno and Pawpaw Formations of the Texas Comanchean* (Univ. Texas Bull. 1856, 172 p., 1918); **W. S. Adkins** and **W. M. Winton,** *Paleontological Correlation of the Fredericksburg and Washita Formations in North Texas* (Univ. Texas Bull. 1945, 128 p., 1919); **W. S. Adkins,** *New Rudistids from the Texas and Mexican Cretaceous,* and *Texas Comanchean Echinoids of the Genus* Macraster, in *Contributions to Geology, 1930* (Univ. Texas Bull. 3001, 207 p., 1930); **W. S. Adkins,** *Some Upper Cretaceous Ammonites in Western Texas,* in *Contributions to Geology, 1931* (Univ. Texas Bull. 3101, 239 p., 1931).

89. **Emil Böse,** *On a New* Exogyra *from the Del Rio Clay and Some Observations on the Evolution of* Exogyra *in the Texas Cretaceous* (Univ. Texas Bull. 1902, 22 p., 1919); **Emil Böse,** *Cretaceous Ammonites from Texas and Northern Mexico (in* Univ. Texas Bull. 2748, p. 143-212, 1927); **Emil Böse** and **O. A. Cavins,** *The Cretaceous and Tertiary of Southern Texas and Northern Mexico (in* Univ. Texas Bull. 2748, p. 7-142, 1927).

90. **F. B. Plummer** and **R. C. Moore,** *Stratigraphy of the Pennsylvanian Formations of North-Central Texas, op. cit.* (45). For the surveys of the Pennsylvanian formations of the Osage Plains done by the Dumble Survey, *see* **W. K. Ferguson,** *Geology and Politics in Frontier Texas, 1845-1909, op. cit.* (3), p. 137-140. **N. F. Drake's** *Report on the Colorado Coal Field of Texas,* which appeared in the Fourth Annual Report of the Dumble Survey, was republished as Univ. Texas Bull. 1755, 75 p., 1917. In 1921, Udden wrote that "the publications [of Dumble and Cummins] on Texas geology are yet, after 30 years, standard works to be consulted by all who work on geological problems in this state" (**J. A. Udden** to **George Crossland**, Mar. 21, 1921, Bureau of Economic Geology).

91. **J. A. Udden**, assisted by **D. M. Phillips,** *A Reconnaissance Report on the Geology of the Oil and Gas Fields of Wichita and Clay Counties, op. cit.* (23).

92. **W. E. Galloway** and **L. F. Brown, Jr.,** *Depositional Systems and Shelf-Slope Relationships in Upper Pennsylvanian Rocks:* Univ. Texas, Bur. Econ. Geology Rept. Inv. 75, 62 p., 1972.

93. **F. B. Plummer** and **R. C. Moore,** *Stratigraphy of the Pennsylvanian Formations of North-Central Texas, op. cit.* (45), title page. **E. H. Sellards'** important pioneer work, *On the Underground Position of the Ellenburger Formation in North-Central Texas, With a Preliminary Contour Map, op. cit.* (54), consolidated all of the information gained through drilling in this area, and was an early indicator of the subsurface features that have controlled Pennsylvanian and Permian oil production in West and North-Central Texas.

94. *See* **W. K. Ferguson,** *Geology and Politics in Frontier Texas, 1845-1909, op. cit.* (3), p. 137-140.

95. **C. L. Baker,** *Origin of Texas Red Beds:* Univ. Texas Bull. 29, 8 p., 1916.

96. *Ibid.*, p. 8.

97. **J. W. Beede,** *The Geology of Runnels County* (Univ. Texas Bull. 1816, 64 p., 1918); **J. W. Beede** and **W. P. Bentley,** *The Geology of Coke County* (Univ. Texas Bull. 1850, 81 p., 1918).

98. *Ibid.*, p. 51.

99. **J. W. Beede** and **D. D. Christner,** *The San Angelo Formation and the Geology of Foard County:* Univ. Texas Bull. 2607, 57 p., 1926.

100. **C. N. Gould,** *The Correlation of the Permian of Kansas, Oklahoma, and Northern Texas:* Am. Assoc. Petroleum Geologists Bull., v. 5, no. 2 (Feb. 1926), p. 144-153. A fine earlier work is **C. H. Gordon,** *Geology and Underground Waters of Wichita Region, North-Central Texas:* U.S. Geol. Survey Water-Supply Paper 317, 88 p., 1913.

101. For a list of the publications on the vertebrate paleontology of Texas by **S. W. Williston,** *see* Univ. Texas Bull. 3232, p. 945-946. The list for **E. C. Case** is even more impressive *(ibid.,* p. 835-837). Of especial interest is his *The Red Beds Between Wichita Falls, Texas, and Las Vegas, New Mexico, in Relation to Their Vertebrate Fauna:* Jour. Geology, v. 22, p. 243-259, 1914. *See also* **C. D. White,**

Permian of Western America from the Paleontological Standpoint: Pan-Pacific Scientific Congress, Australia, 1923, Pr. 2, p. 1050-1077, 1924. The outstanding Bureau publication by **A. S. Romer,** *Vertebrate Faunal Horizons in the Texas Permo-Carboniferous Red Beds,* in *Contributions to Geology, 1928:* Univ. Texas Bull. 2801, 202 p., 1928, includes a good history of vertebrate collecting in Texas beds. Romer also presents the vertical distribution of the vetebrate remains in tabular form and a map of Baylor and Archer Counties showing the stratigraphy of the vertebrate deposits.

102. **C. N. Gould,** *The Correlation of the Permian of Kansas, Oklahoma, and Northern Texas, op. cit.* (100), p. 151.

103. **C. L. Baker,** *Depositional History of the Red Beds and Saline Residues of the Texas Permian,* in *Contributions to Geology, 1929:* Univ. Texas Bull. 2901, 226 p., 1929, p. 10-13.

104. **M. G. Cheney,** *Stratigraphic and Structural Studies in North-Central Texas:* Univ. Texas Bull. 2913, 29 p., 1929.

105. **E. H. Sellards,** *The Pre-Paleozoic and Paleozoic Systems in Texas,* in *The Geology of Texas, Vol. I, Stratigraphy:* Univ. Texas Bull. 3232 (Aug. 22, 1932), p. 147, 1933.

106. For a description of the reconnaissance work of J. A. Taff and R. S. Tarr of the Dumble Survey and that of R. T. Hill and George Richardson of the U.S. Geological Survey, *see* **W. K. Ferguson,** *Geology and Politics in Frontier Texas, 1845-1909, op. cit.* (3), p. 140-143.

107. **J. A. Udden,** *The Geology of the Shafter Silver Mine District, Presidio County, Texas:* Univ. Texas Bull. 24 (Min. Sur. Ser., Bull. 8), 1904, p. 20.

108. **J. A. Udden, C. L. Baker,** and **Emil Böse,** *Review of the Geology of Texas, op. cit.* (42).

109. *See* the W. S. Adkins Collection, Conversations with C. L. Baker.

110. **E. Böse,** *The Permo-Carboniferous Ammonoids of the Glass Mountains and Their Stratigraphic Significance:* Univ. Texas Bull. 1762, 241 p., 1917.

111. **J. A. Udden,** *Notes on the Geology of the Glass Mountains,* and **C. L. Baker** and **W. F. Bowman,** *Geologic Exploration of the Southeastern Front Range of Trans-Pecos Texas:* Univ. Texas Bull 1753, 177 p., 1917.

112. *Ibid.,* p. 77.

113. **C. L. Baker,** *Exploratory Geology of a Part of Southwestern Trans-Pecos Texas:* Univ. Texas Bull. 2745, 70 p., 1927.

114. **P. B. King,** *The Geology of the Glass Mountains, Texas, Part I, Descriptive Geology, op. cit.* (76), and **R. E. King,** *The Geology of the Glass Mountains, Texas, Part II, Faunal Summary and Correlation of the Permian Formations With Description of Brachiopoda, op. cit.* (76). *See also* **J. A. Udden** to **Charles Schuchert,** supervisor of the Kings' graduate work at Yale, for the financial support given by the Bureau (Mar. 3, 1927).

P. B. King, *Geology of the Marathon Region, Texas* (U.S. Geol. Survey Prof. Paper 187, 148 p., 1938), is an extension of the studies done in the adjoining area. King conducted 8 months of field work in the Marathon area in 1929 and 1930 under the auspices of the U.S. Geological Survey.

115. **P. B. King,** *The Geology of the Glass Mountains, Texas, Part I, Descriptive Geology, ibid.*,p. 12.

116. **J. A. Udden,** *The Deep Boring at Spur [Dickens County, Texas], op. cit.* (17); **J. A. Udden,** *Potash in the Texas Permian:* Univ. Texas Bull. 17, 59 p., 1915.

117. **J. A. Udden,** *Notes on the Geology of the Glass Mountains, op. cit.* (111), p. 57-58.

118. **Martin W. Schwettman,** *Santa Rita:* Univ. Texas Press, Austin, Tex., 43 p., 1943.

119. **H. W. Hoots,** *Geology of a Part of Western Texas and Southeastern New Mexico, With Special Reference to Salt and Potash:* U.S. Geol. Survey Bull. 780, p. 33-126, 1926.

120. *Ibid.*, p. 119.

121. *See* **E. H. Sellards,** *Paleozoic of the Llanoria Geosyncline*, in *The Geology of Texas, Vol. I, Stratigraphy* (Univ. Texas Bull. 3232 (Aug. 22, 1932), p. 127-130, 1933), and **P. T. Flawn, August Goldstein, Jr., P. B. King,** and **C. E. Weaver,** *The Ouachita System* (Univ. Texas Pub. 6120, p. 8-9, 1961) for the history of the recognition of this geosyncline.

E. O. Ulrich discusses the geosyncline in an interesting letter to **C. L. Baker** (Jan. 12, 1922, Bureau of Economic Geology):

> As to the relations of the Marathon formations to those of the Ouachitas, I was already convinced in 1910-11 that the latter extended in a curving direction beneath the cover of Mesozoic rocks to southwest Texas. This extension has been largely demonstrated by deep well records published by Sellards and Miser. This early view [of mine] was based on the probable distribution of Ouachita Ordovician graptolite zones before I had even heard of the Marathon outcrops of lower Paleozoics; and in a paleo-geographic map published on page 38 of the "Revision of the Paleozoic Systems" (Geological Society of America Bull., v. 22, p. 281-680, Aug. 31, 1911) I indicated with a thick, too strongly curved dashed line this extension of the Ouachita geosyncline southwestwardly [this curves southwestwardly to San Antonio area, and then veers south and southeast along the Gulf Coast of Mexico].

122. **J. A. Udden,** *Observations on Two Deep Borings Near the Balcones Faults (With Discussion):* Am. Assoc. Petroleum Geologists Bull. 3, p. 124-131, 1919.

123. *Ibid.*, p. 129.

124. **C. L. Baker** and **W. F. Bowman,** *Geologic Exploration of the Southeastern Front Range of Trans-Pecos Texas, op. cit.* (111), p. 111. *See also* **P. T. Flawn,** *Basement Rocks of Texas and Southeast New Mexico* (Univ. Texas Pub. 5605, p. 58-59, 1956) for discussion of early theories.

125. **E. H. Sellards,** *Paleozoic of the Llanoria Geosyncline, op. cit.* (121), p. 128-129.

126. **Sidney Powers**, *Age of the Folding of the Oklahoma Mountains — The Ouachita, Arbuckle, and Wichita Mountains of Oklahoma, and the Llano-Burnet and Marathon Uplifts of Texas:* Geol. Soc. America Bull. 39, p. 1031-1071, 1928.

127. **R. C. Moore**, *Framework of Southeastern North America* (abs.): Pan-Am. Geologist, p. 141, 1928.

128. **E. H. Sellards**, *Underground Position of the Pre-Cambrian Complex in Texas* (abs.): Geol. Soc. America Bull. 40, p. 134, 1929; Pan-Am. Geologist, v. 51, p. 159, 1929.

129. **M. G. Cheney**, *Stratigraphic and Structural Studies in North-Central Texas, op. cit.* (104).

130. **E. H. Sellards**, *Rocks Underlying the Cretaceous in the Balcones Fault Zone of Central Texas:* Am. Assoc. Petroleum Geologists Bull. 15, p. 819, 1931. **W. A. J. M. van Waterschoot van der Gracht** arrived independently at conclusions similar to those developed by the above-mentioned geologists and delivered his paper before the same New Orleans meeting. See *The Permo-Carboniferous Orogeny in the South-Central United States* (abs.): Am. Assoc. Petroleum Geologists Bull. 15, p. 991-1057, 1931, with correction in Am. Assoc. Petroleum Geologists Bull. 16, p. 102, 1932.

There is an excellent bibliography of previous studies on the Llanoria geosyncline in **W. S. Adkins** and **M. B. Arick**, *Geology of Bell County, Texas*: Univ. Texas Bull. 3016, 92 p., 1930. Another interesting report is **Richard A. Jones**, *The Paleozoic of the Pedernales Valley in Gillespie and Blanco Counties*, in *Contributions to Geology, 1929:* Univ. Texas Bull. 2901, p. 95-130, 1929.

131. **P. B. King**, *The Geology of the Glass Mountains, Texas, Part I, Descriptive Geology, op. cit.* (76), p. 114.

132. *Ibid.*, p. 116.

133. The history of geologic investigations in these Pliocene strata of West Texas is found in **F. B. Plummer**, *Pliocene Systems of West Texas*, in *The Geology of Texas, Vol. I, Stratigraphy, op. cit.* (78), p. 763-764. A fine bibliography on the geological literature in this area is in **L. C. Reed** and **O. M. Longnecker, Jr.**, *The Geology of Hemphill County, Texas:* Univ. Texas Bull. 3231, 98 p., 1932.

134. **C. L. Baker**, *Geology and Underground Waters of the Northern Llano Estacado:* Univ. Texas Bull. 57, 115 p., 1915.

135. The results of the work of Udden and Patton were published as the work of **L. T. Patton**, *The Geology of Potter County:* Univ. Texas Bull. 2330, 180 p., 1923.

136. Dr. Sellards campaigned for the establishment of a Geological Research Museum in Texas, and protested against the fact that "the fine [vertebrate] material of Texas is gradually going to the collection of other states or foreign countries or unfortunately is being lost when uncovered or otherwise brought to light" (mimeographed paper entitled *The Need of a Foundation for Geological Research in Texas, 1929,* Bureau of Economic Geology). *See also* **E. H. Sellards** to **C. L. Baker**, Dec. 11, 1928, Bureau of Economic Geology: "When will the time

come when this kind of work in vertebrate paleontology will be done in Texas and for Texas?"

137. **C. L. Baker,** *Geology and Underground Waters of the Northern Llano Estacado, op. cit.* (134), p. 7.

138. *Ibid.*, p. 8.

139. *Ibid.*, p. 24-32.

140. **L. C. Reed** and **O. M. Longnecker, Jr.**, *The Geology of Hemphill County, Texas, op. cit.* (133), p. 9.

141. *Ibid.*, p. 5.
A bibliography of W. D. Matthew's works on Texas vertebrates is found in *The Geology of Texas, Vol. I, Stratigraphy* (Univ. Texas Bull. 3232, Aug. 22, 1932, p. 894, 1933). Also, on p. 764, **F. B. Plummer** notes that Matthew, "just before his untimely death, prepared a manuscript setting forth his observations of the Tertiary geology of the northern Staked Plains in Texas." This manuscript, "Observations on the Tertiary of the Staked Plains," was submitted to the American Museum of Natural History in 1924.

142. **L. C. Reed** and **O. M. Longnecker, Jr.**, *The Geology of Hemphill County, Texas, op. cit.* (133), p. 5.

143. *Ibid.*, p. 5-6.

144. *Ibid.*, p. 6.

145. **H. X. Bay,** *A Study of Certain Pennsylvanian Conglomerates of Texas*, in *Contributions to Geology, 1932:* Univ. Texas Bull. 3201, p. 149-188, 1932. His bibliography includes all of the early classics on sedimentation.
Other pioneer works published by the Bureau which include sedimentary analyses include **T. L. Bailey,** *The Geology and Mineral Resources of Colorado County* (Univ. Texas Bull. 2333, 259 p., 1923); *The Gueydan, A New Middle Tertiary Formation from the Southwestern Coastal Plain of Texas* (Univ. Texas Bull. 2645, 187 p., 1926); and **L. T. Patton,** *The Geology of Potter County, op. cit.* (135).

146. **J. A. Udden,** *The Mechanical Composition of Clastic Sediments:* Geol. Soc. America Bull., v. 25 (Dec. 14, 1914), p. 655-744. An even earlier work is *The Mechanical Composition of Wind Deposits*, Augustana Library Publications No. 1, 1898.

147. **J. A. Udden,** *A Neglected Field in Stratigraphy,* in *Contributions to Geology, 1928:* Univ. Texas Bull. 2801, p. 55-66, 1928.

148. *Ibid.*, p. 58-59.

149. **E. H. Sellards** to **R. A. F. Penrose,** Jan. 26, 1927, Bureau of Economic Geology.

150. H. P. Bybee noted that in 1930 there were over 900 geologists in the state who were members of the A.A.P.G. (**Bybee** to **E. H. Sellards,** Apr. 19, 1930, Bureau of Economic Geology).

151. He also looked to avoid any duplication of effort. In 1917, he wrote that there was so much oil company work being done in the Gulf Coast region that the Bureau should expend its funds elsewhere (**J. A. Udden** to **W. B. Phillips**, Nov. 22, 1917, Bureau of Economic Geology).

152. **E. L. DeGolyer**, *The Development of the Art of Prospecting:* Princeton Univ. Press, Princeton, New Jersey, 38 p., 1940.

153. **E. W. Owen**, *Trek of the Oil Finder: The History of Petroleum Exploration, op. cit.* (18), p. 749.

154. *Ibid.*, p. 750.

155. *See* the well-researched *History of HO&RC, A Study in Industrial Growth*, by **Henrietta Larson** and **Kenneth W. Porter**: Harper & Bros., New York, 769 p., 1959.

Alva Ellisor presents a fine discussion of Pratt's early background in *Rock-hounds of Houston, op. cit.* (31), p. 24-26. **Everett DeGolyer's** *Wallace Everette Pratt, First Sidney Powers Memorial Medalist, An Appreciation* (Am. Assoc. Petroleum Geologists Bull., v. 19, no. 4 [May 1945], p. 478-490) is an outstanding tribute to this great man and brilliant geologist.

156. **Wallace Pratt**, *Geological Structure and Producing Areas in North Texas Petroleum Fields:* Am. Assoc. Petroleum Geologists Bull., v. 3, p. 52, 1919.

Larson and **Porter** (*ibid.*, p. 114) view the rise of geologists to positions of paramount importance in Humble's exploration organization as the most significant new development in the company's producing operations until the late 1920's. Early in 1919, the number of geologists under Pratt increased from 3 to 10, and in 1920, a geological research laboratory was set up in Houston.

See also **Wallace Pratt** to **W. K. Ferguson**, Nov. 14, 1973.

> The American oil industry when I entered it in 1916 was all but devoid of technology outside its refining department. Its production and exploration activities were completely dominated by so-called "practical oil men." Very few geologists were on the job and these were new employees. Petroleum engineers were even rarer. Moreover, management in the production departments was very disdainful — often suspicious — of technology. When I was employed as a geologist by Texaco in 1915 a senior Vice President (a former driller) in my presence spoke contemptuously of geologists as "those wise sons of bitches from Judea" ("Judeah," as he rendered the last word). When I first reported to my boss, the Division Superintendent of Production (a former driller) he said: "I think I ought to tell you, Mr. Pratt, we don't think much of geologists around here." A few months later, an Executive Vice President of another company telephoned a request that I call at this office. I did so, and he offered me a job: "Mr. Pratt," he began, "you are a geologist. My company needs a geologist. I don't know why in the world we need a geologist but they say we do and we want you to come work for us."

Director G. O. Smith of the U.S. Geological Survey commented on the growth of private research in 1916 (Thirty-Seventh Annual Report, p. 7):

> The increased extent to which the large industrial corporations of the country now support their own departments of scientific research is a gratifying development of the times. . . . Never before, perhaps, has the demand been so keen for research that is no less than even scientific in spirit and in method but that now has a definite purpose and yields definite results.

157. Udden proudly observed that the Bureau had played a leading role in introducing "a scientific attitude among the people interested in oil production" (**J. A. Udden** to **D. T. MacDonald,** June 13, 1924, Bureau of Economic Geology).

158. Udden's successor, Dr. E. H. Sellards, defined the "strategic" information exchanged between the Bureau and commercial geologists (**Sellards** to **D. C. Barton,** Dec. 10, 1928, Bureau of Economic Geology):

. . . that which is not imminently significant with regard to prospective drilling sites. Broad structural features and pure stratigraphy are freely exchanged types of information. The only useful exploratory information released by the Bureau are county reports. Bureau personnel are available for private contract work during their vacations. . . . Well samples are given to the Bureau in return for their description, and such information obtained is available to the public except probably where specifically held tight.

However, this cooperative attitude of the oil companies stopped short of monetary contributions to the Bureau (**Wallace Pratt** to **E. H. Sellards,** June 13, 1927, Bureau of Economic Geology).

159. The sharing of geological information between Federal, State, and private geologists that so creditably characterized the geological research scene in Texas is perhaps best described in **Julia Gardner's** "Acknowledgments" in her paper on *The Midway Group of Texas* (Univ. Texas Bull. 3301, p. 5-6, 1933).

160. **J. Elmer Thomas,** *A.A.P.G.'s Growth Parallels the Expansion of Petroleum Geology:* Oil and Gas Jour., v. 38, no. 48 (Apr. 11, 1940), p. 30-40. *See also* **Sidney Powers,** *The History of the A.A.P.G.* (Am. Assoc. Petroleum Geologists Bull., v. 8, no. 2, Feb. 1929, p. 153-170), and the more recent and comprehensive article by **Harold T. Morley,** *A History of the American Association of Petroleum Geologists, The First Fifty Years* (Am. Assoc. Petroleum Geologists Bull., v. 50, no. 4, Apr. 1966, p. 669-820).

161. **J. A. Udden,** *Hints to Prospective Oil Geologists:* Am. Assoc. Petroleum Geologists Bull., v. 1, p. 127-130, 1917.

162. **F. B. Plummer** and **R. C. Moore,** *Stratigraphy of the Pennsylvanian Formations of North-Central Texas, op. cit.* (45). *See* Chapter I for information on the micropaleontological studies published by the Bureau.

163. **K. C. Heald,** *Research and the A.A.P.G.:* Am. Assoc. Petroleum Geologists Bull., v. 12, no. 9 (Sept. 1928), p. 939-948. *See also* **G. M. Knebel,** *The History and Activities of the A.A.P.G. Research Committee, 1923-1955:* Shale Shaker, v. 6, no. 6 (Feb. 1956), p. 5-19.

164. *See* **Walter H. Bucher,** *National Research Council and Co-operation in Geologic Research:* Geol. Soc. America Bull., v. 53, no. 9 (Sept. 1942), p. 1331-1354. This study has an excellent bibliography. *See also* **W. H. Twenhofel,** *A History of the National Research Council, 1919-1933, Division of Geology and Geography:* Science, v. 66, no. 2009 (June 30, 1933), p. 618-620. The work of this division was regularly discussed in *Proceedings* of the Geological Society of America.

165. The confidential minutes of the meeting of the Executive Committee of the Division of Geology and Geography for February 3, 1923 are in the Bureau files.

Dr. Udden wrote an interesting reply to an inquiry sent out by Professor N. M. Fenneman regarding the gaps in the science of geology and the lines of possible future advance (**J. A. Udden** to **N. M. Fenneman**, Feb. 24, 1923, Bureau of Economic Geology). Udden stressed the lack of information on the mineralogy and petrology of the clastic elements making up different stratigraphic units in different regions, and also the elements of bedding in sedimentary rocks.

166. The bureaucratic penchant for weighty committee titles was not a New Deal development. *See* **Arthur C. Lawson** to **J. A. Udden**, June 30, 1924, Bureau of Economic Geology.

167. The Geological Society of America became a very important source of research funds in 1931 when R. A. F. Penrose bequeathed to the Society more than $3.5 million. The Penrose Bequest grants-in-aid of research are discussed in note 432.

168. An example of the scope which the National Research Council added to research possibilities is described in a letter from **E. H. Sellards** to **W. H. Twenhofel** (Sept. 18, 1931, Bureau of Economic Geology):

> In understanding our stratigraphic problems of the west margin of the Gulf of Mexico, it is of course very important that studies of the formations in Mexico be included. Naturally our state and governmental funds cannot very well be applied to investigations in foreign territory. I should therefore like to speak especially for this particular investigation which aids in giving correlations around the west margin of the Gulf of Mexico.

Sellards submitted the outline of a project submitted by Julia Gardner:

> . . . the specific determination of a large collection of Eocene fossils from northern Mexico, their correlation with other Eocene faunas of the Gulf Province and the interpretation of the Eocene history of the western Gulf Province in light of these faunas.

Miss Gardner was eventually able to carry out these studies by virtue of a grant from the R. A. F. Penrose, Jr., Bequest of the Geological Society of America. Some of the results were published as *Mollusca of the Tertiary Formations of Northeastern Mexico:* Geol. Soc. America, Mem. 11 (Dec. 15, 1945), 332 p.

169. One of the committees of the Division of Geology and Geography of the National Research Committee is the Committee on Studies in Petroleum Geology, and the members of that committee were also members of the A.A.P.G. (**K. C. Heald**, *Research and the A.A.P.G., op. cit.* [163], p. 943).

170. **E. Böse**, *Cretaceous Ammonites from Texas and Northern Mexico, op. cit.* (89); **F. B. Plummer** and **R. C. Moore**, *Stratigraphy of the Pennsylvanian Formations of North-Central Texas, op. cit.* (45); **P. B. King**, *The Geology of the Glass Mountains, Texas, Part I, Descriptive Geology, op. cit.* (76); **R. E. King**, *The Geology of the Glass Mountains, Texas, Part II, Faunal Summary and Correlation of the Permian Formations with Description of Brachiopoda, op. cit.* (76); **D. O. Carsey**, *Foraminifera of the Cretaceous of Central Texas* (Univ. Texas Bull. 2612, 56 p., 1926); **H. J. Plummer**, *Foraminifera of the Midway Formation in Texas, op. cit.* (30); and **J. Gardner**, *The Midway Group of Texas, op. cit.* (159).

171. Udden was a member of the Committee on Sedimentation of the National Resource Council (*see* **Arthur C. Trowbridge** to **J. A. Udden,** Nov. 13, 1922, Bureau of Economic Geology). Sellards also became a committee member under the Division of Geology and Geography of the N.R.C.

172. **Raymond C. Moore,** *Society of Economic Paleontologists and Mineralogists Firmly Established:* Oil and Gas Weekly, v. 38, no. 48 (Apr. 11, 1940), p. 42, 59.

173. Proceedings of the Conference to Discuss the Problem of the Pennsylvanian of the Western Interior Region held at Norman, Oklahoma, May 19, 1928 (Bureau of Economic Geology). C. N. Gould cited the purpose of the conference to be "to get together a vast amount of information that is now being done by oil geologists and tied up in the confidential files of the oil companies."

174. Proceedings of the Meeting of Committees for Texas to Further Cooperative Investigation of the Permo-Carboniferous held at Eastland, Texas, July 14, 1928 (part of the Cooperative Research of the Permo-Carboniferous Formations of the Mid-Continent under the auspices of the Society of Economic Paleontologists and Mineralogists, Bureau of Economic Geology).

175. Research notes — Research Committee Report 1929-1930: Am. Assoc. Petroleum Geologists Bull., v. 14, no. 6 (June 1930), p. 809-812.

176. The results of this cooperative mapping program include the following Mineral Resource Pamphlets of the Bureau of Economic Geology: **W. S. Adkins,** *Mineral Resources of Bell County* (p. 1-16, 1929); **E. H. Sellards,** *Mineral Resources of Bastrop County* (p. 17-40, 1929); **E. H. Sellards,** *Mineral Resources of Travis County* (p. 41-69, 1930); **E. H. Sellards,** *Mineral Resources of Williamson County* (p. 70-92, 1930); **F. M. Getzendaner,** *Mineral Resources of Uvalde, Zavala, and Maverick Counties* (p. 93-140, 1931).

L. T. Patton comments on this cooperative work in *The Geology of Stonewall County, Texas* (Univ. Texas Bull. 3027, p. 6-7, 1930):

> Subsequent to the field work done by the author, the Cooperative Mapping Committee of the A.A.P.G. and the B.E.G. carried on mapping work, including extensive plane table surveys in this and adjacent counties. Data from this work have been added to the geologic map mainly in details of boundary lines obtained from plane table surveys in this and adjacent counties, which were not possible for a geologist working without assistance to obtain.

177. For further information on this project, *see* **E. H. Sellards** to **K. C. Heald,** Jan. 13, 1928, and **K. C. Heald** to **E. H. Sellards,** Apr. 30, 1928, and rough progress reports of Project 25B by **E. M. Hawtof** for 1928 and 1930, Bureau of Economic Geology. The final report was published as *Earth Temperatures in Oil Fields,* Part I, "Results of Deep Well Temperature Measurements in Texas": Am. Petroleum Inst., Production Bull. 205, p. 62-108, 1930.

178. **F. B. Plummer** and **E. C. Sargent,** *Underground Waters and Subsurface Temperatures of the Woodbine Sand in Northeast Texas:* Univ. Texas Bull. 3138, 178 p., 1931.

This is Part II of A.P.I. Project 25B. This bulletin contains an excellent bibliography on the geology of East Texas.

179. In June 1930, Virgil Barnes took over the A.P.I. Project 25B for its terminal year. His work was concentrated in North Texas where he found the temperature gradients were higher over the Red River Uplift, where Precambrian rocks were nearer the surface than they were in the basin to the south. Briefs of the results of Barnes' work were published as *Earth Temperatures of North-Central Texas* (Am. Assoc. Petroleum Geologists Bull., v. 16, no. 4 [Apr. 1932], p. 413-416) and *Oil Field Waters of North-Central Texas, ibid.*, p. 409-411. A comprehensive report was published several years later as "Earth Temperatures and Oil Field Waters of North-Central Texas," in *Texas Mineral Resources:* Univ. Texas Pub. 4301, p. 323-357, 1943 (1946).

180. **E. H. Sellards** to **H. Y. Benedict**, Sept. 1, 1928, Bureau of Economic Geology. Sellards added:

> In September 1927, in a general letter sent out to the geologists of Texas, I invited the cooperation of all geologists in the state in assembling data on the areal, stratigraphic and structural geology. The immediate response to this invitation is [overwhelming]. In addition, the general, broad idea of cooperative work in developing the knowledge of the geology has taken form to such extent that there has been organized a group of geologists for cooperative research in the Mid-Continent region, [the A.A.P.G.], and, for the present at least, most actively in the Texas section . . . Under the direction of the District Chairman approximately 100 local geologists are enrolled in the effort to assemble and put in usable shape the available information on Texas geology.

E. W. Owen observed that "the A.A.P.G. is the instrument by which you got away from the old backbiting, destructive tactics of an earlier generation, and arrived at a really cooperative spirit" (conversation with **W. K. Ferguson**, 1972).

181. The local geological societies that began to emerge in the late 1920's were vital keys to the cooperative programs spearheaded by the A.A.P.G. John Reeves, Director of the Kansas Geological Society, speaking at the 1928 Conference to Discuss the Problem of the Pennsylvanian of the Western Interior Region, felt that the local geological societies could act as the key local clearinghouses for gathering information from company geologists for such cooperative research projects as the one they were discussing (173). **J. Elmer Thomas**, first president of the Southwestern Association of Petroleum Geologists, wrote that as the A.A.P.G. had become so large as to be cumbersome, the formation of local voluntary study groups within local geological societies should be encouraged. Research papers given by these groups at the meetings of the geological societies would facilitate the exchange of geological thought on local problems. Thomas also noted that the society guidebooks were compendiums of important research information (**Thomas**, *A.A.P.G.'s Growth Parallels the Expansion of Petroleum Geology, op. cit.* [160], p. 32).

One example of the valuable activities of these local societies is the following note from the "At Home and Abroad" section of the Bulletin of the American Association of Petroleum Geologists (v. 11, no. 2 [Feb. 1927], p. 105):

> The West Texas Geological Society met in San Angelo, and adopted a committee report presented by Edgar Kraus [of the Atlantic Producing Company] providing for a series of cooperative field investigations with the students of the University of Texas and E. H. Sellards of the Bureau of Economic Geology . . . Eventually this work will be carried from all directions toward the center of the great salt basin.

182. Thirty-Fifth Annual Report, U.S. Geological Survey, 1914, p. 10. The introduction to this report is directed toward a proposed amendment to restrict the geological work of the Geological Survey to public lands.

183. **R. T. Hill**, *Geography and Geology of the Black and Grand Prairies, Texas:* U.S. Geol. Survey, Twenty-First Ann. Rept., pt. 7, 1901, 666 p.

R. T. Hill and **T. W. Vaughan**, *Description of the Nueces Quadrangle:* U.S. Geol. Survey, Geol. Atlas, Nueces folio (no. 42), 4 p., 1898.

R. T. Hill and **T. W. Vaughan**, *Description of the Austin Quadrangle:* U.S. Geol. Survey, Geol. Atlas, Austin folio (no. 76), 8 p., 1902.

T. W. Vaughan, *Description of the Uvalde Quadrangle:* U.S. Geol. Survey, Geol. Atlas, Uvalde folio (no. 64), 7 p., 1900.

G. B. Richardson, *Description of the El Paso District:* U.S. Geol. Survey, Geol. Atlas, El Paso folio (no. 166), 11 p., 1909.

G. B. Richardson, *Description of the Van Horn Quadrangle:* U.S. Geol. Survey, Geol. Atlas, Van Horn folio (no. 194), 9 p., 1914.

Sidney Paige, *Description of the Llano and Burnet Quadrangles:* U.S. Geol. Survey, Geol. Atlas, Llano-Burnet folio (no. 183), 16 p., 1912.

Charles Gould, *The Geology and Water Resources of the Eastern Portion of the Panhandle of Texas:* U.S. Geol. Survey Water-Supply Paper 154, 64 p., 1906.

Charles Gould, *The Geology and Water Resources of the Western Portion of the Panhandle of Texas:* U.S. Geol. Survey Water-Supply Paper 191, 70 p., 1907.

C. H. Gordon, *Geology and Underground Waters of the Wichita Region, North-Central Texas:* U.S. Geol. Survey Water-Supply Paper 317, 88 p., 1913.

184. For further discussion, *see* **W. K. Ferguson**, *Geology and Politics in Frontier Texas, 1845-1909, op. cit.* (3), Appendix D, p. 173-177.

185. **J. A. Udden** to **W. C. Mendenhall**, July 21, 1924, Bureau of Economic Geology. *See also* **Mendenhall** to **Udden**, Aug. 21, 1924, Bureau of Economic Geology.

186. **R. T. Hill** to **J. A. Udden**, May 1, 1925, Udden Papers.

187. **W. B. Phillips** to **J. A. Udden**, Mar. 31, 1914, Bureau of Economic Geology. Phillips also wrote to Udden (May 30, 1914, Bureau of Economic Geology):

> Cipher is a cipher and figger is a figger. All for the white man. None for the nigger. This is my idea of cooperation; in cooperative work we are mostly the nigger.

188. **E. W. Owen**, conversation with **W. K. Ferguson**, 1972.

189. **E. W. Owen**, conversation with **W. K. Ferguson**, 1972. The trend toward more open cooperation with state surveys may have been given its main impetus during the First World War. The increased amount of cooperative work between the U.S. Geological Survey and the state surveys is described in the Thirty-Eighth Annual Report (1917).

190. U.S. Geological Survey, Twenty-Ninth Annual Report, 1909, p. 22.

191. *See* Chapter 9 for a discussion of the cooperation between the Bureau, the A.A.P.G., the N.R.C., and the G.S.A. on studies of the Permian deposits of Texas during the 1930's.

192. The most important publications based on the Texas Coastal Plain research by the U.S. Geological Survey include:

Alexander Deussen, *Geology and Underground Waters of the Southeastern Part of the Texas Coastal Plain:* U.S. Geol. Survey Water-Supply Paper 335, 365 p., 1914.

L. W. Stephenson, *The Cretaceous-Eocene Contact in the Atlantic and Gulf Coastal Plain:* U.S. Geol. Survey Prof. Paper 90-J, p. 155-182, 1915. Unfortunately, he ignored the work of J. A. Taff, who knew more about this than all the rest put together.

A. C. Trowbridge, *A Geologic Reconnaissance in the Gulf Coastal Plain of Texas, near the Rio Grande:* U.S. Geol. Survey Prof. Paper 131D, p. 85-107, 1923.

Alexander Deussen, *Geology of the Coastal Plain of Texas West of the Brazos River:* U.S. Geol. Survey Prof. Paper 126, 145 p., 1924.

In her fine "Historical Background of the Midway Group" (*op. cit.*, **159**), which is in fact a brilliant concise "case history" of geological work in the state, **Julia Gardner** observes that Deussen's *Geology of the Coastal Plain of Texas West of the Brazos River* was a distinguished contribution to the geology of that time in which it was written, but the delay in publication of almost a decade of rapidly accumulating geologic knowledge had lent an historic interest to a paper distinctly in advance of its time. Trowbridge's *Geologic Reconnaissance in the Gulf Coastal Plain of Texas near the Rio Grande,* begun much later than Deussen's work, was published a year earlier, and completed the areal reconnaissance studies of the Coastal Plain (p. 14-15). Miss Gardner's work represented the paleontological aspect of the Coastal Plain studies. The best description of the overall scope of these studies is found in the Thirty-Seventh Annual Report of the U.S. Geological Survey, p. 45-46.

193. Director **G. O. Smith** informed Director **W. B. Phillips** of the course of the Coastal Plain investigations (**Smith** to **Phillips,** Dec. 15, 1909, Bureau of Economic Geology) and Phillips replied, telling him of the work of the new Bureau of Economic Geology, and stating that no duplication of effort was likely (**Phillips** to **Smith,** Dec. 31, 1909, Bureau of Economic Geology).

194. Thirty-Fourth Annual Report, U.S. Geol. Survey, p. 67, 1913. Also, L. W. Stephenson spoke highly of the tremendous assistance he had received and of the great benefit he had derived from his association with the men of the Bureau (*Notes on the Stratigraphy of the Upper Cretaceous Formations of Texas and Arkansas:* Am. Assoc. Petroleum Geologists Bull., v. 11, no. 1 (Jan. 1927), p. 2.

In turn, the Geological Survey made all of Deussen's research available for use in compiling the first complete geological map of the state which was included in Bull. 44 of the Bureau, *Review of the Geology of Texas,* by **J. A. Udden, C. L. Baker,** and **Emil Böse,** *op. cit.* (42).

195. **W. C. Mendenhall** to **J. A. Udden,** July 15, 1924, Bureau of Economic Geology.

196. The map legend reads as follows:

> [compiled] between the years 1924 and 1932 by the United States Geological Survey, in co-operation with the Bureau of Economic Geology of the University of Texas, the geologists of Texas and the oil companies of Texas. . . .

197. This is noted in the Forty-Sixth Annual Report of the U.S. Geological Survey, 1925, p. 43. Also, W. B. Land was working in the Mesozoic rocks of western Texas. It is also noted that Stanton held a conference with the geologists of the Bureau (p. 44).

198. *See* **W. C. Mendenhall** to **E. H. Sellards**, July 14, 1929, Bureau of Economic Geology.

199. **Julia Gardner**, *The Midway Group of Texas, op. cit.* (159). The contact is described in a letter from **F. W. Graff**, Assistant to the University President, to **E. H. Sellards**, July 6, 1922, Bureau of Economic Geology, and also in **Sellards** to **L. W. Stephenson**, May 25, 1925, B.E.G. The actual "Cooperative Agreement Between the Texas Bureau of Economic Geology and the U.S. Geological Survey, With Reference to the Completion of A Report on the Geology and Paleontology of the Midway Formation in Texas" signed by both directors is found in the Bureau files in the U.S.G.S. folder, and dated Oct. 5, 1926.

200. A complete survey of Udden's potash investigations is found in Udden's preface to **H. W. Hoots**, *Geology of a Part of Western Texas and Southeastern New Mexico with Special Reference to Salt and Potash, op. cit.* (119), p. 33-36.

For the overall scope of the U.S. Geological Survey potash investigations, *see* **E. H. Sellards**, *Economic Geology of Texas — The Search for Potash in the United States*, in *The Geology of Texas, Vol. II, Structural and Economic Geology:* Univ. Texas Bull. 3401, p. 653-663, 1934 (1935). *See also* **E. H. Sellards** and **E. P. Schoch**, *Core Drill Tests for Potash in Midland County, Texas*, in *Contributions to Geology, 1928:* Univ. Texas Bull. 2801, p. 159-201, 1928.

201. **J. A. Udden**, *The Deep Boring at Spur [Dickens County, Texas], op. cit.* (17), p. 87-88.

202. **Hoyt Gale**, *The Search for Potash in the Desert Basin Region:* U.S. Geol. Survey Bull. 530A, p. 295, 1913.

203. **David White** to **J. A. Udden**, Nov. 19, 1918, Bureau of Economic Geology.

204. **Walter Lang**, *Potash Investigations in 1924:* U.S. Geol. Survey Bull. 785B, p. 29-43, 1926.

205. There is a manuscript of this act in the Bureau files which was sent to Dr. Udden on April 19, 1924.

206. Outstanding examples of the U.S. Geological Survey topographic work in Texas are found in the Thirty-Ninth Annual Report, 1918, p. 105, and Forty-Third Annual Report, 1922, p. 43. The former was somewhat of an exceptional year as there were extensive U.S.G.S. topographic mapping efforts concentrated along the Mexican border and the Gulf Coast as a result of the war scare. The war also stimulated the development of aerial photographic mapping techniques.

207. **J. A. Udden** to University President **R. E. Vinson**, Jan. 22, 1921, Bureau of Economic Geology. According to Elizabeth Stiles, Udden also arranged for the State Reclamation Department to ask for an additional $30,000 in their budget to be used to match Federal topographic mapping funds. This plan failed also (**Stiles** to **G. O. Smith**, Feb. 18, 1921, Bureau of Economic Geology).

208. For a thorough discussion of the Temple Act, *see* the Journal of the Association of American State Geologists, v. 2, no. 3 (July 1, 1921), p. 9-14. No appropriations were ever made for the implementation of this act.

209. **Udden** wrote to Texas Representative **James R. Buchanan** (Aug. 11, 1919, Bureau of Economic Geology) that "the Mexican Scare led the government to do quite a little work along the Rio Grande last year and the year before." Also, two Bureau men were apprehended by the Texas Rangers on suspicion of spy activities (**Udden** to Adj. General **Henry Hutchings**, May 25, 1918, Bureau of Economic Geology).

210. *See* **Udden** to **M. R. Gutsch,** Director of the Texas War Records Collection, Nov. 20, 1920, Bureau of Economic Geology, and Major General **W. M. Black,** Chief of the Corps of Engineers, to **Udden,** Oct. 9, 1917, Bureau of Economic Geology. There is an excellent description of the accelerated wartime activities of the Geological Survey in the Thirty-Eighth Annual Report, 1917, p. 7-9.

211. **D. D. Christener,** *The Geology of Terrell County:* Univ. Texas Bull. 1819, p. 5-32, 1918; **O. C. Wheeler,** *Roads of Terrell County, ibid.,* p. 33-49; **John R. Roberts** and **J. P. Nash,** *The Geology of Val Verde County:* Univ. Texas Bull. 1803, 51 p., 1918; **James P. Nash, C. L. Baker, E. L. Porch, Jr.,** and **R. G. Tyler,** *Road Building Materials in Texas:* Univ. Texas Bull. 1839, 159 p., 1918.

212. A total review of the geological work of the Federal survey in Texas is beyond the scope of this paper, but the Fiftieth Annual Report (1929) includes a full representative description of such activities.

213. See note 194, and the following letters exchanged between **E. H. Sellards** and **W. C. Mendenhall:** July 3, 1929, July 14, 1929, and Aug. 14, 1929, Bureau of Economic Geology.

214. **C. N. Gould** to **E. H. Sellards,** Feb. 28, 1930, Bureau of Economic Geology.

215. **Julia Gardner,** *The Midway Group of Texas, op. cit.* (159), p. 5-6.

216. **P. T. Flawn,** *Geology in the State Government of Texas:* Univ. Texas, Bur. Econ. Geology Geol. Circ. 65-5, p. 10, Nov. 1965.

217. **Arthur Duggan,** *Texas Ground Water Law,* in *Proceedings,* Texas Water Law Conferences, Nov. 20-21, 1952, June 10-11, 1952: School of Law, Univ. Texas.

218. **P. T. Flawn,** *Geology in the State Government of Texas, op. cit.* (216). Also, *Mineral Resources and Conservation in Texas*: Univ. Texas, Bur. Econ. Geology Geol. Circ. 70-1, 20 p., Feb. 1970.

219. **P. T. Flawn,** *Geology in the State Government of Texas, op. cit.* (216), p. 7.

220. For a very readable and perceptive study of the history of the conservation movement in America, *see* **Frank E. Smith,** *The Politics of Conservation:* Pantheon Publ., New York, 338 p., 1966. For further information on the subject covered here, *see* Chapters 6 and 7, "T. R." and "The First Crusade."

221. **P. T. Flawn,** *Mineral Resources and Conservation in Texas, op. cit.* (218), p. 7.

222. **F. E. Smith,** *The Politics of Conservation, op. cit.* (220), p. 88.

223. Thirtieth Annual Report of the U.S. Geological Survey, p. 7-8, 1909. For a more economically directed discussion of conservation of natural resources by Director **G. O. Smith,** see *Brakes for the Mineral Industry,* an address to the Section of Mineral Law of the American Bar Association (Washington, p. 7, 1913).

224. **G. O. Smith,** *The Classification of Public Lands:* U.S. Geol. Survey Bull. 537, p. 7, 1913.

225. In the Thirty-Fourth Annual Report (1913) the director reported a 37 percent net growth in personnel since 1907. He attributed this growth to the increase in resource management services performed by the Geological Survey.

226. For a thorough and lively account of the Federal survey's concern with the oil reserves under the public lands in California, *see* **E. W. Owen,** *The Trek of the Oil Finder: The History of Petroleum Exploration, op. cit.* (18), Chapter 6, "Inception of Professional Practice, 1891-1910."

227. Gifford Pinchot corresponded with Texas Governor T. M. Campbell about the affairs of the National Conservation Commission, and encouraged Campbell to establish a Texas Conservation Commission. Campbell did establish such a commission, of which W. B. Phillips was a member. However, according to Phillips, the commission met only a few times, accomplished nothing, and died a quiet death (**W. B. Phillips** to **J. B. White,** President of National Conservation Congress, Sept. 10, 1912). *See also* Governors' Correspondence, Texas State Archives, **Pinchot** to **Campbell,** Nov. 21, 1908, 4-25/86.

228. **Ralph Arnold,** *Two Decades of Petroleum Geology, 1903-1922* (Am. Assoc. Petroleum Geologists Bull., v. 7, no. 6, Nov. 1923, p. 603-624) presents a contemporary and personal view of the issue of withdrawal of public oil lands.

The Geological Survey's call for the withdrawal of all public petroleum lands appeared in **D. T. Day,** "Petroleum Resources of the United States," in *Papers on the Conservation of Natural Resources:* U.S. Geol. Survey Bull. 394, 1909. This is the collection of papers by the Geological Survey on the mineral resources and reserves of the nation that were compiled for the National Conservation Commission.

229. Thirty-First Annual Report, 1910, p. 10. Public lands suspected of having potash and phosphate reserves were also withdrawn from entry (*see* **G. O. Smith,** *The Classification of Public Lands, op. cit.,* 224).

230. Twenty-Fourth Annual Report, U.S. Geol. Survey, 1902, p. 213.

231. **M. O. Leighton,** *Sewage Pollution in the Metropolitan Area Near New York City and Its Effect on Inland Water Resources:* U.S. Geol. Survey Water Supply and Irrigation Paper No. 71.

232. *Ibid.,* p. 11.

233. *Ibid.,* p. 61. In a supplementary note in **Lemuel Sackett** and **Isiah Bowman,** *The Disposal of Strawboard and Oil-Well Wastes* (U.S. Geol. Survey

Water Supply and Irrigation Paper No. 113, 1905), M. O. Leighton comments that where oil deposits were developed, stream pollution was inevitable. He said that it would be wiser to let the oil run down creek beds than to allow it to stand in tanks and seep down into the ground. He said that such pollution is "part of the price paid for the accumulation of national wealth" (p. 50).

234. **M. O. Leighton,** *Pecuniary Damage to Municipal Water Supply Occasioned by City and Industrial Wastes:* U.S. Geol. Survey Water Supply and Irrigation Paper No. 79, p. 11, 1903.

235. *Ibid.*, p. 186.

236. **Edwin Goodell,** *A Review of the Laws Forbidding Pollution of Inland Waters in the United States:* U.S. Geol. Survey Water Supply and Irrigation Paper No. 103, 1904. For a discussion of Texas anti-water pollution laws, see p. 50.

237. The Thirty-third Legislature of Texas passed *An Act to Prevent Pollution of Water Courses or Other Public Bodies of Water of the State of Texas* (General Laws, Chap. 47, p. 90-91, March 17, 1913). This act stipulated that cities would have 3 years to comply with the provisions of the act. The author has not searched for the record of enforcement of these Texas pollution laws.

238. The best contemporary overview on river pollution, its biological and chemical aspects, its general and special sanitary significance, and legal ramifications was by **M. O. Leighton,** *Pollution of the Illinois and Mississippi Rivers by Chicago Sewage, A Digest of the Testimony Taken in the Case of the State of Missouri and the Sanitary District of Chicago* (U.S. Geol. Survey Water Supply and Irrigation Paper No. 194, 1907). The State of Missouri claimed that there had been an increase in typhoid fever in St. Louis, which the State attributed to the sewage dumped into the Illinois River by the Sanitary District of Chicago. The defendants were able to prove that the fault lay with the much greater pollution entering the Mississippi River in the State of Missouri itself.

A classic early study of geologic investigations in support of managed use of water resources is **Grove Karl Gilbert,** *Hydraulic Mining Debris in the Sierra Nevada* (U.S. Geol. Survey Prof. Paper 105, 1917). Gilbert studied the geologic and physiographic phases of the problem of detrital waste due to agriculture, grazing, road cuts, and mining in the rivers and streams of the Sierra Nevada. He concluded that there was nearly as much detrital debris entering the water courses due to erosion caused by agriculture and logging as there was as a result of hydraulic mining. Gilbert's suggested solution was the construction of large storage reservoirs outside of the regular river channels. The debris would settle in these reservoirs and not be carried down and deposited in the bays, hindering navigation.

239. Forty-First Annual Report, U.S. Geol. Survey, 1920, p. 8.

240. For further information on the Irrigation Survey, *see* **E. W. Sterling,** *The Powell Irrigation Survey, 1888-1892:* Mississippi Valley Historical Review, v. 27, no. 12, p. 421-434, Dec. 1940.

241. **M. O. Leighton** and **Rene Tavernier**, *The Public Utility of Water Powers and Their Governmental Regulation:* U.S. Geol. Survey Water Supply Paper 238, 1910.

242. **W. S. Murray and others**, *A Super Power System for the Region Between Boston and Washington:* U.S. Geol. Survey Prof. Paper 123, p. 9-10, 1921.

243. Fortieth Annual Report, 1919. The Director's analysis of the Super Power Project was in keeping with the economic imperialism of the day (*ibid.*, p. 17):

> Only by increased economy in the production and distribution of power will it be possible for our manufacturers to decrease their product expenses and compete successfully in the world's markets, maintaining at the same time the American standard of wages and living.

244. **Marius R. Campbell** and others, "The Northern Pacific Route, with a Side Trip to Yellowstone Park," Part A, *Guidebook of the United States:* U.S. Geol. Survey Bull. 611, 1915.
Willis T. Lee, Ralph W. Stone, Hoyt Gale, and others, "The Overland Route with a Side Trip to Yellowstone Park," Part B, *Guidebook of the United States:* U.S. Geol. Survey Bull. 612, 1915.
N. H. Darton and others, "The Santa Fe Route, with a Side Trip to the Grand Canyon of the Colorado," Part C, *Guidebook of the United States:* U.S. Geol. Survey Bull. 613, 1915.
J. S. Diller and others, "The Shasta Route and Coast Line," Part D, *Guidebook of the United States:* U.S. Geol. Survey Bull. 614, 1915.

245. **N. H. Darton** and **P. B. King**, *Western Texas and Carlsbad Caverns:* 16th Internat. Geol. Cong., Guidebook 13, Excursion C-1, 38 p., 1933.
W. E. Wrather and others, *Oklahoma and Texas:* 16th Internat. Geol. Cong., Guidebook 6, Excursion A-6, 91 p., 1933.

246. "An Act to encourage irrigation and to provide for the acquisition of the right to the use of water . . . for irrigation, and for mining, milling, and stock-raising in the arid districts of Texas" (Twenty-First Legislature, General Laws, Chap. 88, p. 100-103, March 19, 1889).
Arthur Duggan, "Texas Ground Water Law," in *Proceedings*, Texas Water Law Conferences, Nov. 20-21, 1952, June 10-11, 1954 (School of Law, Univ. Texas) presents a very concise survey of legislation dealing with State management of water resources.

247. Chap. 171, General Laws of the Thirty-Third Regular Session, p. 358-379, Apr. 9, 1913. An act creating the State Levee and Drainage Board to construct and maintain dams, levees, canals to prevent flooding and procure drainage was passed in 1909 (Chap. 81, Thirty-First Regular Session, p. 136-137, Mar. 19, 1909).

248. This Conservation Amendment is Sec. 59a, Art. SVI of the Texas Constitution.

249. Thirty-Fifth Legislature, Fourth Called Session, Senate Jour., p. 2, Feb. 26, 1918.

250. Thirty-Fifth Legislature, Fourth Called Session, General Laws, Chap. 25, p. 40-43, Mar. 21, 1918. In 1931, legislation was passed giving the Board regulatory powers over underground waters also: "An Act for the conservation, preservation, protection and distribution of the underground water supply of the State; defining the duties of the Board of Water Engineers for said purpose; providing for the plugging, casing or capping of wells to prevent contamination of the fresh water supply" (Forty-Second Legislature, Regular Session, Chap. 261, p. 432-433, May 28, 1931).

However, the courts did not uphold these regulatory powers. In a memorandum to the Board of Water Engineers on Sept. 14, 1931, W. N. White, Director of the Groundwater Studies of the U.S. Geological Survey in Texas, commented on the danger that the flow of the large springs near Balmorhea and Fort Stockton would decrease if many wells were drilled in the Toyah Basin for irrigation purposes (U.S. Geol. Survey groundwater correspondence files, 1930-39, Texas Water Commission). In a suit brought to curtail irrigation pumping, the State court reaffirmed the doctrine of absolute ownership of ground waters. However, this does not stand in the way of proper management of ground water in the prevention of waste, as provided for in the Conservation Amendment.

Finally, in 1949, the Legislature passed the Texas Underground Water Conservation Law (Fifty-First Legislature, Regular Session, General Laws, Chap. 306, p. 559-564, June 2, 1949): "An Act to provide for the creation and organization of underground water conservation districts to provide for the conservation, preservation, protection, and recharging and prevention of waste of underground waters." This law treats underground water simply as an extractive mineral, such as oil. It authorizes prevention of waste by injunction, provides for the spacing of wells, and provides for limitation of production to prevent waste. The basic approach to the problem of maintaining ground water reserves is artificial recharge activity. Texas is ahead of other states in this regenerative, constructive approach to the solution of ground water supply problems (**Arthur Duggan**, "Texas Ground Water Law," *op. cit.* [217], p. 18). The recharge principle was not mentioned in pre-1933 materials investigated.

251. "An Act to conserve the oil and gas resources of the State of Texas; to define waste in the production of said minerals; to invest the Railroad Commission of Texas with the authority to make and enforce all needful rules and regulations with respect to the conservation of said resources" (Thirty-Sixth Legislature, Regular Session, General Laws, Chap. 155, p. 285-288, Mar. 31, 1919).

252. See *George Charles Butte*, in **W. P. Webb**, ed., *The Handbook of Texas, Vol. I:* The Texas Historical Assoc., Austin, Tex., p. 258, 1952. *See also* "The Butte Papers" in The University of Texas Archives.

For a comprehensive discussion of the legislative history of the oil conservation program of the Railroad Commission, *see* **Jack C. Rothwell**, *The Conservation Program of the Railroad Commission and the Structure of Crude Oil Prices in Texas* (Univ. Texas, Ph.D. dissert., 258 p., 1958), especially Chap. 4, "History of Oil Conservation in Texas," p. 90-149. *See also* "The East Texas Ordeal and the Establishment of Proration, 1931-35," in *History of HO&RC, A Study in Industrial Growth, op. cit.* (155).

253. The operators of the Yates Field had accepted a voluntary proration scheme in 1927.

254. Forty-Second Legislature, First Called Session, Senate Jour., p. 31, July 22, 1931. *See also* the *Austin American*, July 23, 1932, p. 1.

255. Forty-Second Legislature, Fourth Called Session, General Laws, Chap. 2, p. 3-10, Nov. 12, 1932.

256. **Herbert Davenport**, "Riparian versus Appropriative Rights: The Texas Experience," *Proceedings*, Texas Water Law Conferences, Nov. 20-21, 1952, June 10-11, 1954 (School of Law, Univ. Texas). For information on irrigation laws, *see* "The Arid Region Doctrine of Appropriative Rights," *ibid.*, p. 155-169.

257. Governor Pat Neff was very interested in the reclamation of overflow and arid lands. The Legislature under Neff passed "an act requiring the Board of Water Engineers and the State Reclamation Engineer . . . to make an adequate topographic and hydrographic survey of the stream watersheds of the State" (Thirty-Eighth Legislature, Regular Session, General Laws, Chap. 11, p. 19-20, Jan. 21, 1923). Other acts dealing with reclamation — power to incur debts and issue bonds, and to fix liens and collect taxes — were also passed during Neff's administration.

His interest in conservation also led him in another direction. Upon his urging, the State Parks Board was established (Thirty-Eighth Legislature, Second Called Session, Chap. 25, p. 58-60, May 21, 1923). For further information, *see* **Emma Morrill Shirley**, *The Administration of Pat M. Neff, Governor of Texas, 1921-1925:* Baylor Bull., v. 41, no. 4, Dec. 1938.

258. **W. K. Ferguson**, *Geology and Politics in Frontier Texas, 1845-1909, op. cit.* (3).

259. *See* **W. K. Ferguson**, *Geology and Politics in Frontier Texas, 1845-1909, op. cit.* (3), Chap. 6, "The Diffusion of Public Resources," p. 98-113.

260. *Ibid.*, p. 98-100.

261. It is interesting that there is no mention of the Bureau of Economic Geology or any resource management scheme in the annual reports of the Texas General Land Office from 1909-1932. This points out the lack of coordination between the work of the Land Office and the Bureau. Previously, the close cooperation between the Dumble Survey and the Land Office had preceded the passage of legislation that made the actual discovery of minerals the basis for the classification of land as mineral bearing (*see* **W. K. Ferguson**, *Geology and Politics in Frontier Texas, 1845-1909, op. cit.* [3], p. 123-124).

262. In a letter **W. B. Phillips** wrote to **John Kerr** (Dec. 16, 1909, Bureau of Economic Geology), he stated:

> It is a very serious question that confronts us, the conservation of our material resources without impinging upon the rights of the individuals or the corporations and without restricting the development of the state. Waste is waste, only if it cannot profitably be used. . . . I am around the state a good deal and find that most of our people are far more concerned over questions of development than over questions of conservation. This was distinctly the dominant note at the West Texas Development Congress which met last week in San Angelo.

Phillips also commented on this frontier spirit of unlimited exploitation in a letter to Gifford Pinchot. The Bureau director noted that his "own business is confined almost entirely to questions of development, rather than of conservation" (**W. B. Phillips** to **Gifford Pinchot**, Dec. 29, 1909, Bureau of Economic Geology). Phillips had been Secretary of the Texas Conservation Commission but resigned in 1910.

263. C. L. **Baker**, *Geology and Underground Waters of the Northern Llano Estacado, op. cit.* (134).

264. **Walter Webb**, *The Great Plains:* Ginn & Co., Boston, Mass., 345 p., 1931.

265. C. L. **Baker**, *Geology and Underground Waters of the Northern Llano Estacado, op. cit.* (134). The democratic concern with the creation of single-family ranching and farming units out of the public domain was reexpressed in a resolution endorsing reclamation and irrigation projects in 1921:

> Whereas, The future growth and prosperity of Texas and the entire West and Southwest is of vital importance in connection with the maintenance of our national strength and of inestimable value in connection with the future defense of the Nation, and is almost wholly dependent upon the development and utilization of natural resources, particularly the reclamation by irrigation and drainage of the vast areas of now unproductive desert and swamp lands . . .
>
> Whereas, There is an urgent necessity for stimulating activity in that greatest of all American industries — home building — and for the lending of National support to the back to the farm movement in order that our discontented floating population might be permanently cemented to the soil and its loyalty as citizens quarenteed [guaranteed] to the nation (Thirty-Seventh Legislature, Regular Session, House Journal, p. 302, March 7, 1921).

This Concurrent Resolution was introduced by Mr. W. W. Steward of Balmorhea, Reeves County, in the Trans-Pecos.

266. W. B. Phillips was interested in water resource management, especially in the Trans-Pecos region, which was of particular interest to him as the only potentially precious mineral-rich area in the state:

> I believe that the future of [the Davis Mountains] region will depend upon the storing of storm waters in reservoirs constructed at the foot of the mountains and distributing irrigation water as it may be needed . . . To my mind, this is the biggest problem we have in Texas today, the use of storm waters (**W. B. Phillips** to **Hervey Hicks**, Jan. 18, 1910, Bureau of Economic Geology).

267. **E. H. Sellards** to **H. Y. Benedict**, Jan. 28, 1929, Bureau of Economic Geology.

Sellards had recommended in the March 1928 budget that one associate and four assistant geologists be added to the staff of the Bureau to study the problems of underground water. The Regents rejected these additions.

In 1922, Governor Neff had asked Sellards to attend a meeting with the State Reclamation Department and the Board of Water Engineers to consider the problems of storm water control (**P. M. Neff** to **E. H. Sellards**, July 15, 1922, Bureau of Economic Geology). Sellards also was a member of the Governor's Advisory Council of Engineers on Conservation in 1924 (Minutes of Council of Engineers on Conservation, Jan. 8, 1924, Bureau of Economic Geology).

268. "An act authorizing the State Board of Water Engineers to have made a scientific and sanitary study and investigation and report on the sources, amount, and quality of the underground water supply in Texas, and the conservation, maintenance and supplementing of the same" (Forty-First Legislature, Second Called Session, General Laws, Chap. 37, p. 63-64, July 8, 1929). Of course, the U.S. Geological Survey had matching funds, so the Board of Water Engineers contracted with them rather than with the Bureau. The appropriation could have gone to the Bureau rather than the Board, as Sellards had requested. Then the Bureau would have arranged cooperative work with the U.S. Geological Survey.

Walter N. White was the hydraulic engineer in charge of the ground-water investigations in Texas, and O. E. Meinzer was in charge of the Division of Ground Water of the U.S. Geological Survey at the time the cooperative agreement was made. John A. Norris was the Chairman of the Board of Water Engineers. The first cooperative projects undertaken were investigations of the ground waters of certain South Texas counties that form the Winter Garden district, with special reference to the supply available for irrigation. These counties were developing extensive winter truck farming. The first work began in the summer of 1929, and the results were published as Water-Supply Papers of the U.S. Geological Survey: **John T. Lonsdale,** *Geology and Ground Water Resources of Atascosa and Frio Counties, Texas* (Water-Supply Paper 676, 1935); **Albert Sayre,** *Geology and Ground Water Resources of Uvalde and Medina Counties, Texas* (Water-Supply Paper 678, 1936); and **John T. Lonsdale** and **James R. Day,** *Geology and Ground Water Resources of Webb County, Texas* (Water-Supply Paper 778, 1937). The works by Lonsdale were a cooperative project between the Engineering Experimental Station of the Agricultural and Mechanical College of Texas, the Texas Board of Water Engineers, and the U.S. Geological Survey.

269. The Board of Water Engineers had contracted with the Bureau for geological surveys of proposed irrigation projects and damsites (for examples of contract, *see* Memorandum of Agreement, March 19, 1920, Bureau of Economic Geology). In 1920, R. A. Liddle investigated the geology of a proposed reservoir site on the San Saba River in the western part of San Saba County under this contract (**J. A. Udden** to **John A. Norris,** Chairman of the Board of Water Engineers, June 1, 1920, Bureau of Economic Geology). Leroy Patton conducted several similar surveys of proposed reservoir sites in West Texas along the upper forks of the Brazos River, the Pease River, and the Canadian River. A superficial summary of these studies is **Patton,** *Geology and the Location of Dams in West Texas:* Econ. Geology, v. 19, no. 8, p. 756-761, Dec. 1924. *See also* **J. A. Udden** to University President **W. S. Sutton,** Nov. 2, 1923, Bureau of Economic Geology.

270. In **Sellards**' very comprehensive report on *Water Supplies in Texas*, in *The Geology of Texas, Vol. II, Structural and Economic Geology* (Univ. Texas Bull. 3401 (Jan. 1, 1934), p. 351-402, Dec. 1935), he fails to discuss the questions of recharge. On the whole, this is a very concise review. Sellards concentrates on the underground water resources of the semi-arid Trans-Pecos, the western Edwards Plateau region, and the High Plains. He also presents a bibliography of underground water resource studies for other parts of the state, including mimeographed preliminary reports by the U.S. Geological Survey done in cooperation with the Board of Water Engineers.

271. In a letter to **W. W. Scott** of the U.S. Bureau of Mines, **J. A. Udden** commented that (Nov. 6, 1923, Bureau of Economic Geology):

> . . . some weeks ago this department arranged with the Oil and Gas Division of the State Railway [Railroad] Commission to make a study of the Luling Field.

Udden added that the Commission geologist had quit, and that he did not know what further work was intended by the Oil and Gas Division. He then described the extensive core analyses that the Bureau was doing in connection with the Luling Field study (**Udden** to **W. W. Scott,** Nov. 6, 1923, Bureau of Economic Geology).

272. **P. T. Flawn,** *Geology in the State Government of Texas, op. cit.* (216).

273. **R. T. Hill** to **Wallace Pratt,** Apr. 30, 1920, Bureau of Economic Geology.

274. The lack of cooperation between the Oil and Gas Division of the Railroad Commission and the Bureau is illustrated in a letter from Udden in 1920 (**Udden** to **W. F. Bowman,** Feb. 28, 1920, Bureau of Economic Geology):

> By the way, I saw in a newspaper item the other day that two more appointments are going to be made by Major G. C. Butte of deputies to enforce regulations with regard to the preservation of oil and gas in this State.

275. Udden answered a request made by Acting University President W. S. Sutton in 1923 as to what extent the work of the Bureau had been helpful in bringing about the making of the oil test at Santa Rita in Reagan County as follows:

> In June 1916, the Board of Regents of the University asked the undersigned if preliminary surveys for oil or other minerals should in my judgement be made on the University lands. I replied to the inquiry on June 22, 1916 (**E. H. Sellards** to Major **G. W. Littlefield,** Chairman, Board of Regents, Bureau of Economic Geology), calling attention to the existence of sulphur under the University lands in Culberson County, and to the existence of salt beds and possible potash salts in the region of the Panhandle and the Llano Estacado. . . . I also made a statement . . . as follows: "There exists an old axis of uplift extending northeast of the northern end of Brewster County . . . [and] it is not entirely unreasonable to suppose that at some point along this axis there may be a structure into which oil may have gathered from the lower rocks."

Udden goes on to state that in Univ. Texas Bull. 1857 (**R. A. Liddle** and **T. M. Prettyman,** *Geology and Mineral Resources of Crockett County, op. cit.* (87), the axis of the Marathon fold was shown to cut the southeast part of Reagan County. He finally added that "Mr. Pickerell of the Santa Rita Oil Company" informed Udden that their leases had been secured after they had been made aware of the course of the Marathon fold by the report in Univ. Texas Bull. 1857 (**J. A. Udden** to **W. S. Sutton,** Oct. 15, 1923, Bureau of Economic Geology). *See also* Udden's correspondence in Appendix IV.

276. Minutes of the Board of Regents, vol. G, p. 391, Apr. 29, 1929; **E. H. Sellards** to **H. Y. Benedict,** May 3, 1929, Bureau of Economic Geology.

277. The Forty-First Legislature passed the act creating the Board for Lease of University Lands on March 29, 1929 (Regular Session, General Laws, Chap. 282, p. 616-621). This act authorized the Board to employ a geologist and mineralogist "who shall keep informed with reference to the minerals in the University lands

and all activities under this and previous leases and sales and shall report to the Board all information gained with reference hereto" (p. 621). For a concise history of the administration of the University lands, *see* the manuscript by Berte R. Haigh in the Bybee Papers, The University of Texas Archives.

278. The Bureau directors could have assumed the responsibility of overseeing the development of minerals on University lands, and provided regular reports to the Regents. Also, they might have published the county ground-water reports separately and advertised them more. There is no logical reason why the Bureau should not have assumed the supervision of the ground-water resource studies for the Board of Water Engineers.

Keith Young makes the following observation on the emergence of "mission-oriented" geological staffs in State agencies, such as the Railroad Commission, which are engaged in technical service rather than research (**Keith Young**, *History of Geology at The University of Texas, op. cit.* [31], p. 17):

> Sellards is supposed to have avoided participation by the Bureau of Economic Geology in the State regulation of the petroleum industry [or water resources]. . . . Probably one should consider what effect on the Bureau's research program [which was outstanding during Sellards' directorship] the assumption of large scale supervisory, regulatory, and technical services should have had. The remarkable research contributions of the Bureau, with a small staff and limited budget, may justify Sellards' decisions.

279. It should be noted that mineral inventories and the study of the geology of any area are in fact basic foundations for any program of resource management.

280. Udden, as a representative of the American Association of State Geologists, encouraged the officers of the American Association of Petroleum Geologists to rally support for the passage of the Super Power Project bill in Congress. He called the project a truly national economic enterprise of great importance, and added: "In time, I believe that similar surveys should be made looking to the utilization of water power in the West, as well as in the East" (**J. A. Udden** to **Officers of the A.A.P.G.**, Apr. 7, 1920, Bureau of Economic Geology). There is no record that he endorsed any similar project in Texas.

One interesting note with regard to the public services provided by the Bureau is a request by the Rehabilitation Training Officer with the Federal Board for Vocational Education that the Bureau provide geological exhibits for the training center at Ft. Bayard, New Mexico: "We have a thousand men located at this place who are ex-servicemen, many of them from the State of Texas, and we are trying to interest a large class of them in geology and mining" (**Wallace M. Fox** to **J. A. Udden**, Nov. 23, 1920, Bureau of Economic Geology).

281. The following excerpts from a speech given by Governor James E. Ferguson at an Old Settler's Picnic in Valley Mills typifies the nature of his attack on the University:

> Parents of college students should not have to pay for tailor made clothes, ten dollar bouquets, automobiles, silk stockings, golf balls and high balls, poker chips, fraternity dues, fraternity pins and mandolins and a hundred other foolish and extravagant things which the rich crowd says is necessary to carry on this belshazzar revel down at the State University. I say that not only are too many people going hog wild over higher education, but that some people have become plain fools over the idea that we have to have an army of educated fools to run the government (**Ralph W. Steen**, *The Texas News*, Austin, Tex., p. 154, 1955).

See also Chap. 6, "The War on the University," *in* **Ralph W. Steen,** *The Political Career of James E. Ferguson, 1913-1917, op. cit.* (1), p. 112-140.

282. **Sellards** suggested to University President **W. M. W. Splawn** that the University research division be renamed the Division for the Conservation and Development of Natural Resources of Texas (**Sellards** to **Splawn,** June 16, 1925, Bureau of Economic Geology). Sellards said that this title would recognize more fully the actual function of the research departments — the Bureau of Economic Geology, the Bureau of Industrial Chemistry, and the Bureau of Engineering Research — which was "to lay the foundation for the continued and enlarged development of the natural resources and the related economic interests of the State." His failure to mention that this development should be as efficient as possible to prevent waste points up a lack of concern with the concept of conservation.

The creation of the Conservation Branch of the U.S. Geological Survey in July of 1925 was officially announced in the Forty-Seventh Annual Report (p. 72). This Branch combined the functions of the Land Classification Branch of the Geological Survey and the Mineral Leasing Branch of the Bureau of Mines: to classify the public lands with respect to minerals, water power, and agricultural value, and to supervise the mineral and power development on such lands and the mineral development on Indian lands.

283. During the Sellards years, Secretary Josephine Casey assumed more and more of the administrative duties of the director. Her name was synonymous with the management of the Bureau.

284. *See* Governor **Miriam Ferguson** to **E. H. Sellards,** June 7, 1934, Bureau of Economic Geology. *See also* mimeographed paper on *The Origin and Purposes of the Texas Planning Board,* in Bureau of Economic Geology for 1934.

285. Governor **James Allred** to **E. H. Sellards,** Dec. 12, 1923, Bureau of Economic Geology.

286. **E. H. Sellards** to **Everett L. Looney,** Dec. 31, 1934, Bureau of Economic Geology.

287. "Activities of the T.P.B.," by **E. A. Wood,** Director, address delivered to the Southwestern Social Science Association, Apr. 10, 1936, Bureau of Economic Geology. In this address, Wood reported that the T.P.B. had approved statewide projects estimated to provide work for 57,890 men at an estimated cost of $36,029,498. The projects included a Brazos River project, a statewide mineral survey, an underground water survey, a tax survey, an educational survey, and a price index survey. These were to be Federally funded.

288. **M. M. Leighton,** *A Model State Resource Survey,* in *Quarter-Centennial Memorial Volume of the Division of Natural Resources:* Univ. Texas Bull. 3501, p. 165, 1935.

289. For example, in the minutes of the meetings of the Mineral Resource Committee of the T.P.B., which was chaired by E. H. Sellards and included C. L. Baker, the only discussions relative to planned resource development are discussions of ways to change the mining laws so as to encourage more prospecting.

290. *See* **E. H. Sellards** to **Henry B. Ward**, Permanent Secretary of the American Association for the Advancement of Science, Dec. 18, 1933, and University President **H. Y. Benedict** to the **State Comptroller**, Dec. 26, 1933, Bureau of Economic Geology. For the role of the American Association of State Geologists, *see* Appendix V.

291. **E. H. Sellards** to **Leonard Orynski**, President of the Dallas Geological Society, Jan. 6, 1934, Bureau of Economic Geology.

292. Minutes of meeting of the Mineral Resource Council, T.P.B. folder, Oct. 1, 1935, Bureau of Economic Geology.

293. "Mineral Resource Survey of Texas," typescript included in letter from **E. H. Sellards** to University President **H. Y. Benedict**, Nov. 14, 1935, Bureau of Economic Geology.

294. **E. H. Sellards** to University Acting President **J. C. Calhoun**, Dec. 22, 1927, Bureau of Economic Geology.

295. "Texas Mineral Drive Started," *Austin Statesman*, Sept. 19, 1939.

296. **E. H. Sellards** to University President **H. P. Rainey**, June 30, 1941, Bureau of Economic Geology.

297. From 1936 to 1943, the Bureau issued 57 Mineral Resource Survey Circulars based on the results of the work under the W.P.A. project sponsored by the Bureau. The complete series of these mimeographed circulars is found at the Bureau.

298. The Federal Connally Act in 1937 prohibiting interstate and foreign shipment of any petroleum and petroleum products produced in violation of state laws added teeth to the Texas Railroad Commission proration orders by virtually destroying the market for contraband oil. *See* **F. E. Smith**, *The Politics of Conservation, op. cit.* (220), p. 285.

299. In T.P.B. folder for 1925, Bureau of Economic Geology.

300. **E. H. Sellards** to **E. A. Wood**, Mar. 20, 1935, Bureau of Economic Geology.

301. During the Roosevelt years the Federal conservation movement was directed toward reclamation projects, such as the Grand Coulee Dam, soil conservation programs, more rigid regulation of grazing use of public lands, and increasing the acreage in national parks. For the best description of the tremendous impact of the Tennessee Valley Authority program, *see* "Built for the People of the United States," *in* **F. E. Smith**, *The Politics of Conservation, op. cit.* (220). This program was the first experience anywhere in coordinated control of a watershed. For the story of President Roosevelt's role in these innovative conservation schemes of the 1930's, *see* the chapter, "F.D.R."

302. Governor **James Allred** to **E. H. Sellards**, July 2, 1938, Bureau of Economic Geology.

303. Mimeographed pamphlet, "The Need for Administrative Control of Ground Water in Texas," in T.P.B. folder for 1938, Bureau of Economic Geology.

304. C. S. Clark, who was Chairman of the Board of Water Engineers, also attended a meeting of the Subcommittee on Pollution of the Water Resource Committee of the T.P.B. at which the members discussed the pollution of streams in East Texas where oil well wastes were polluting the waters of the Sabine and Neches Rivers (**C. S. Clark** to **Walter N. White**, Sept. 1, 1938, Bureau of Economic Geology).

305. **Phil Goodrum** to **E. H. Sellards**, Nov. 19, 1943, Bureau of Economic Geology.

306. **L. W. Woods**, State Superintendent of Education, to **E. H. Sellards**, Nov. 22, 1943, Bureau of Economic Geology.

The wildlife and nature organizations were just voices in the wilderness in the thirties, when the public conscience was ignorant of the harmful effects of pollution, and pollution itself was just beginning its great leap forward. At the annual meeting of the Texas Academy of Science in 1940, the Texas Federation of Nature Clubs hosted a meeting on conservation. Two of the principal talks were "What White Man Has Done to American Wildlife," and "Present Status of Stream and Fresh Water Pollution in Texas" (Texas Academy of Science folder for 1940, Bureau of Economic Geology). Such talks are more likely to be heard in legislative halls today.

307. Texas Industrial and Commercial Research folder, May 4, 1938, Bureau of Economic Geology.

308. *Ibid.*, Apr. 17, 1938, Bureau of Economic Geology.

309. *Ibid.*, May 21, 1938, Bureau of Economic Geology.

310. With the expiration of the Texas Planning Board, the Governor recommended that the University of Texas, A&M College, and Texas Technological College pool their facilities and participate in a coordinated movement to further the development of the state's resources. Their joint effort was labeled the Texas Industrial Program, with William McGill as the Governor's coordinator. Apparently, it was completely overshadowed by the Texas Industrial and Commercial Research Council (Texas Industrial Program folder, Nov. 15, 1940, Bureau of Economic Geology).

311. **E. H. Sellards** to **D. B. Casteel**, May 9, 1935, Bureau of Economic Geology.

312. **F. B. Plummer**, "Texas Water Resources," in *Texas Mineral Resources:* Univ. Texas Bull. 4301, p. 301-312, 1943 (1946).

313. Forty-First Legislature, Second Called Session, Local and Special Laws, Chap. 37, p. 63-64, July 8, 1929.

314. *See* manuscript, **C. L. Baker**, "Geology of Dam Sites in Comanchean Rocks on the Brazos River in Johnson, Hood, Bosque, and Hill Counties, Texas," in open files of the Bureau.

315. **E. H. Sellards** to University President **H. Y. Benedict**, Mar. 16, 1936, Bureau of Economic Geology.

The law creating the Brazos River Conservation and Reclamation District had marked the first time that any state bestowed administrative authority to a district for the control of a river and its watershed. The watershed of the Brazos covers 65 counties and includes one-fourth of the state's population. The district, with the aid of W.P.A. funds, completed Morris Sheppard dam, creating Possum Kingdom Lake, in 1941. Twelve more dams were to be constructed under the master reclamation program for the district.

John A. Norris resigned as Chairman of the Board of Water Engineers in February 1936 to become General Manager of the Brazos River Conservation Reclamation District. Prior to this change, Norris, as Chairman of the Water Conservation Committee of the Texas Planning Board, had appointed a strong committee including Dr. Leroy Patton of Texas Tech and C. L. Baker of the Bureau to study the problems of power development, water supply, and sewage disposal (Texas Planning Board Bull., July 24, 1935, Bureau of Economic Geology). No record of a comprehensive report by this committee has been found.

316. *See* Colorado Valley Authority folder for 1934, Bureau of Economic Geology. *See also* **Raymond Brooks,** Secretary of the C.V.A., to **E. H. Sellards,** Sept. 1, 1934, Bureau of Economic Geology

317. Forty-Third Legislature, Fourth Called Session, General Laws, Chap. 7, p. 19-32, Nov. 13, 1934). The Lower Colorado River Authority administrative district includes the counties of San Saba, Llano, Burnet, Blanco, Travis, Bastrop, Fayette, Colorado, Wharton, and Matagorda. The Central Colorado River Authority in Coleman County and the Upper Colorado River Authority in Coke and Tom Green Counties were established in 1935. The former pioneered in the construction of small farm and ranch dams. C. L. Baker examined the geology of a proposed dam site in Coke County in that same year (**E. H. Sellards** to **Penrose Metcalf,** Chairman, Upper Colorado River Authority, June 5, 1935, Bureau of Economic Geology). In 1941, H. B. Stenzel made an examination of a proposed dam site near Port Arthur for the Lower Neches Valley Authority (**E. H. Sellards** to **A. N. Peckham,** President, Lower Neches Valley Authority, July 7, 1941, Bureau of Economic Geology).

318. *See* **Tate Dalrymple** and others, *Major Texas Floods of 1936* (U.S. Geol. Survey Water-Supply Paper 816, 1937), and **Seth Breeding** and **Tate Dalrymple,** *Texas Floods of 1938 and 1939* (U.S. Geol. Survey Water-Supply Paper 914, 1944).

319. *See* **Samuel F. Turner,** "Report of the W.P.A. Ground Water Projects Sponsored by the T.B.W.E.," August 1, 1936 (in U.S.G.S. folder, 1936, Bureau of Economic Geology). The final report was **W. N. White, W. L. Broadhurst,** and **J. W. Lang,** *Groundwater in the High Plains of Texas:* U.S. Geol. Survey Water-Supply Paper 889-F, 1946.

320. *Ibid.*, p. 406.

321. Annual Report of the Secretary of the Interior for 1935 (1936), p. 183-185.

322. Annual Report of the Secretary of the Interior for 1940 (1941), p. 167.

323. For an excellent concise survey of the history of the State parks of Texas, *see* **Ross A. Maxwell,** "Introduction" to *Geologic and Historic Guide to the State Parks of Texas:* Univ. Texas, Bur. Econ. Geology Guidebook 10, 197 p., 1970.

324. *See* **C. N. Gould** to **E. H. Sellards,** Dec. 3, 1935, Bureau of Economic Geology, concerning the establishment of Longhorn Caverns State Park. In 1936, the National Park Service appointed 21 geologists to preserve the geological features of parks and prepare descriptive material thereon.

325. *Centennial News,* v. 1, no. 23, March 16, 1936.

326. Bureau of Economic Geology files.

327. **Glen L. Evans,** *Memorial to E. H. Sellards. See* Appendix II.
See also **Wallace Pratt** to **W. K. Ferguson,** Aug. 11, 1973:

> I knew Dr. Sellards, and, I think Dr. Beede, too, at Kansas University Geological Survey, under Erasmus Haworth when I was an assistant there in 1908. Both were scholarly, serious-minded and very industrious geologists. Later, when Dr. Sellards came to the Bureau from Florida, he had demonstrated his competency as a practical, systematic organizer and administrator. But I think he lacked personality. I wanted to help him with his problems in Texas but they came at a time when I was absorbed with my own responsibilities and I contributed no useful service to him.

See also **Edgar Owens** to **W. K. Ferguson,** personal communication:

> Sellards had very close contact with the other State geologists and got along well with them. He was a peculiar character. I used to go up there quite often and I never was at the Bureau when Udden was still there. But Sellards was typical of the academic people in his generation. Ollie and I went to a tea at Sellards' house one time and it was the stiffest thing that I ever attended in my life. The whole Bureau and the Department of Geology were all that way at the time. And not until Mrs. DeFord came down there [in 1948] did things change. From that time on, things got to be a lot of fun. I enjoyed going there then. But in spite of Sellards' good will and cooperative spirit and everything, it was still cold.

328. **Alva Ellisor,** *Rockhounds of Houston, op. cit.* (31), p. 27. The Houston Geological Society made Sellards an Honorary Member in 1945, and the A.A.P.G. followed suit in 1946.

329. **E. H. Sellards,** *Early Man in America — A Study in Prehistory:* Univ. Texas Press, Austin, Tex., XVI, 221 p., 1952.

330. **E. H. Sellards** to University Acting President **J. C. Calhoun,** Oct. 18, 1937. Judge Meade and Cyrus Ray of Abilene, President of the Texas Archaeological and Paleontological Society, were the outstanding private vertebrate collectors in the state.
Père Teilhard de Chardin, consulting paleontologist of the National Geological Survey of China, was a participant in this symposium.

331. For the story of this discovery, *see* **J. A. Meade** to **E. H. Sellards,** July 13, 1945, Bureau of Economic Geology. Meade received his B.A. (1935) and M.A. (1937) in geology from the University of Nebraska.

332. **E. H. Sellards,** *Artifacts Associated with Fossil Elephant (Roberts County):* Geol. Soc. America Bull., v. 49, no. 7 (July 2, 1938), p. 999-1010.

E. H. Sellards, *Pleistocene Artifacts and Associated Fossils from Bee County, Texas,* and **Thomas N. Campbell,** *Notes on Artifacts,* and **Glen Evans,** *Notes on Terrace Deposits:* Geol. Soc. America Bull., v. 51, no. 11 (Nov. 1, 1940), p. 1627-1657.

E. H. Sellards, Glen Evans, and **Grayson Meade,** *Fossil Bison and Associated Artifacts from Plainview, Texas:* Geol. Soc. America Bull., v. 56, no. 12 (Oct. 1947), p. 917-954.

Another important contribution by **E. H. Sellards** was his *Early Man in America; Index to Localities and Selected Bibliography* (Geol. Soc. America Bull., v. 51, no. 3, Mar. 1, 1940, p. 373-431) and the similar paper covering the years 1940-1945 in Geol. Soc. America Bull., v. 58, no. 10 (Oct. 1947), p. 955-977. Loren Eisley reviewed the first paper in Am. Antiquity, v. 7, p. 84, 1941.

Another interesting find was the discovery of Pleistocene stone images in a gravel pit in Henderson County, which Sellards felt were the oldest known carvings in the western world (**E. H. Sellards,** "Stone Images from Henderson County, Texas": Am. Antiquity, v. 7, no. 1 [July 1941], p. 29-38). Sellards corresponded with **A. V. Kidder** of the Division of Historical Research of the Carnegie Institute of Washington about these images in September 1938 (Bureau of Economic Geology).

333. Sellards had done excellent work in historical physiography in connection with the Red River boundary dispute between Oklahoma and Texas back in 1923.

334. Sellards' first paper on this subject was in connection with Dr. J. E. Pearce's discovery of the first evidences of early man in Texas, *Stream Terrace Building in Connection with Human Occupancy in Central Texas,* Proclamations, Geol. Soc. America for 1934, p. 106 (abs.), June 1935.

335. **E. H. Sellards** to **Edson Bastin,** May 22, 1936, Bureau of Economic Geology.

336. Along with Alexander Deussen, A. C. Trowbridge, and Albert W. Weeks of the U.S. Geological Survey, **W. Armstrong Price** had been a pioneer investigator of the terraces of Texas. One of his earliest important papers was *The Reynosa Problem of South Texas and the Origin of Caliche* (Am. Assoc. Petroleum Geologists Bull., v. 17, no. 4, May 1933, p. 488-522). For a discussion of the proposed study, *see* **E. H. Sellards** to **Charles P. Berkey,** Secretary of the Geological Society of America, Aug. 31, 1936, Bureau of Economic Geology.

337. **E. H. Sellards,** *Artifacts Associated with Fossil Elephant (Roberts County), op. cit.* (332). Concurrently, Sellards corresponded with **Edgar B. Howard** of the Academy of Natural Sciences of Philadelphia about the possibility of a cooperative project for making a study of the terraces of the Brazos River (**Sellards** to **Howard,** Mar. 21, 1938, Bureau of Economic Geology).

338. **E. H. Sellards, Thomas N. Campbell,** and **Glen Evans,** *Pleistocene Artifacts and Associated Fossils from Bee County, Texas, op. cit.* (332).

339. **E. H. Sellards,** *Terrace Deposits as an Aid to Age Determinations of Early Man* (abs.): Geol. Soc. American Bull., v. 52, no. 23 (Dec. 1, 1941), p. 2007-2008. Another significant study that Sellards sponsored was the fine paper on *Heavy Minerals of Colorado River Terraces of Texas,* Jour. Sed. Petrography, v. 14, no. 2 (Aug. 1944), p. 86-93, by **Robert W. Mathis.**

340. An interesting outgrowth of Sellards' interest in terrace studies was the cooperation between the Bureau and the State Highway Department in the development of an "electrical resistivity machine" to locate and delineate gravel deposits. The Highway Department came up with information on the character, depth, and quantity of gravel by using this device, and then W.P.A. laborers under Bureau supervision proved the accuracy of the machine's results by digging pits. The results were good. *See* **Julian Montgomery,** State Highway Engineer, to **J. M. Isbell,** District Highway Engineer for Wichita Falls, June 24, 1935, Bureau of Economic Geology, *and* **E. H. Sellards** to **Herman Gunter,** Director of the Florida Geological Survey, Jan. 11, 1938. Also, some results of the studies are found in the mimeograph folder for 1938, Bureau of Economic Geology.

341. Sellards carried on a voluminous correspondence with the leading vertebrate paleontologists of the country: Dr. Alfred Romer of the Museum of Comparative Zoology at Harvard; C. W. Gilmore, Curator of Vertebrate Paleontology of the Smithsonian Institution; J. T. Gregory, Curator of Vertebrate Paleontology at the University of Michigan; and Dr. Edward Howard, Curator of the University Museum at the University of Pennsylvania. Interest in the rich vertebrate faunas in Texas was stirred up by Sellards, and in an effort to halt the outflow of specimens, in 1939 the Legislature passed a law authorizing the Secretary of State to establish regulations for explorations in or on any prehistoric ruins or archaeological or vertebrate paleontological sites in Texas (Forty-Second Legislature, Regular Session, Chap. 1, p. 60-61, May 18, 1939).

The law also called for a $50 collecting permit for nonresidents. This law occasioned a sharp exchange between Dr. Romer and Dr. Sellards, when Romer cited the law as an attempt on the part of Sellards to appropriate to himself the results of the work of others (**Romer** to **Sellards,** June 11, 1941, Bureau of Economic Geology).

342. **E. H. Sellards** to **L. T. Patton,** Aug. 18, 1941, Bureau of Economic Geology.

343. C. J. Hesse had been an assistant at the Museum of Paleontology at the University of California at Berkeley.

W. S. Strain, Curator of the Museum of the College of Mines and Metallurgy at El Paso, was also an active vertebrate collector. He supervised a W.P.A. collecting expedition in Brewster and Presidio Counties in 1940 for the Bureau (**E. H. Sellards** to **W. S. Strain,** Jan. 18, 1940, Bureau of Economic Geology).

344. In a 1945 report to the President of the University, Dr. Sellards noted that H. J. Sawin was working on two problems: (1) the Triassic vertebrate fauna and its application to the subdivision of the Triassic deposits of Texas; and (2) the Permian and late Carboniferous vertebrates and their application to the subdivision of the Permian and Carboniferous formations of Texas. For the years 1944 and 1945, Sawin's work was supported in part by two grants, one of $865.00 from the University Research Institute and one of $600.00 made by the Geological Society of America.

One of Sawin's outstanding reports for the Bureau was "Amphibians from the Dockum Triassic of Howard County, Texas" in *Contributions to Geology, 1944:* Univ. Texas Bull. 4401, p. 361-484, 1944 [Dec. 1945]. He resigned to join the faculty of the University of Houston on August 31, 1945.

345. File of mimeographed materials for 1928, Bureau of Economic Geology.

346. **E. H. Sellards** to Dr. **Henry Pritchett**, March 26, 1934, Bureau of Economic Geology. He also wrote **Wallace Pratt** about the same proposal on November 10, 1933, Bureau of Economic Geology. On March 27, 1937, he again wrote Pratt about the possibility of private contributions for endowing vertebrate collecting in Texas, Bureau of Economic Geology.

347. *See* Appendix VI, *Proposed Institute for Geological Research at the University of Texas.*

Dr. Stenzel, who was in charge of the geological displays in the Museum, commented that it was well known that the most noted Texas localities had been gutted by eastern museums. He also noted that "the collection of the University of Munich excels by far all the material that all the scientific institutions of Texas have collected in the red beds of west Texas . . . and the University of Oklahoma has far better vertebrate material from Texas than the University of Texas" (Bureau of Economic Geology).

348. **W. P. Webb**, *Handbook of Texas History, Vol. II, op. cit.* (252), p. 747.

349. Sellards corresponded often with his close friend Professor Charles Schuchert about the drive for an "Institute of Geologic Research" and later the Museum. A letter from **Schuchert** to **Sellards** in 1938 illustrates the role played by this pair during the successful drive for the Museum (April 12, 1938, Bureau of Economic Geology):

> Glory to you and your new Museum! You and I are the boys that got the propaganda started, and I will never forget our first talks and how I started the engine going by giving $100 as the first donation. Keep it up, but get some one started to dig up the Permian monsters for your Museum.

350. In 1940, Dr. **John A. Wilson** of the University of Michigan wrote Dr. **John Lonsdale** about the possibility of creating a vertebrate paleontology laboratory under the Bureau, the Museum, and the Department of Geology at the University (Sept. 30, 1940, Geology Department correspondence). The vertebrate collection established for the W.P.A. mineralogical-paleontological project had been moved from San Antonio to Austin in 1939. Dr. Wilson later joined the University after the war, and became Director of the Vertebrate Paleontology Laboratory of the Bureau at the Balcones Research Center.

351. Sellards also kept extensive files on earthquakes, and contributed papers on "The Valentine, Texas, Earthquake" in *Contributions to Geology, 1932* (Univ. Texas Bull. 3201, p. 113-138, 1932), "The Texas-Oklahoma Earthquake of April 11, 1934" in *Quarter-Centennial Memorial Volume of the Division of Natural Resources* (Univ. Texas Bull. 3501, p. 259-261, 1935), and "The Borger, Texas, Earthquake of June 19, 1936" in *Contributions to Geology, 1939* (Univ. Texas Bull. 3945, p. 699-704, 1939). He also delivered a paper at the 1935 meeting of the American Association for the Advancement of Science (A.A.A.S.) on the relations between earthquakes, the remains of Early Man north of Round Rock, and recent movement along the Balcones Fault Zone (**E. H. Sellards** to **James B. Macelwane**, Chairman of the Geology and Geography Section of the A.A.A.S., in A.A.A.S. folder for 1934, Bureau of Economic Geology).

352. **E. H. Sellards** to **Berte Haigh**, University Lands Office, Midland, Texas, Mar. 3, 1938, Bureau of Economic Geology.

353. Mimeograph folder for 1941, Bureau of Economic Geology. The Geological Society of America had turned down a request for funds from the Penrose Bequest for constructing shafts in the crater (Dr. **H. R. Aldrich**, Secretary of the G.S.A., to **E. H. Sellards**, Nov. 3, 1941, Bureau of Economic Geology).

354. *See* **Glen Evans** to **E. H. Sellards**, Apr. 18, 1942, and **Sellards** to **H. W. Merritt**, of Humble Oil & Refining Company, Sept. 16, 1942, Bureau of Economic Geology. *See also* **Sellards** and **Glen Evans**, "Statement of Progress of Investigation at Odessa Meteor Craters," Univ. Texas, Bur. Econ. Geology, 12 p., Sept. 1, 1941.
In a 1961 paper, Glen Evans discussed all of the investigations of the crater. This paper is in the open files of the Bureau.

355. **Edgar W. Owen**, Review of *The Geology of Texas, Vol. I, Stratigraphy*, by **E. H. Sellards, W. S. Adkins**, and **F. B. Plummer**, Univ. Texas Bull. 3232 [1933]: Am. Assoc. Petroleum Geologists Bull., v. 18, no. 4, p. 554-556, Apr. 1934.

356. *See* **Frank E. Lozo**, *Memorial, Walter Scott Adkins*, in Appendix II.

357. **C. L. Baker** to **H. Y. Benedict**, Sept. 1, 1935, Bureau of Economic Geology.

358. **C. L. Baker** to **W. S. Adkins**, Dec. 2, 1943, Bureau of Economic Geology. The range of Baker's geological knowledge as evidenced in his correspondence attests to the brilliance of his intellect. He handled inquiries ranging from detailed mineralogical analyses to the complex problems of the structural and stratigraphic geology of the Trans-Pecos to the homesteading possibilities in the Marathon Basin.

359. Press release, May 24, 1938, Bureau of Economic Geology.

360. **V. E. Barnes** and **Mildred Barnes**, eds., *Tektites:* Dowden, Hutchinson & Ross, Stroudsburg, Pa., 445 p., 1973.

361. **V. E. Barnes**, "North American Tektites," in *Contributions to Geology, 1939:* Univ. Texas Bull. 3945, p. 477-582, June 1940.
This paper demonstrates Barnes' competence and thoroughness as a scientist, and includes a fine 14-page bibliography. Barnes also compiled the "Catalogue of Texas Meteorites," which appeared in the same Bulletin, p. 583-608, along with several other descriptions of meteorite sightings and impacts.

362. Leo Hendricks left Shasta Oil Company to join the Bureau as supervising geologist for management of well samples in 1937.

363. **E. H. Sellards** to **W. E. Wrather**, Sept. 10, 1943, Bureau of Economic Geology. In this letter, Sellards stated that a "better understanding of the character and development in the State of the units of this group will be distinctly useful to those who are prospecting for oil and gas."

364. The origin and scope of this cooperative project is best described in a "Memorandum for Messrs. Cloud and Crowley, November 17, 1943" from **Hugh Miser** (U.S. Geological Survey folder for 1943, Bureau of Economic Geology).

365. These Ellenburger and Central Mineral Region studies firmly demonstrated Barnes' genius for stratigraphic observation in the field.

366. Stenzel was also Hans Closs' first Ph.D. in petrofabrics.

367. **H. B. Stenzel,** *The Geology of Leon County:* Univ. Texas Bull. 3818, 61 p., 1938.

368. Stenzel's passion for detail often precluded the completion of his studies. He spent many years studying the geology of Houston County, but the only published result was "Faulting in Northwestern Houston County," in *Texas Mineral Resources:* Univ. Texas Pub. 4301 [1946], p. 19-27.

369. For the bibliographies of these geologists, *see* **R. M. Girard,** *Bibliography and Index of Texas Geology, 1933-1950:* Univ. Texas Pub. 5910, 238 p., 1959. Stenzel's extensive paleontological publications are found there.

370. **H. B. Stenzel, F. E. Turner,** and **C. J. Hess,** *Brackish and Non-Marine Miocene Sediments in Southeastern Texas:* Am. Assoc. Petroleum Geologists Bull., v. 28, no. 7 (July 1944), p. 977-1011.

371. **F. B. Plummer** and **Gayle Scott,** *Upper Paleozoic Ammonites in Texas, Pt. I,* in *The Geology of Texas, Vol. III, Upper Paleozoic Ammonites and Fusulinids:* Univ. Texas Bull. 3701, p. 13-516, 1937.

372. *Geology of Palo Pinto County* by **F. B. Plummer** and **Joseph Hornberger, Jr.,** is a classic county geological report (Univ. Texas Bull. 3534, 240 p., 1935).

373. **F. B. Plummer,** *The Carboniferous Rocks of the Llano Region of Central Texas:* Univ. Texas Pub. 4329 [1950], 170 p.

374. **E. H. Sellards, C. L. Baker,** and others, *The Geology of Texas, Vol. II, Structural and Economic Geology (exclusive of petroleum):* Univ. Texas Bull. 3401 [1935], 884 p.

375. The procedures followed in compiling the map, the acknowledgments, and the various contour surfaces used are described in Univ. Texas Bull. 3401, *ibid.,* p. 133-136. Leo Hendricks was co-author of this map. Sellards prepared the first two editions. The third edition was largely prepared by Hendricks.

An example of the interchange of ideas on the major structural features is found in this excerpt from a letter from **Archie R. Kautz** of Empire Gas and Fuel to Dr. **Sellards** (June 19, 1933, Bureau of Economic Geology):

> Concerning Circular No. 3 concerning the Amarillo Uplift, I wish to state that we had a discussion . . . at our June meeting of the Panhandle Geologic Society. Shortly you may expect a letter concerning the paper, which will embody a majority opinion of the Society. It must be explained that we cannot say unanimous because of company policy some of the members are not in a position to express an opinion.

376. **E. H. Sellards** wrote on the "Structural Geology of Texas East of Pecos River" (*ibid.,* p. 11-133), and **C. L. Baker** described the "Structural Geology of Trans-Pecos Texas" (*ibid.,* p. 137, 214).

377. *Ibid.,* p. 482-502, 641-832.

378. **E. H. Sellards** and **Leo Hendricks,** "Occurrence of Oil and Gas in Texas," in *Texas Mineral Resources:* Univ. Texas Pub. 4301, 1943 [1946], p. 179-190.

Virgil E. Barnes, "Earth Temperatures and Oil Field Waters of North-Central Texas," in *Texas Mineral Resources:* Univ. Texas Pub. 4301, 1943 [1946], p. 313-357.

379. **V. E. Barnes, R. F. Dawson,** and **G. A. Parkinson,** *Building Stones of Central Texas:* Univ. Texas Pub. 4246, 1942 [1947], 198 p.

380. **F. M. Bullard** and **R. H. Cuyler,** "The Upper Pennsylvanian and Lower Permian Section of the Colorado River Valley, Texas," in *Quarter-Centennial Memorial Volume of the Division of Natural Resources:* Univ. Texas Bull. 3501, 1935, p. 191-258. The authors did their field work in connection with the course in field geology offered by the University of Texas during the summers of 1930, 1932, and 1934.

381. **F. B. Plummer** and **J. Hornberger, Jr.,** *Geology of Palo Pinto County, Texas, op. cit.* (372).

382. The results of this work were published in *Stratigraphic and Paleontologic Studies of the Pennsylvanian and Permian Rocks in North-Central Texas* (Univ. Texas Bull. 3801, 1938 [1939], 252 p.

W. Lee, *Stratigraphy of the Cisco Group of the Brazos Basin, Texas;*

C. O. Nickell, *Stratigraphy of the Canyon and Cisco Groups on Colorado River in Brown and Coleman Counties, Texas;*

W. Lee, *Comparison of Brazos and Colorado River Sections;*

J. S. Williams, *Carboniferous Invertebrate Fossils (Except Fusulinids) from North-Central Texas;*

L. G. Henbest, *Notes on the Ranges of Fusulinidae in the Cisco Group (Restricted) of the Brazos River Region, North-Central Texas.*

383. **F. M. Bullard** and **R. H. Cuyler** published their report in Univ. Texas Bull. 3501 *op. cit.* (380) after Nickell had finished his field work. A footnote on the bottom of page 91 describes the various discrepancies between the measured sections of Bullard and Cuyler, and those of Nickell.

384. Univ. Texas Bull. 3801, *op. cit.* (382), p. 95. The reference is to **M. G. Cheney,** *Stratigraphic and Structural Studies in North-Central Texas, op. cit.* (104).

385. **F. B. Plummer,** *The Carboniferous Rocks of the Llano Region of Central Texas, op. cit.* (373). For an up-to-date discussion of the various stratigraphic classifications for the Pennsylvanian and Lower Permian in the Brazos basin, *see* **L. F. Brown, Jr.,** *Stratigraphy of the Blach Ranch-Crystal Falls Section (Upper Pennsylvanian), Northern Stephens County:* Univ. Texas, Bur. Econ. Geology Rept. Inv. 41, 1960, 45 p.

386. **F. B. Plummer** and **Gayle Scott,** *Upper Paleozoic Ammonites in Texas, Pt. I, op. cit.* (371). Dr. Stewart Weller had first proposed this study in 1925, when Plummer was lecturing at the University of Chicago. Scott joined Plummer the following year, and from 1929 on, Dr. Sellards encouraged the continuation of this work. The Geological Society of America made a contribution from the

Penrose Bequest to be applied toward the completion of the project, under the general grant for "Group Study to Establish More Satisfactory Criteria for the Stratigraphic Correlation of the Mississippi, Pennsylvanian and Permian Sediments in the Texas Region."

387. **C. O. Dunbar** and **J. W. Skinner**, *Permian Fusulinidae of Texas, Pt. II*, in *The Geology of Texas, Vol. III, Upper Paleozoic Ammonites and Fusulinidae:* Univ. Texas Bull. 3701, 1937, p. 523-822. In their work on the stratigraphy of the Glass Mountains, the King brothers had found that the extensive faunas they had collected would take several years to study. So they turned the fusulines over to Dr. Dunbar, who, along with Robert King, was to study these microfossils. King's oil company work consumed his time so he refused to accept the authorship of the paper. Dunbar was subsequently joined by Skinner.

388. Other significant paleontological works include **R. C. Moore** and **F. B. Plummer**, "Crinoids from the Upper Carboniferous and Permian Strata in Texas," in *Contributions to Geology, 1939* (Univ. Texas Bull. 3945, 1939, p. 9-468); **R. C. Moore** and **R. M. Jeffords**, "Description of Lower Pennsylvanian Corals from Texas and Adjacent States," in *Contributions to Geology, 1944* (Univ. Texas Pub. 4401, 1944 [1945], p. 77-208); and **H. J. Plummer**, "Smaller Foraminifera in the Marble Falls, Smithwick, and Lower Strawn Around the Llano Uplift, in Texas," *ibid.*, p. 209-271.

389. **R. C. Moore** and others, *Report on the Correlation of Pennsylvanian Formation of North America:* Geol. Soc. America Bull., v. 55, no. 6 (June 1944), p. 657-706.
C. O. Dunbar and others, *Correlation of the Permian Formation of North America:* Geol. Soc. America Bull., v. 71, no. 12 (Dec. 1960), p. 1763-1805.

390. **H. B. Stenzel**, *The Geology of Leon County, op. cit.* (367).

391. *Ibid.*, p. 218-224. **David Crow** wrote a report on "Lignite Mining in Leon County" (p. 230-245).

392. **H. B. Stenzel**, "The Yegua Problem," in *Contributions to Geology, 1939:* Univ. Texas Bull. 3945, 1939, p. 847-910.

393. **Dan Feray** was able to use some of Stenzel's information in his doctoral dissertation at the University of Wisconsin, *Relation of the Foraminifera to the Sedimentary Characteristics of the Weches Formation* (1948).

394. **H. B. Stenzel**, "Faulting in Northwestern Houston County, Texas," in *Texas Mineral Resources:* Univ. Texas Pub. 4301, 1943 [1946], p. 19-27.

395. **E. H. Sellards** to **T. W. Painter**, Jan. 15, 1945, Bureau of Economic Geology.

396. **R. A. Liddle**, *The Van Oil Field, Van Zandt County, Texas:* Univ. Texas Bull. 3601, 1936, 79 p.
B. C. Renick, *The Jackson Group and the Catahoula and Oakville Formations in a Part of the Texas Gulf Coastal Plain:* Univ. Texas Bull. 3619, 1936, 101 p.

397. **D. C. Barton, C. H. Ritz** and **Maude Hickey**, *Gulf Coast Geosyncline:* Am. Assoc. Petroleum Geologists Bull., v. 17, no. 12 (Dec. 1933), p. 1446-1458.

In the section "Cenozoic Systems" in Univ. Texas Bull. 3232, Plummer observed that the periodic transgressions and regressions of the Cenozoic seas "have been explained by some geologists as due to periodic isostatic adjustments in a geosyncline or embayment following heavy loading with sediments and corresponding removal of material from adjacent land masses; and by others as due to periodic compaction of clays by dehydration of colloids at depth following long epochs of deposition . . ." (p. 529). However, there is no discussion of Barton's paper in this bulletin, or even in Bulletin 3401, *Structural and Economic Geology.*

Willis G. Meyer, *Stratigraphy and Historical Geology of Gulf Coastal Plain in the Vicinity of Harris County, Texas* (Am. Assoc. Petroleum Geologists Bull., v. 23, no. 2, Feb. 1939, p. 145-211) is an excellent pioneer discussion of the Gulf Coast Geosyncline and a review of the literature thereon.

398. **D. C. Barton**, *Deltaic Coastal Plain of Southeastern Texas:* Geol. Soc. America Bull., v. 41, 1930, p. 359-382.

In Bulletin 3232, Mr. Plummer noted that "Barton in his excellent account of recent deltaic sedimentation along the coastal plain has given the best description of the building-out process" (p. 526).

399. **R. J. Russell, H. V. W. Howe,** and **J. H. McGuirt**, "Physiography of Coastal Southwest Louisiana," in *Geology of Cameron and Vermilion Parishes:* Louisiana Geol. Survey Geol. Bull. No. 6, 1935, p. 1-72.

R. J. Russell, "Physiography of the Lower Mississippi Delta," in *Geology of Plaquemines and St. Bernard Parishes:* Louisiana Geol. Survey Geol. Bull. No. 8, 1936, p. 3-199.

H. V. W. Howe, *Stratigraphic Evidence for Gulf Coast Geosyncline* (abs.): Geol. Soc. America Proc. 1935, 1936, p. 82.

R. J. Russell, *Quarternary History of Louisiana:* Geol. Soc. America Bull., v. 55, 1940, p. 1199-1234.

400. **L. W. Storm**, *Resume of Facts and Opinions on Sedimentation in Gulf Coast Region of Texas and Louisiana:* (Am. Assoc. Petroleum Geologists Bull., v. 29, no. 9, Sept. 1945, p. 1304-1335). This is a thorough summary which includes discussion of the effects of glacial periods on post-Tertiary shoreline deposition. He also notes the extension of this concept to Tertiary deposition, during which there were changes of sea level unrelated to the accumulation or melting of ice caps. Storm says that his paper was an attempt to round out the work of the Corpus Christi Geologists Study Group. He described the Texas Gulf Coast as the best area in the world for the study of clastic sediments.

401. **Sidney Paige**, *Description of the Llano and Burnet Quadrangles:* U.S. Geol. Survey Geol. Atlas, Llano-Burnet Folio (no. 183), 1912, 16 p.

402. **Josiah Bridge**, *The Correlation of the Upper Cambrian Sections of Missouri and Texas with the Section in the Upper Mississippi Valley:* U.S. Geol. Survey Prof. Paper 186L, 1937, p. 233-237.

403. **Josiah Bridge** and **V. E. Barnes**, "Report of Facies Change in the Cambrian of Central Texas," National Resources Council, Report of Committee on Marine Ecology as Related to Paleontology, 1941-1942 (Dec. 1942), p. 11-12.

404. **P. E. Cloud, Jr., V. E. Barnes,** and **Josiah Bridge,** "Stratigraphy of the Ellenburger Group in Central Texas, A Progress Report," in *Texas Mineral Resources:* Univ. Texas Pub. 4301, 1943 [1946], p. 133-161.

P. E. Cloud, Jr., and **V. E. Barnes,** *The Ellenburger Group of Central Texas:* Univ. Texas Pub. 4621, 1946 [1948], 473 p.

Also, **V. E. Barnes, P. E. Cloud, Jr.,** and **L. E. Warren** authored a paper describing the first Devonian strata discovered in Texas in "The Devonian of Central Texas," in *Texas Mineral Resources:* Univ. Texas Pub. 4301, 1943 [1946], p. 163-177.

405. **Leo Hendricks,** "Subsurface Division of the Ellenburger Formation in North-Central Texas," in *Contributions to Geology, 1939:* Univ. Texas Bull. 3945, 1939, p. 923-968.

406. **Leo Hendricks** and **A. J. Crowley,** *Lower Ordovician and Upper Cambrian Subsurface Division in North-Central Texas:* Am. Assoc. Petroleum Geologists, v. 29, no. 4 (Apr. 1945), p. 413-425.

407. **A. J. Crowley,** "Subdivisions of the Ellenburger Limestone in the Southwest Part of the K.M.A. Oil Field, North-Central Texas," in *Texas Mineral Resources:* Univ. Texas Pub. 4301, 1943 [1946], p. 191-194.

408. **J. T. Gregory,** "Osteology and Relationships of *Trilophosaurus*" (p. 273-330), and **H. J. Sawin,** "Amphibians from the Dockum Triassic of Howard County" (p. 361-399), in *Contributions to Geology, 1944:* Univ. Texas Pub. 4401, 1944 [1945].

409. **Glen Evans** and **Grayson Meade,** "Quaternary of the Texas High Plains" (p. 485-507), and **Grayson Meade,** "The Blanco Fauna" (p. 509-540), in *Contributions to Geology, 1944:* Univ. Texas Pub. 4401, 1944 [1945].

410. **C. L. Baker,** *Geology and Underground Waters of the Northern Llano Estacado, op. cit.* (134).

W. D. Matthew, "Observations on the Tertiary of the Staked Plains," manuscript (dated 1924) in open files of Bureau of Economic Geology.

L. T. Patton, *Some Observations on the So-Called Lakes of the Llano Estacado of Texas* (abs.): Geol. Soc. America Proc. 1934 (June 1935), p. 451.

411. **Glen Evans** and **Grayson Meade,** "Quaternary of the Texas High Plains," *op. cit.* (409), p. 485.

412. **Grayson Meade,** "The Blanco Fauna," *op. cit.* (409), p. 519. Meade includes a superior survey of previous work.

Mention should be made of **C. J. Hesse**'s paper, "A Pliocene Vertebrate Fauna from Higgins, Lipscomb County, Texas," in *Contributions to Geology, 1939* (Univ. Texas Bull. 3945, 1939, p. 671-698). This contains descriptions of the vertebrate fossils collected by L. C. Reed and O. M. Longnecker, Jr., and sent to the University of California by C. L. Baker in 1928.

413. **L. W. Stephenson,** *The Larger Invertebrate Fossils of the Navarro Group of Texas (Exclusive of Corals and Crustaceans and Exclusive of the Fauna of the Escondido Formation):* Univ. Texas Pub. 4101, 1941, 641 p.

See also **E. H. Sellards** to Dean **A. P. Brogan,** Feb. 23, 1940, Bureau of Economic Geology.

In the autumn of 1935, Stephenson collected material for his paper on the *Stratigraphic Relations of the Austin, Taylor, and Equivalent Formations in Texas:* U.S. Geol. Survey Prof. Paper 186G, 1937, p. 133-146.

Also, Stephenson was chairman of the Subcommittee on the Cretaceous of the Atlantic and Gulf Coastal Plains of the National Resource Council. His committee finished *Correlations of the Outcropping Cretaceous Formations of the Atlantic and Gulf Coastal Plain and Trans-Pecos Texas* in 1941 (Geol. Soc. America Bull., v. 53, no. 3, Mar. 1, 1942, p. 435-448).

414. An example of Stephenson's neo-Lamarckian prose that defies all studies in modern evolution follows (**L. W. Stephenson,** *The Larger Invertebrate Fossils of the Navarro Group of Texas (Exclusive of Corals and Crustaceans and Exclusive of the Fauna of the Escondido Formation), ibid.*, p. 33):

> The sea bottom suitable for the existence of the multiplicity of organisms that had been maintaining a lazy existence in the salubrious waters of a broad continental shelf was thus reduced to a narrow strip. In this drastically restricted zone competition was severe, and the forms that were unable to adapt themselves to the cramped living conditions perished in the struggle, [which] continued for a considerable stretch of geologic time. The competition stimulated evolution and the forms that survived the struggle were changed creatures when, with the advance of the Midway sea, they finally returned to their ancestral home. So greatly were they changed in form and surface features that they must be classed as new species. Along with these changed forms came also immigrant strangers from other seas.

Also, **H. B. Stenzel** contributed many papers on Cretaceous crabs, the most significant of which was "Decapod Crustaceans from the Cretaceous of Texas," in *Texas Mineral Resources:* Univ. Texas Pub. 4301, 1943 [1946], p. 401-476.

415. In 1939, George Branner, Chairman of the Committee to Consider Reorganization of the Federal Departments Connected with Mineral Development of the American Association of State Geologists (Dr. Sellards was a member of this committee), composed a report outlining what the committee believed the primary functions of the U.S. Geological Survey should be. They included mineral exploration and inventory, and improvement of the "exploratory methods now in use, so that they may become more effective aids in the location of new deposits." The committee also recommended that the Survey decentralize operations to facilitate cooperative, coordinative, and advisory work with respect to local geological agencies. Branner also recommended that the topographic mapping and the measurement of stream flow services be separated from the main work of the Survey — discovery, evaluation, and conservation of mineral resources. These extraneous services had overshadowed in importance and appropriations the work of the Geologic Branch over the years (A.A.S.G. folder, Nov. 27, 1939, Bureau of Economic Geology).

416. In his "History of the Bureau of Economic Geology," which is found in the open files of the Bureau (H-La, 1946), Everett DeGolyer comments that the greatest weakness of the Bureau is its failure to make itself as useful to all industry as it has to the petroleum industry. He notes that the very great strength of the Illinois Geological Society lies in its strong position with all phases of industry (p. 7-8).

417. **C. P. Ross**, "Preliminary Report on the Terlingua Quicksilver District, Brewster County, Texas," in *The Geology of Texas, Vol. II, Structural and Economic Geology:* Univ. Texas Bull. 3401, 1934 [1935], p. 558-568.

C. P. Ross and **W. E. Cartwright**, "Preliminary Report on the Shafter Mining District, Presidio County, Texas," in *The Geology of Texas, Vol. II, Structural and Economic Geology:* Univ. Texas Bull. 3401, *ibid.*, p. 573-608.

C. P. Ross, *The Geology and Ore Deposits of the Shafter Mining District, Presidio County, Texas:* U.S. Geol. Survey Bull. 928-B, 1943, p. 45-125.

418. **E. B. Eckel** and **P. E. M. Purcell**, "The Iron Ores of East Texas," in *The Geology of Texas, Vol. II, Structural and Economic Geology:* Univ. Texas Bull. 3401, 1934 [1935], p. 482-503.

E. B. Eckel, *The Brown Iron Ores of Eastern Texas:* U.S. Geol. Survey Bull. 902, 1938, 157 p.

419. **P. B. King**, "Clay Deposits of the San Antonio Area and Morris County, Texas," in *Clay Investigations in the Southern States, 1934-1935:* U.S. Geol. Survey Bull. 901, 1940, p. 93-188.

420. **P. B. King**, *Geology of the Southern Guadalupe Mountains, Texas:* U.S. Geol. Survey Prof. Paper 215, 1948, 183 p. King noted that the main economic interest of his work was indirect — the stratigraphic value to petroleum geologists and Permian Basin producers.

421. Julia Gardner conducted extensive investigations of the Eocene and other Tertiary faunas of northeastern Mexico for the Survey. Her studies were supported by grants from the National Research Council and the Penrose Bequest of the Geological Society of America, and the resultant publications included **Julia Gardner** and **E. O. Bowles**, *The* Venericardia planicosta *Group in the Gulf Province* (U.S. Geol. Survey Prof. Paper 189-F, 1939, p. 143-215); **Julia Gardner**, *Notes on Fossils from the Eocene of the Gulf Province: 1, The annelid genus* Tubulostium; *2, The gastropod families Cassididae, Ficidae, and Buccinidae* (U.S. Geol. Survey Prof. Paper 193-B, 1939, p. 17-44); and **Julia Gardner**, *Mollusca of the Tertiary Formations of Northeastern Mexico* (Geol. Soc. America Mem. 11, Dec. 15, 1945, 332 p.).

Also of interest are two popular geologic guidebooks which **N. H. Darton** prepared: *Guidebook of the Western United States, Part F, The Southern Pacific Lines, New Orleans to Los Angeles* (U.S. Geol. Survey Bull. 845, 1933, 304 p.) and **N. H. Darton** and **P. B. King**, *Western Texas and Carlsbad Caverns, op. cit.* (245).

422. **N. H. Darton, L. W. Stephenson,** and **J. A. Gardner,** *Geologic Map of Texas:* U.S. Geol. Survey, map, 4 sheets, scale 1:500,000, 1937. Abs., Geol. Soc. America Bull., v. 44, pt. 1 (Feb. 28, 1933), p. 82.

In his Memorial to N. H. Darton, **J. R. Pemberton** reports (Am. Assoc. Petroleum Geologists Bull., v. 33, no. 1, Jan. 1949, p. 116-124):

> He reviewed and checked in the field all previously published information, and petroleum companies freely gave him what information they had. He soon found however that much of this vast area was unexplored. Fortunately adequate topographic maps were available for much of the rugged Big Bend country, but in other critical parts of Trans-Pecos Texas he made his own topographic maps as bases for the geology. These maps were made in his

geological notebook by plotting a running compass traverse, metered by his automobile odometer, and taking compass bearings on all topographic and cultural landmarks. Darton's amazing ability to judge relative altitudes by eye was checked from time to time by barometer readings. These field plats were gradually built together into 30-minute sheets showing both topography and geology, using for control all triangulation stations and benchmarks. The resulting field sheets were in Darton's words meant to present a "valid rather than accurate map of the area."

Pemberton found it incredible that these field topographic and geologic maps grew at the rate of 500 mi² a week. Darton completed the major task of mapping in six field sessions.

In spite of the fact that extensive W.P.A. funds were granted to the U.S. Geological Survey and the U.S. Coast and Geodetic Survey for making special purpose maps, no comprehensive national topographic mapping program was established (no appropriations for carrying out the Temple Act had ever been passed). In the absence of any such program, it was inevitable that the map-using agencies make their own maps. William Bowie, Chairman of the Committee on Policy, Advisory Council to the Federal Board of Surveys and Maps, sent out a paper on "Status of National Mapping" on July 19, 1939, in which he called for support for Senate Resolution No. 87 which recommended that appropriations be made to cover a 15-year mapping program to complete the mapping of the nation (Texas Planning Board folder, 1939, Bureau of Economic Geology).

423. **E. H. Sellards** to **J. S. Williams,** U.S. Geological Survey, June 3, 1943, Bureau of Economic Geology.

424. **Glen Evans,** *Fluorspar Deposits in the Eagle Mountains of Hudspeth County, Texas:* Univ. Texas, Bur. Econ. Geology Min. Res. Circ. 26 (June 21, 1943), 12 p.

Glen Evans, "Fluorspar in Trans-Pecos Texas," in *Texas Mineral Resources:* Univ. Texas Pub. 4301, 1943 [1946], p. 227-238.

For an up-to-date study, *see* **W. N. McAnulty, Sr.**, *Fluorspar in Brewster County, Texas:* Univ. Texas, Bur. Econ. Geology Geol. Circ. 67-2, 1967, 16 p.

425. **Harry Ladd,** Regional Geologist for the U.S. Geological Survey, to **E. H. Sellards,** May 18, 1943, Bureau of Economic Geology.

426. **V. E. Barnes, S. S. Goldich,** and **F. Romberg,** *Iron Ore in the Llano Region, Central Texas:* Univ. Texas, Bur. Econ. Geology Rept. Inv. 5 (Apr. 1949), 50 p.

427. For a good brief history of the iron industry in Texas, *see* **W. P. Webb** and **H. Bailey Carroll,** *The Handbook of Texas, Vol. I,* "The Iron and Steel Industry," p. 894.

An interesting side note is a report by the Texas Industrial and Commercial Research Council, "Brief Concerning the Proposal That Our Federal Government Finance an Iron and Steel Manufacturing Plant in East Texas," Aug. 23, 1941, in Texas Industrial and Commercial Research Council folder for 1941, Bureau of Economic Geology.

428. **E. H. Sellards** to **L. B. Johnson,** Oct. 24, 1944, Bureau of Economic Geology.

429. For the history of the Balcones Research Center, see Chapter XI.

430. **F. B. Plummer** and **Bruce Grant** of the Petroleum Engineering Department of the University of Texas published a paper on *Oil Shale of Central Texas* (Oil and Gas Jour., v. 43, no. 22, Oct. 7, 1944, p. 66-70), in which they discussed tests on the petroleum content of the Barnett shale which Grant had conducted, and then the outcrops of the Barnett in San Saba County that showed the most potential.

431. Some typical reports that were produced under this program include **Carl Fries, Jr.**, *Optical Calcite Deposits of the Republic of Mexico* (U.S. Geol. Survey Bull. 954-D, 1948) and **Carl Fries** and **W. F. Foshag**, *Tin Deposits of the Republic of Mexico* (U.S. Geol. Survey Bull. 935-C, 1943, p. 99-176). Bulletins 935, 953, and 954 were issued only as separate chapters under the general title *Geologic Investigations in the American Republics for the Years 1941-43, 1945, and 1946,* respectively.

432. **Helen R. Fairbanks** and **Charles P. Berkey,** *Life and Letters of R. A. F. Penrose, Jr.:* Geol. Soc. America, 1952, 765 p.

The market value of the Penrose Fund of the Geological Society of America on Oct. 31, 1935, was $4,314,403.75 (G.S.A. Proc. for 1935 [1936], p. 17).

433. **H. R. Fairbanks** and **C. P. Berkey,** *Life and Letters of R. A. F. Penrose, Jr., ibid.,* p. 760.

434. In connection with the Paricutin study, **V. E. Barnes** and **F. Romberg** made a gravity survey of the area, the results of which were published in the Geological Society of America Bulletin for 1948.

Grants-in-aid from the Penrose Bequest that dealt specifically with geological research in Texas from 1933-1944 are listed below, along with the amount of the grant:

1933.	Philip King, "Permian Rocks of Trans-Pecos Texas"	$1,400
1933.	Josiah Bridge, "Cambrian and Ordovician Strata of the Central Mineral Region of Texas"	$ 450
1933.	E. H. Sellards, et al., "Stratigraphic Correlation of Mississippian, Pennsylvanian, and Permian Sediments in the Texas Region"	$3,000
1933.	Alfred S. Romer, "The Early History of the 'Red Beds' Vertebrate Fauna"	$ 326
1937.	John T. Lonsdale, "Petrography and Petrology of the Igneous Rocks of the Terlingua Quadrangle, Brewster and Presidio Counties, Texas"	$1,104
1938.	C. C. Albritton, Jr., and Kirk Bryan, "Investigation of Late Pleistocene Deposits Near Alpine, Texas, for the Purpose of Establishing the Chronology of Human Remains and Fossils of Extinct Animals Contained Therein"	$ 500
1938.	H. G. Richards, "Pleistocene Terraces of the Gulf Coastal Plain"	$ 200
1940.	John T. Lonsdale and Ross A. Maxwell, "Geology of the Chisos Mountains Area, Texas"	$1,338
1940.	S. S. Goldich and M. A. Elms, "Structural Geology and Petrology of the Buck Hill Quadrangle, Texas"	$ 350
1944.	H. B. Stenzel, "Investigation of Fossil Decapod Crustaceans of North America"	$ 600

1944.	H. J. Sawin, "Skeleton of the Triassic Reptile, Desmatosuchus"......	$ 600
1944.	F. M. Bullard, "Study of the Space Relations and Products of Paricutin" (Bullard was a member of the Geology Department of the University of Texas.)	$ 800

Over the years 1935 to 1940, R. T. Hill received a total of $8,400 in grants from the Penrose Bequest. His initial project, "History of Geologic Investigations in the Southwest," 1935-1938, $6,000, did not develop into a published report. However, his studies on "Identification and Interpretation of the Routes of Cabeza de Vaca and Coronado" led to the publication of a series of articles in the *Dallas Morning News*.

435. Geol. Soc. America Proc. 1934 [1935], p. 45-47.

436. **Parker Trask** edited two outstanding works under the direction of his committee: *Recent Marine Sediments: A Symposium* (Am. Assoc. Petroleum Geologists, Tulsa, Okla., 1939, 736 p.) and *Applied Sedimentation* (Wiley & Co., New York, 1950, 707 p.). The report of W. H. Twenhofel's committee was published only by the National Resource Council, "Report of the Committee on Paleoecology, 1935-36" (Washington, 1936).

437. Geol. Soc. America Proc. for 1935 [1936], p. 30-33.

438. Dr. Sellards was President of the S.E.P.M. in 1938 and Dr. Stenzel was Secretary-Treasurer in 1939.

439. **R. K. DeFord**, ed., *The Permian System of the United States West of the Mississippi River:* West Texas-New Mexico Symposium, Pt. 1, Am. Assoc. Petroleum Geologists Bull., v. 24, no. 1 (Jan. 1940), and Pt. 2, A.A.P.G. Bull., v. 26, no. 4 (Apr. 1942).

440. **Chester R. Longwell** edited this *Tectonic Map of the United States* (Am. Assoc. Petroleum Geologists Bull., v. 28, no. 12 [Dec. 1941], p. 1767-1774). The Division of Geology and Geography of the National Resources Council had initiated the project in 1923, and it soon became a first rate illustration of the scope of cooperation in geological research. The committee members worked as volunteers. About one half of the members were from the U.S. Geological Survey, which donated a great amount of research energies. The National Research Council paid for clerical and technical assistants, the American Association of Petroleum Geologists paid all costs of drafting and preparation, and the A.A.P.G. and the Geological Society of America paid for the costs of the plates.

441. In 1934, M. G. Cheney, Chairman of the Research Committee, observed that the main research role of the A.A.P.G. must be to suggest "well thought out, attractive, definite research projects in petroleum geology" that would attract research funds from the government, oil companies, or private institutions. The A.A.P.G. should not attempt to raise and administer large sums of money for research projects, but it can formulate and sponsor such projects (A.A.P.G. Bull., v. 28, no. 10, Oct. 1944, p. 1543-1544). Cheney also noted that most of the broad research studies in the area of petroleum geology in the past had been under the auspices of the American Petroleum Institute, which administered research funds from the Rockefeller Foundation. However, the only geological investiga-

tion then being sponsored by the A.P.I. was a project on "Transformation of Organic Matter into Petroleum and Identification of Source Beds."

It is interesting to note that in 1945, Professor E. J. Lund of the University of Texas first suggested the establishment of an Institute of Marine Science in a letter to the Regents of the University (**M. G. Cheney,** Chairman of the Research Committee of the A.A.P.G., to **E. J. Lund,** Feb. 16, 1945, Bureau of Economic Geology). In this letter, Cheney commented that "we will have to turn to a greatly expanded fundamental research program to meet the increased difficulties in successful exploration for maintenance of oil production and oil reserves."

442. **D. C. Barton** to **J. M. Dawson,** June 9, 1945, Bureau of Economic Geology.

443. **J. M. Dawson** to **D. C. Barton,** July 17, 1945, Bureau of Economic Geology.

444. **J. R. Sandidge** to **H. P. Rainey,** May 7, 1943, Bureau of Economic Geology.

445. **E. H. Sellards** to **W. E. Wrather,** Sept. 10, 1943, Bureau of Economic Geology.

446. **W. E. Wrather** to **E. H. Sellards,** Oct. 18, 1943, Bureau of Economic Geology.

447. **E. H. Sellards** to the **Committee on Projects, Geol. Soc. America,** Aug. 22, 1933, Bureau of Economic Geology.

448. *See* G.S.A. folder for 1939, Project Grant 44-33, Bureau of Economic Geology.

449. **E. H. Sellards** to **C. O. Dunbar,** Chairman of the Committee on Stratigraphy and Chairman of the Subcommittee on the Permian of National Research Council, Mar. 12, 1934, Bureau of Economic Geology.

450. *Ibid.*

451. See *Correlation Charts Prepared by the Committee on Stratigraphy of the N.R.C.:* Geol. Soc. America Bull., v. 53, no. 3 (Mar. 1, 1942), p. 425-434.

452. Permian Subcommittee of the National Research Council's Committee on Stratigraphy (Carl O. Dunbar, Chairman), "Correlation of the Permian Formations of North America" (Geol. Soc. America Bull., v. 71, no. 12, Dec. 1960, p. 1763-1805). Sellards was also a member of this subcommittee. The bibliography for this paper contains 485 entries.

The list of subcommittees and chairmen is as follows: Cambrian, B. F. Howell; Ordovician, W. H. Twenhofel; Silurian, C. K. Schwartz; Devonian, G. Arthur Cooper; Mississippian, Carey Croneis; Pennsylvanian, Raymond C. Moore; Permian, Carl Dunbar; Triassic and Jurassic, John B. Reeside, Jr.; Cretaceous of the Atlantic and Gulf Coastal Plain, Lloyd V. Stephenson; Cretaceous of Western North America, Charles E. Weaver; and Cenozoic of the Atlantic and Gulf Coastal Plains and the Caribbean Region, C. Wythe Cooke. *See also* notes 392 and 396.

453. *See* Memorial to John T. Lonsdale, Appendix II. Samuel P. Ellison was Chairman of the Faculty Committee which authored this memorial.

It is interesting to note that Monroe G. Cheney, Pennsylvanian stratigrapher and independent geologist from Coleman, was also given serious consideration as a candidate for the post of director (**E. H. Sellards** to Acting University President **T. H. Painter,** Mar. 31, 1934, Bureau of Economic Geology). Lonsdale assumed his post in September.

454. **H. B. Stenzel, A. E. Weissenborn,** et al., *Geological Resources of the Trinity River Tributary Area in Oklahoma and Texas:* Univ. Texas Pub. 4824, 1948, 252 p.

455. This mimeographed report is in the Arkansas, White and Red River Basins Interagency Committee folder in the Bureau files for 1955.

456. **J. M. Perkins** and **J. T. Lonsdale,** *Mineral Resources of the Texas Coastal Plain (Preliminary Report):* Univ. Texas, Bur. Econ. Geology Min. Res. Circ. 38, 1955, 50 p.

457. **J. W. Dietrich** and **J. T. Lonsdale,** *Mineral Resources of the Colorado River Industrial Development Association Area:* Univ. Texas, Bur. Econ. Geology Rept. Inv. 37, 1958, 84 p.

458. **R. A. Maxwell,** *Mineral Resources of South Texas: Region Served Through the Port of Corpus Christi:* Univ. Texas, Bur. Econ. Geology Rept. Inv. 43, 1962, 140 p.

459. Folders containing the bulletins of the Texas Water Conservation Association are in the Bureau files.

460. **Walter P. Webb,** *More Water for Texas: The Problem and the Plan:* Univ. Texas Press, Austin, Tex., 1954, 69 p.

461. For a good discussion of the historical evolution of Texas water laws, *see* **Arthur P. Duggan,** "Texas Ground Water Law" (p. 11-30) and other articles in *Proceedings of the Water Law Conferences.* The Fifty-Third Regular Session of the Texas Legislature passed an "act declaring the public policy of this state with reference to the establishment of a more comprehensive program for the prevention, abatement, and control of stream pollution; creating the five member Water Pollution Advisory Council" (Chap. 353, p. 868-869, June 8, 1953). At least the thought was there.

462. Fifty-Fifth Legislature, First Called Session, Chap. 11, p. 23-30, Dec. 2, 1957.

463. **P. T. Flawn,** *Geology in the State Government of Texas:* Univ. Texas, Bur. Econ. Geology Geol. Circ. 65-5, 1965, 11 p.

464. Support for pollution controls increased under Presidents Kennedy and Johnson and culminated in the passage of the 1965 New Water Quality Act which greatly increased the pollution enforcement powers of the Federal government. In 1966, President Johnson placed all anti-pollution activities in the Interior Department (*see* **F. E. Smith,** *The Politics of Conservation, op. cit.* [220], p. 290-291).

465. *See* Annual Report of the Secretary of the Interior, p. xix.

466. An example of the work of the Geological Survey in this area was the geological mapping work done by W. R. Hansen in the Del Rio area in 1958-1959 as preparation for water resources studies in support of the National Water Resources Program (Annual Report of the Secretary of the Interior for 1958-1959).

467. Annual Report for 1951, p. iii. In the 1958 report, Secretary Fred Seaton still spoke of the unprecedented defense needs.

468. *Ibid.*, p. xvii. In the Annual Report for 1948, Secretary Krug noted that in spite of the abundance of coal reserves, the nation was getting a smaller and smaller percentage of its energy from coal, and increasing its use of oil and gas which were our most exhaustible energy resources (p. 3).

469. **J. T. Lonsdale** to **Harold Wanless,** Mar. 25, 1948, Bureau of Economic Geology.

470. **Watson Monroe** to **J. T. Lonsdale**, June 9, 1949, Bureau of Economic Geology.

471. **W. A. Heck, K. A. Yenne,** and **L. G. Henbest,** *Boundary of the Pennsylvanian and Permian (?) in the Subsurface Scurry Reef, Scurry County, Texas:* Univ. Texas, Bur. Econ. Geology Rept. Inv. 13, 1952, 17 p.

472. **D. A. Myers, P. T. Stafford,** and **R. J. Burnside,** *Geology of the Late Paleozoic Horseshoe Atoll in West Texas:* Univ. Texas Pub. 5607, 1956, 113 p.

473. **J. A. Reinemund** to **J. T. Lonsdale,** Oct. 21, 1955, Bureau of Economic Geology.

474. The full scope of the activities of the geologists of the U.S. Geological Survey in Texas in 1960 is illustrated in a letter from **Charles A. Anderson,** Chief Geologist, to **J. T. Lonsdale** on Apr. 5, 1960 (Bureau of Economic Geology), and he describes the continuing work on the Texas Pennsylvanian oil and gas investigations as follows:

> *Texas Pennsylvanian Oil and Gas Investigations.* **D. A. Myers** in charge. Field studies for this work have been completed, and the project has been recessed while the project chief is on another assignment, but work on a report on the geology of the Wayland quadrangle is being done as time permits.
>
> A report by **R. T. Terriere,** *Petrography and Environmental Analysis of Some Pennsylvanian Limestones from Central Texas,* is in review.
>
> U.S.G.S. Professional Paper 315-A, *Geology of Part of the Horseshoe Atoll in Scurry and Kent Counties, Texas,* by **P. T. Stafford,** and Professional Paper 315-B, *Geology of Part of the Horseshoe Atoll in Borden and Howard Counties,* by **R. J. Burnside,** were published in September 1959.
>
> Reports *Stratigraphic Distribution of Some Pennsylvanian Fusulinidae from Brown and Coleman Counties,* by **D. A. Myers,** and *Stratigraphy of the Outcropping Pennsylvanian and Lower Permian Rocks in Brown and Coleman Counties,* by **D. H. Eargle,** are in press as Chapters C and D, respectively, of Professional Paper 315.
>
> A report, *Geology of Grosvenor Quadrangle,* by **R. T. Terriere,** is in press as U.S.G.S. Bulletin 1096-A and another report, *Geology of the Cross Plains Quadrangle, Brown, Callahan, Coleman, and Eastland Counties,* by **P. T. Stafford,** is being edited for U.S.G.S. Bulletin publication.

Stratigraphy of the Cisco Group, Wayland Quadrangle, Stephens and Eastland Counties, Texas, by **D. A. Myers**, has been approved for publication in the Permian Basin Section-SEPM Field Trip Guidebook.

475. There is another letter in the Bureau of Economic Geology files from **Arthur Baker**, Acting Director of the U.S. Geological Survey, to Senator **Lyndon B. Johnson**, July 1, 1955, in which Baker cites the need for revision of the Texas geological map. He notes that the State government had failed to appropriate any funds for mapping work by the Federal bureau. The State of Oklahoma contributed $200,000 toward the compilation of a new geologic map.

476. **Ralph Miller**, Chief of Fuels Branch, to **J. T. Lonsdale**, Feb. 16, 1956, Bureau of Economic Geology.

Actually, Virgil Barnes and Preston Cloud brought up the idea of preparing a new state geologic map following their Ellenburger work, but received no encouragement from Lonsdale.

477. In a letter to **K. C. Heald** on December 21, 1960, **P. T. Flawn** said that the Bureau was making tentative plans to undertake the compilation and publication of a new state geologic map. The funding for this project is explained in the Bureau's Annual Report for 1961, p. 6:

> To finance this project, oil companies and professional geological groups contributed more than $41,000 to the Bureau's Petroleum Industry Fund, and the Bureau pledged its own special endowment funds to bring the total to over $50,000. Moreover, oil companies are contributing geologic data which have enormous dollar value. . . . Monies from the sale of the first published sheets will be used to finance additional compilation.

This was strictly a State project and quite unique in that aspect.

478. The Atomic Energy Commission also contracted the U.S. Geological Survey to conduct studies of the lignite deposits of East Texas (*see* **Maurice Deul**, *The Lignite Project*, U.S.G.S. Report TEI-218, Feb. 1952, p. 217-220). The Alcoa plant in Rockdale, Texas, began operation in 1952 using electrical power generated by burning lignite for fuel. The A.E.C. wanted to investigate the possibilities of powering uranium mills in the same way.

479. Eleventh Semi-Annual Report of the Balcones Research Center (Dec. 1954-June 1955), p. 4.

480. **D. H. Eargle**, *Some Uranium Occurrences in West Texas:* Univ. Texas, Bur. Econ. Geology Rept. Inv. 27, 1956, 23 p.

481. **D. H. Eargle** and **J. L. Snider**, *A Preliminary Report on the Stratigraphy of the Uranium-Bearing Rocks of the Karnes County Area, South-Central Texas:* Univ. Texas, Bur. Econ. Geology Rept. Inv. 30, 1957, 30 p.

482. *Ibid.*, p. 26.

483. Fourteenth Semi-Annual Report of the Balcones Research Center (June 1956-Dec. 1956).

484. For the history of uranium production in Texas up to 1970, *see* **D. H. Eargle, G. W. Hinds**, and **A. M. D. Weeks**, *Uranium Geology and Mines, South Texas:* Univ. Texas, Bur. Econ. Geology Guidebook 12, 1971, p. 7-11.

485. **R. M. Moxham,** Staff Geologist of the Geophysical Branch of the U.S. Geological Survey, to **J. T. Lonsdale,** Apr. 4, 1956, Bureau of Economic Geology.

486. *See* **A. D. Weeks** and **D. H. Eargle,** *Texas Coastal Plain Geophysical and Geologic Studies:* U.S. Geol. Survey Rept. TEI-752, Dec. 1959, p. 33-51; and **D. H. Eargle** and **R. M. Moxham,** *Airborne Radioactivity and Geologic Map of the Coastal Plain Area, Southeast Texas:* U.S. Geol. Survey Geophys. Inv. 198, 1961.

487. *See* Twenty-Second Semi-Annual Report of the Balcones Research Center, 1960, p. 7, Report of the U.S. Geological Survey Laboratory, D. Hoye Eargle, Project Chief.

> During 1960 a comprehensive report on the uranium-bearing rocks of the southeast Texas Coastal Plain has been prepared and submitted to the Geological Survey for review. These studies concerned geophysical and geological aspects of that region obtained from (1) airborne geophysical survey and compilation of a geologic map of an area of about 12,000 square miles of southeast Texas from the Balcones Escarpment in the vicinity of San Antonio to the Gulf Coast in the vicinity of Corpus Christi, and (2) detailed surveys and studies of a smaller area centered in northwestern Karnes County. . . . Preliminary maps showing the distribution of radioactivity patterns are being prepared in the Washington office for publication in the Geophysical Map Series on geologic base maps which were compiled in the laboratory.

The following year, Eargle reported that the functions of this project had been transferred in October 1960 to the Branch of Special Projects, Office of Inter-Agency Programs and Supporting Activities of the Geological Survey (Twelfth Annual Report for 1961, p. 8).

488. **P. T. Flawn** and **G. H. Anderson,** *Prospecting for Uranium in Texas:* Univ. Texas, Bur. Econ. Geology Min. Res. Circ. 37, 1955, 22 p.

489. *See* **J. T. Lonsdale** to **Charles Anderson,** Chief Geologist, U.S. Geological Survey, Apr. 12, 1960.

490. The main monographic publications of the Bureau were called University of Texas Bulletins up to No. 3701. The next number, 3801, was labeled University of Texas Publication. The University of Texas series has been limited to administrative or other official publications. The first important change in method of publication by the Bureau was the introduction of the Reports of Investigations in 1946 to furnish a medium for quick publication for shorter papers of current interest.

491. **H. R. Aldrich,** Secretary of the Geological Society of America, to **J. T. Lonsdale,** Jan. 6, 1947, and May 2, 1947, Bureau of Economic Geology, in "West Texas Project" folder for that year (Box 4Y212). Some of the published results of research supported in part by funds from the Penrose Bequest include:

S. S. Goldich and **M. A. Elms,** *Stratigraphy and Petrology of Buck Hill Quadrangle, Texas:* Univ. Texas, Bur. Econ. Geology Rept. Inv. 6, 1949, 50 p.

G. K. Eifler, Jr., *Geology of the Barrilla Mountains, Texas:* Univ. Texas, Bur. Econ. Geology Rept. Inv. 8, 1951, 15 p.

C. G. Moon, *Geology of the Agua Fria Quadrangle, Brewster County, Texas:* Univ. Texas, Bur. Econ. Geology Rept. Inv. 15, 1953, 45 p.

R. L. Erickson, *Stratigraphy and Petrology of the Tascotal Mesa Quadrangle, Texas:* Univ. Texas, Bur. Econ. Geology Rept. Inv. 18, 1953, 34 p.

W. N. McAnulty, *Geology of Cathedral Mountain Quadrangle, Brewster County, Texas:* Univ. Texas, Bur. Econ. Geology Rept. Inv. 25, 1955, 48 p.

The last three of these were University of Texas doctoral dissertations, along with the unpublished dissertation of Richard R. Bloomer, *Geology of the Christmas and Rosillos Mountains, Brewster County, Texas*, which was funded in part by the Geological Society of America and in part by the Bureau. Also, Moon and Erickson had research scholarships from the Humble Oil & Refining Company and the Standard Oil Company of Texas, respectively. *See also* note 576.

492. **J. T. Lonsdale** to Acting University President **T. S. Painter**, Mar. 20, 1947, Bureau of Economic Geology files. Goldich took a 1-year leave of absence in 1947 to undertake for the U.S. Geological Survey research on bauxite deposits on some of the islands captured from the Japanese. He then returned to join the faculty at the University of Minnesota.

493. These two led the Big Bend field trip at the 1949 annual meeting of the Geological Society of America in El Paso. Also, the Bureau files contain a fascinating letter from **Ross Maxwell** to **O. C. Wallmo** of the Texas Agricultural Experiment Station in reply to an inquiry about the ecological changes in the Big Bend region since the 1930's (Apr. 12, 1950, Bureau of Economic Geology).

494. The results of this joint project initiated in 1936 were published 31 years later by the Bureau as *Geology of Big Bend National Park, Brewster County, Texas* (Univ. Texas Pub. 6711, 1967, 320 p.), by **R. A. Maxwell, J. T. Lonsdale, R. T. Hazzard,** and **J. A. Wilson.**

495. **J. T. Lonsdale** to **H. R. Aldrich,** Aug. 1952, Bureau of Economic Geology.

496. 1952.

497. *See* **R. K. DeFord,** *Tertiary Formations of Rim Rock Country, Presidio County, Trans-Pecos Texas* (Univ. Texas, Bur. Econ. Geology Rept. Inv. 36, 1958, p. 6), for the students involved in this project.

498. Also, such geologists as P. B. King of the U.S. Geological Survey did extensive work in the Trans-Pecos. The International Boundary and Water Commission of the United States and Mexico contributed copies of nine maps displaying the areal geology of a 7-mile strip straddling the Rio Grande from the confluence of Fresno Creek above Lajitas to Del Rio (**L. W. Hewitt,** Commissioner, to **J. T. Lonsdale,** Oct. 20, 1955, Bureau of Economic Geology).

499. **J. T. Lonsdale** to **P. T. Flawn,** Feb. 12, 1949, Bureau of Economic Geology.

500. 1953.

501. The cooperative efforts are described in the letter from **J. T. Lonsdale** to **P. B. King,** Dec. 21, 1951, Bureau of Economic Geology.

502. **J. T. Lonsdale** to **Charles Milton,** U.S. Geological Survey, May 24, 1951, Bureau of Economic Geology.

503. **P. T. Flawn,** *Basement Rocks of Texas and Southeast New Mexico:* Univ. Texas Pub. 5605, 1956, 261 p.

504. *Ibid.*, p. 58-61.

505. **H. J. Morgan, Jr.**, *Paleozoic Beds South and East of the Ouachita Folded Belt:* Am. Assoc. Petroleum Geologists Bull., v. 36, no. 12 (Dec. 1952), p. 2226.

506. Publicity folder, 1954, Bureau of Economic Geology.

507. **J. B. Carsey** to **J. T. Lonsdale**, July 27, 1954, Bureau of Economic Geology.

508. **J. T. Lonsdale** to **J. B. Carsey**, July 30, 1954, Bureau of Economic Geology.

509. This cooperation is demonstrated by Flawn's coauthors — August Goldstein, Jr., of Stanolind Oil Company, P. B. King of the U.S. Geological Survey, and C. E. Weaver of Continental Oil Company. *The Ouachita System* (Univ. Texas Pub. 6120, 1961) is also indicative of the very basic and sophisticated nature of oil company research, and their willingness to support such research is shown by the grant of $20,000 by the Standard Oil Company of New Jersey to the Basement Rock Project of the American Association of Petroleum Geologists in 1959. Flawn was Chairman of this committee which was to study the nature and structure of the ancient crystalline rocks of North America.

510. *Ibid.*
See also **P. T. Flawn** to **W. K. Ferguson**, Jan. 24, 1975.

In thinking about the work I did at the Bureau, I think that if I did anything new or original it was the extension of the surface mapping of Precambrian rocks into the subsurface. Before the publication of *Basement Rocks of Texas and Southeast New Mexico*, I do not believe there had been an attempt at regional geologic mapping of the concealed basement. Bill [W. R.] Muehlberger [of the Geology Department of the University of Texas] later expanded this work. I used the same techniques in the Ouachita study.

511. Later, Humble, Gulf, and Shell authorized that the amounts contributed by them in excess of the original pre-Simpson budget be used to establish a fund in the Bureau called the Petroleum Industry Research and Publication Fund (**J. T. Lonsdale** to **C. P. Boner**, Vice President of the University of Texas, Oct. 21, 1955, Bureau of Economic Geology).

512. **P. E. Cloud, Jr.**, and **V. E. Barnes**, *The Ellenburger Group of Central Texas:* Univ. Texas Pub. 4621, 1946 [1948], 473 p.

513. **J. T. Lonsdale** to **T. S. Painter**, Apr. 18, 1952, Bureau of Economic Geology.

514. **J. T. Lonsdale** to **R. E. McAdams**, Shell Petroleum Company, July 18, 1951, Bureau of Economic Geology.

515. **V. E. Barnes** and others, *Stratigraphy of the Pre-Simpson Paleozoic Subsurface Rocks of Texas and Southeast New Mexico:* Univ. Texas Pub. 5924, 1959, 836 p.

Barnes notes that he should have stated flatly that the only way these rocks can be correlated is by careful sample descriptions. When all properties that can be seen through a microscope are plotted on logs and the logs are compared, correlations can be made.

516. **V. E. Barnes, S. S. Goldich,** and **F. Romberg,** *Iron Ore in the Llano Region, Central Texas* (Univ. Texas, Bur. Econ. Geology Rept. Inv. 5, 1939, 40 p.); **V. E. Barnes,** *High Purity Marble Falls Limestone, Burnet County, Texas* (Univ. Texas, Bur. Econ. Geology Rept. Inv. 17, 1952, 17 p.).

517. **J. T. Lonsdale** to Lt. Commander **David Rex,** Apr. 10, 1946, Bureau of Economic Geology.

518. One published result of this work was **F. Romberg** and **V. E. Barnes,** *Correlation of Gravity Observations with the Geology of the Coal Creek Serpentine Mass, Blanco and Gillespie Counties:* Univ. Texas, Bur. Econ. Geology Rept. Inv. 4, 1949, 11 p.

519. **V. E. Barnes** to **J. T. Lonsdale,** Jan. 15, 1960, Bureau of Economic Geology.

520. Dr. Barnes and his wife, Mildred, edited this volume.

521. The igneous rocks of the Central Mineral Region provide a great laboratory for the study of the structure and petrology of intrusives and extrusives. **R. M. Hutchinson** spent several years studying the Enchanted Rock pluton under a grant from the Geological Society of America. The G.S.A. published his *Structure and Petrology of Enchanted Rock Batholith, Llano and Gillespie Counties, Texas,* in its Bulletin, v. 67, no. 6 (June 1956), p. 763-805.
See also **R. M. Hutchinson** to **J. T. Lonsdale,** Jan. 28, 1954, Bureau of Economic Geology.

522. **D. E. Feray** to **J. H. Russell,** June 12, 1950, Bureau of Economic Geology.

523. **J. T. Lonsdale** to **T. S. Painter,** Feb. 22, 1951, Bureau of Economic Geology.

524. **Henry F. Nelson,** "Deposition and Alteration of the Edwards Limestone, Central Texas," in *Symposium on Edwards Limestone in Central Texas:* Univ. Texas Pub. 5905, 1959, p. 21-95.

525. **F. A. Herald,** ed., *Occurrence of Oil and Gas in Northeast Texas:* Univ. Texas Pub. 5116, 1951, 463 p.

526. April 28, 1947.

527. **F. A. Herald,** ed., *Occurrence of Oil and Gas in West Texas:* Univ. Texas Pub. 5716, 1957, 456 p.
Charles Vertrees is writing the history of the West Texas Geological Society.

528. **J. H. Quinn,** *Miocene Equidae of the Texas Gulf Coastal Plain:* Univ. Texas, Ph.D. dissert., 1954. The Bureau published this report, with the same title, as Univ. Texas Pub. 5516, 1955, 102 p.

529. **J. C. Frye** and **A. B. Leonard,** *Studies of Cenozoic Geology Along Eastern Margin of Texas High Plains, Armstrong to Howard Counties:* Univ. Texas, Bur. Econ. Geology Rept. Inv. 32, 1957, 62 p.

These two subsequently published the following Reports of Investigations: *Pleistocene Molluscan Faunas and Physiographic History of Pecos Valley in Texas* (Univ. Texas, Bur. Econ. Geology Rept. Inv. 45, 1962, 42 p.); *Pleistocene Geology of the Red River Basin in Texas* (Univ. Texas, Bur. Econ. Geology Rept. Inv. 49, 1963, 48 p.); *Relation of Ogallala Formation to the Southern High Plains in Texas* (Univ. Texas, Bur. Econ. Geology Rept. Inv. 51, 1964, 25 p.). Their work has established them as the foremost authorities on the noncoastal Pleistocene deposits of Texas.

530. **J. H. Quinn,** *Pleistocene Equidae of Texas:* Univ. Texas, Bur. Econ. Geology Rept. Inv. 33, 1957, 51 p.

531. **J. H. Quinn** to **J. T. Lonsdale,** Dec. 20, 1956, Bureau of Economic Geology.

532. **Dr. Lonsdale** wrote to **Quinn** on Mar. 26, 1957, advising him that he ought to stick to paleontology.

W. Armstrong Price, consulting geologist in Corpus Christi, continued to be the most active researcher on the Pleistocene geology of the Coastal Plain. He wrote many letters bemoaning the fact that the Bureau did not initiate Pleistocene studies, and he tried to stimulate Director Lonsdale's interest (**Price** to **Lonsdale,** Apr. 7, 1949; **Price** to **H. B. Stenzel,** Aug. 29, 1949; and **Price** to **W. H. Bradley,** Chief Geologist, U.S. Geological Survey, Sept. 19, 1939; Bureau of Economic Geology). Price also wrote a very extensive critique of Quinn's provocative article on *Paired River Terraces and Pleistocene Glaciation* for Dr. Lonsdale (*see* Box 4Y306, Bureau of Economic Geology).

533. **Glen Evans** to **J. T. Lonsdale,** Dec. 14, 1956, Bureau of Economic Geology.

534. **A. B. Frye** to **J. T. Lonsdale,** Feb. 5, 1956, Bureau of Economic Geology.

535. **J. T. Lonsdale** to **A. B. Frye,** Feb. 11, 1956, Bureau of Economic Geology.

536. **A. H. Deen** to **J. T. Lonsdale,** Feb. 1, 1946, Bureau of Economic Geology.

537. **J. T. Lonsdale** to **L. V. Doss,** June 4, 1959, Bureau of Economic Geology.

538. **J. Neils Thompson** to **J. T. Lonsdale,** Apr. 1, 1957, Bureau of Economic Geology.

539. **J. T. Lonsdale** and others to **J. Neils Thompson,** May 12, 1938, Bureau of Economic Geology.

540. Off-Campus [Balcones] Research Center, Semi-Annual Report for 1950.

541. **J. A. Wilson,** *Miocene Formations and Vertebrate Biostratigraphic Units, Texas Coastal Plain:* Am. Assoc. Petroleum Geologists Bull., v. 40, no. 9 (Sept. 1956), p. 2233-2246.

Texas continued to be one of the nation's top vertebrate collecting regions, and was visited often by noted vertebrate paleontologists (*see* **J. T. Gregory,** Department of Vertebrate Paleontology, Yale University, to **J. T. Lonsdale,** Dec. 2,

1958, *and* **A. S. Romer**, Museum of Comparative Zoology, Harvard University, to **J. T. Lonsdale**, Mar. 10, 1952, Bureau of Economic Geology).

542. E. H. Sellards retained an office at the Little Campus quarters of the Bureau and carried on extensive correspondence with archeologists in preparing for a new edition of his *Early Man.*

543. **J. T. Lonsdale** to **J. P. D. Hull**, Business Manager of the American Association of Petroleum Geologists, Sept. 22, 1948, Bureau of Economic Geology. Lonsdale was Chairman of the A.A.P.G. Committee on Conservation of Well Samples for 1952-1953. *See* the A.A.P.G. folder for 1953 in Bureau of Economic Geology files.

544. **D. E. Feray** and **J. L. Starnes**, *Index to Well Samples:* Univ. Texas Pub. 5015, 1950, 148 p.
P. S. Morey, *Index to Well Samples and Cores* (revised through 1962): Univ. Texas, Bur. Econ. Geology Spec. Pub., 1963, 683 p.
M. L. Morrow and **E. T. Moore**, *Addendum to Index to Well Samples and Cores, 1963-1973:* Univ. Texas, Bur. Econ. Geology Spec. Pub., 1976, 196 p.

545. Another interesting organization quartered at the Balcones Research Center was the Texas Archeological Salvage Project. At the end of World War II, the Federal government inaugurated an intensive schedule of dam and reservoir construction, and the National Park Service and the Smithsonian Institution organized the River Basin Surveys at that time. This was a project designed to salvage archeological, paleontological, and historical data endangered by the proposed dams and reservoirs. The Smithsonian administered this project using National Park Service funds. One of their four offices was located in Austin from 1947 to 1958. In the latter year, this office was replaced by the Texas Archeological Salvage Project, a division of the Department of Anthropology at the University of Texas, which at that time established an Archeological Laboratory at the Balcones Research Center. Some of their important initial projects are discussed in the Eighteenth Semi-Annual Report of the Balcones Research Center (June-December 1958).

546. **M. G. Cheney** had tried to encourage Pennsylvanian studies after World War II. He published the results of a symposium on Pennsylvanian problems at the St. Louis meeting of the American Association for the Advancement of Science in 1946 in *Pennsylvanian Classification and Correlation Problems of North-Central Texas:* Jour. Geology, v. 55, no. 3, pt. 2, May 1947, p. 202-219.
Helen Jeanne Plummer, who had become a full-time research scientist for the Bureau in 1948 following her husband's death, died suddenly in January 1951.

547. **L. F. Brown, Jr.**, *Stratigraphy of the Blach Ranch-Crystal Falls Section (Upper Pennsylvanian), Northern Stephens County, Texas:* Univ. Texas, Bur. Econ. Geology Rept. Inv. 41, 1960, 45 p.

548. **H. B. Stenzel**, *Ball Clay of the Troup District, Texas:* Univ. Texas Pub. 5019, 1950, 54 p.; *The Geology of Henrys Chapel Quadrangle, Northeastern Cherokee County, Texas:* Univ. Texas Pub. 5305, 1953, 119 p.

549. **H. B. Stenzel, E. K. Krause,** and **J. T. Twining,** *Pelecypoda from the Type Locality of the Stone City Beds (Middle Eocene) of Texas:* Univ. Texas Pub. 5704, 1957, 237 p.

550. **Josephine Casey** to **W. L. Fisher,** Aug. 9, 1960, Bureau of Economic Geology.

551. **W. L. Fisher,** *Lignites of the Texas Gulf Coastal Plain:* Univ. Texas, Bur. Econ. Geology Rept. Inv. 50, 1963, 164 p.

552. **J. P. Brand,** *Cretaceous of Llano Estacado of Texas:* Univ. Texas, Bur. Econ. Geology Rept. Inv. 20, 1953, 59 p.

J. P. Brand and **R. K. DeFord,** *Comanchean Stratigraphy of Kent Quadrangle, Trans-Pecos Texas:* Univ. Texas, Bur. Econ. Geology Rept. Inv. 34, 1958, 232 p.

D. L. Frizzell, *Handbook of Cretaceous Foraminifera of Texas:* Univ. Texas, Bur. Econ. Geology Rept. Inv. 22, 1954, 232 p.

L. Hendricks, *Geology of Parker County, Texas:* Univ. Texas Pub. 5724, 1957, 67 p.

K. P. Young, *Upper Albian (Cretaceous) Ammonoidea from Texas:* Univ. Texas, Bur. Econ. Geology Rept. Inv. 28, 1957, 33 p.

553. **Dennis Taylor,** exploration geologist of Humble Oil & Refining Company at Tyler, to **W. S. Adkins,** June 6, 1955, Adkins Papers.

554. **R. M. Girard's** excellent *Bibliography and Index of Texas Geology, 1933-1950* (*op. cit.*, 369) continued where the fine bibliography published with Bulletin 3232 ended.

The *Bibliography and Index of Texas Geology, 1951-1960* by **E. T. Moore** and **M. D. Brown** appeared in 1972 to further the outstanding bibliographical work done by the Bureau for the geologists of Texas. Subsequently, Mrs. Moore compiled the *Bibliography and Index of Texas Geology, 1961-1974,* which was published in 1976. Both of these bibliographies were Special Publications of the Bureau.

555. **W. H. Matthews III,** *Texas Fossils: An Amateur Collector's Handbook:* Univ. Texas, Bur. Econ. Geology Guidebook 2, 1960, 123 p.

556. **W. H. Matthews III,** *The Geologic Story of Longhorn Cavern:* Univ. Texas, Bur. Econ. Geology Guidebook 4, 1963, 50 p.

W. H. Matthews III, *The Geologic Story of Palo Duro Canyon:* Univ. Texas, Bur. Econ. Geology Guidebook 8, 1969, 51 p.

R. M. Girard, *Texas Rocks and Minerals: An Amateur's Guide:* Univ. Texas, Bur. Econ. Geology Guidebook 6, 1964, 109 p.

557. **R. A. Maxwell,** *The Big Bend of the Rio Grande, A Guide to the Rocks, Landscape, Geologic History, and Settlers of the Area of Big Bend National Park:* Univ. Texas, Bur. Econ. Geology Guidebook 7, 1968, 138 p.

R. A. Maxwell, J. T. Lonsdale, R. T. Hazzard, and **J. A. Wilson,** *Geology of Big Bend National Park, Brewster County, Texas:* Univ. Texas Pub. 6711, 1967, 320 p.

R. A. Maxwell (with contributions by **L. F. Brown, Jr., G. K. Eifler, Jr.,** and **L. E. Garner),** *Geologic and Historic Guide to the State Parks of Texas:* Univ. Texas, Bur. Econ. Geology Guidebook 10, 1970, 197 p.

558. **J. T. Lonsdale** to **Charles Clark,** Jan. 11, 1950, Bureau of Economic Geology.

559. **Kendall Beaton,** *Enterprise in Oil — A History of Shell in the U.S.:* Appleton-Century-Crofts, New York, N.Y., 1957, p. 684.

560. **L. T. Barrow** wrote the following to **W. K. Ferguson** in relation to research by the oil companies (Dec. 12, 1973):

> Humble was one of the few if not the only major (and minor?) oil companies who thought there were enough oil and gas reserves in the United States to win the Second World War. Therefore, we did not think we could conscientiously ask for deferment of any of our exploratory personnel — research geologists or geophysicists. Many of the latter were taken by University laboratories doing research for the government. As a result, our forces were materially depleted until the war was over. Most of them came back, but the time had come to catch up and expand our efforts. . . . My memory is too poor to recall what was stressed in geological research, and Dr. Harold Fisk is dead, but I do recall that sedimentation was one of his principal interests.

561. **August Goldstein, Jr.,** and **Duane Reno,** *Petrography and Metamorphism of the Sediments of the Ouachita Facies:* Am. Assoc. Petroleum Geologists Bull., v. 36, no. 12 (Dec. 1952), p. 2275-2290.

562. **Shepard W. Lowman** to **J. T. Lonsdale,** Mar. 15, 1938, Bureau of Economic Geology. Later, Lowman observed that the American Petroleum Institute rejected part of the project dealing with ancient sediments "because it contains proposals for a great deal of work that is being done by the companies in their daily search for oil" (Am. Assoc. Petroleum Geologists Bull., v. 32, no. 6, June 1948, p. 1185).

563. **F. P. Sheppard** and **D. G. Moore,** *Central Texas Coast Sediments; Characteristics of Sedimentary Environment, Recent History, and Diagenesis:* Am. Assoc. Petroleum Geologists Bull., v. 39, no. 8 (Aug. 1955), p. 1463-1593. This article points up the fact that the extensive Pleistocene and recent sediments of the High Plains and coast of Texas provide a marvelous laboratory for the study of sedimentation.

564. **University of Texas Geological Society** [Robert L. Folk, leader], *Tertiary Field Trip [guidebook]:* Univ. Texas Geol. Soc., Oct. 1955, 23 p.

565. **Marcus Hanna** to **J. T. Lonsdale,** Aug. 26, 1951, and Feb. 17, 1953, Bureau of Economic Geology.

566. **Grover Murray,** ed., *Sedimentary Volumes in the Gulf Coastal Plain of United States and Mexico:* Geol. Soc. America Bull., v. 63, no. 23 (Dec. 1952), pt. I, p. 1157-1228.

567. **Norman Greenman** and **Rufus LeBlanc,** *Recent Marine Sediments and Environments of the Northwest Gulf of Mexico:* Am. Assoc. Petroleum Geologists Bull., v. 40, no. 5 (May 1956), p. 813-847.

568. For example, **Director Flawn** wrote to **E. S. Larsen, Jr.**, of the Geochemistry and Petrology Branch of the U.S. Geological Survey that "there is a possibility that the Bureau may get a considerable sum of money for geochemical research in Texas by Texas institutions" (Aug. 1954, Bureau of Economic Geology), but there is no record of this grant having been made.

569. **Frank Conselman** to **J. T. Lonsdale,** Feb. 10, 1951, Bureau of Economic Geology.

570. **Keith Young,** *History of Geology at the University of Texas:* Univ. Texas, Bur. Econ. Geology Geol. Circ. 67-3, 1967, p. 29.

571. **L. T. Barrow** to Acting University President **T. S. Painter,** Mar. 6, 1947, Bureau of Economic Geology.

572. **Keith Young,** *History of Geology at the University of Texas, op. cit.* (570), p. 16.

573. Geology Department Staff Meeting, Minutes, September 1942. It is interesting to note that in the correspondence of the Department in the University Archives, the only mention of the Bureau prior to this was in connection with the 1940 meeting of the Geological Society of America in Austin.

574. **J. T. Lonsdale** to Acting University President **T. S. Painter,** Sept. 5, 1947, Bureau of Economic Geology.

575. **J. T. Lonsdale** to University President **Logan Wilson,** July 21, 1953, Bureau of Economic Geology.

576. **R. R. Bloomer,** *Geology of the Christmas and Rosillos Mountains, Brewster County, Texas:* Univ. Texas, Ph.D. dissert., 1949.
R. W. Graves, Jr., *Geology of Hood Spring Quadrangle, Brewster County, Texas:* Univ. Texas, Ph.D. dissert., 1949. Published as Univ. Texas, Bur. Econ. Geology Rept. Inv. 21, 1954, 51 p.
R. L. Erickson, *Stratigraphy and Petrology of the Tascotal Mesa Quadrangle, Texas, op. cit.* (491).

577. **J. T. Lonsdale** to Graduate Dean **Albert P. Brogan,** Jan. 18, 1956, Bureau of Economic Geology.

578. **P. T. Flawn,** *Geology in the State Government of Texas:* Univ. Texas, Bur. Econ. Geology Geol. Circ. 65-5, 1965, 11 p.

579. Examples include: *Map of Oil, Gas and Distillate Fields of Texas as of December 1945,* prepared at the request of the Oil and Gas Division of the Railroad Commission; and a contract between the Bureau and the Texas Water Development Board for the mineral inventory and survey of proposed reservoir sites in southeast (4) and northeast (10) Texas. One of the pitfalls of doing contract work for other agencies is illustrated in the discussions between the Bureau and the Texas Fish and Game Commission on the subject of surveys of oyster shell reefs in Texas bays. Some experiments were started using geophysical techniques, but both the Bureau and the Commission agreed to terminate this project because "private capital should be invested to get this type of information"

(**Lonsdale** to **E. G. Leonardon** of Schlumberger, Feb. 18, 1953, Bureau of Economic Geology).

580. **P. T. Flawn,** *Geology in the State Government of Texas, op. cit.* (578), p. 6.

581. Flawn's reputation in this field is evidenced by the publication of his *Environmental Geology, Conservation, Land Use Planning and Resource Management* (Harper & Row, New York, 1970, 313 p.) in Harper's *Geoscience Series.*

582. Bureau Annual Report for 1961, p. 8.

583. Bureau Annual Report for 1964, p. 2. Flawn also emphasized that conservationists must recognize that an industrial society consumes and alters its environment.

584. Bureau Annual Report for 1970, p. 9, and also p. 9-13 ff.

585. *Ibid.*

586. Bureau Annual Report for 1965, p. 6-7.

587. **L. E. Garner** and **K. P. Young,** *Environmental Geology of the Austin Area: An Aid to Urban Planning:* Univ. Texas, Bur. Econ. Geology Rept. Inv. 86, 1976, 39 p.

588. The U.S. Geological Survey had been classifying Federal lands as to mineral and water resources since its inception, but other than some geological reconnaissance on State lands during Udden's directorship, the Bureau had not done research on the mineral and water resources of these lands.

589. **W. L. Fisher,** *Rock and Mineral Resources of East Texas:* Univ. Texas, Bur. Econ. Geology Rept. Inv. 54, 1965, 439 p.

590. **W. L. Fisher,** *Lignites of the Texas Gulf Coastal Plain:* Univ. Texas, Bur. Econ. Geology Rept. Inv. 50, 1963, 164 p.

591. **P. T. Flawn** to **W. K. Ferguson,** personal correspondence, Jan. 24, 1975.

APPENDIX I

Biographies

VIRGIL E. BARNES

Texas is a state with geologic diversity and immense geologic resources. As a result, Texas has long had a large population of geologists. Some of them, through their work, have become giants in the profession. Among these giants is Virgil Barnes. In his years of service to the Bureau of Economic Geology, The University of Texas, and Texas geology, he has amassed a research record not commonly achieved and very rarely matched. At 71, Virgil Barnes is now technically on modified service. But this is recorded only on paper and his paycheck, not in his activity. He still pursues his work with his long-time dedication, vigor, and intensity.

Virgil Everett Barnes was born in Chehalis, Washington, on June 11, 1903, the only child of Charles and Della Barnes. He grew up on a farm and worked at times in a lumber mill. He went to Washington State to become a civil engineer. But he switched to geology, influenced by a course taught by Olaf Pitt Jenkins, long-time State Mineralogist of California. Barnes completed a Master's degree in geology at Washington State under Harold E. Culver. For his doctoral work, he chose the University of Wisconsin whose faculty at that time boasted such names as Leith, Winchell, Twenhofel, Mead, and Schrock. There Barnes' major interests were hard-rock geology, mining, and metallurgy. He completed his dissertation under Professor A. N. Winchell.

While pursuing his education, Dr. Barnes spent two field seasons with the Washington Geological Survey, one season with the Wisconsin Geological Survey, and one summer prospecting in northern Manitoba for Dominion Explorers, Ltd. He emerged from the Canadian wilderness in October 1929 to witness the beginning of the Great Depression. On finishing his doctorate in 1930 he had two possible jobs, one as a metallurgist in Chicago for U.S. Steel, and the other for the terminal year of a three-year research project of the American Petroleum Institute on oil-field waters and earth temperatures. He took the second job and came to Texas, where the Bureau of Economic Geology provided laboratory and office space.

In Austin, he met a University of Texas student, Mildred Adlof; they were married in 1932. The Barneses have three children: Virgil E. Barnes II, an associate professor of physics and assistant dean of science at Purdue University; Louise Barnes Seeman, a graduate student in developmental psychology at George Peabody College in Nashville; and Elizabeth Barnes Thompson, research biologist at New York University Medical School.

In 1933, Dr. Barnes joined the Topographic Branch of the U.S. Geological Survey as topographic engineer. He worked at Amarillo, Kilgore, and Longview, in Texas, then in Minnesota, and again in East Texas until late in 1936. He left the U.S.G.S. and joined the Bureau of Economic Geology where he began work organizing the jumble of well samples stored on the lower floor of what was known as the Extension Building on the Little Campus.

When Virgil Barnes joined the Bureau of Economic Geology 40 years ago, times were hard and Bureau funds were meager. Barnes, with a Ph.D. in geology and five years of experience, was working for hourly wages. But he worked so many hours a day that soon he was given a regular appointment — for reasons of economy. The ensuing years have not diminished the enthusiasm or his capacity for hard work.

Dr. Barnes' first field work for the Bureau was in the Llano country — the Central Mineral Region of Texas. His first effort resulted in a bulletin on the "Building Stones of Central Texas." In 1939, he began systematic mapping in the Central Mineral Region, and in ensuing years published 24 geologic quadrangle maps and numerous papers and bulletins on mineral deposits, stratigraphy, and geophysical investigations of the area. His works on the Llano area were published by the Bureau and by numerous national and international journals. So thorough has his work been that no one thinks of Central Texas geology without thinking of Virgil Barnes.

During the years of World War II, Dr. Barnes continued work in Central Texas. This period was marked by a splendid example of informal cooperation between two outstanding geologists of two fine organizations — Virgil Barnes of the Bureau and Preston Cloud of the U.S. Geological Survey. This union resulted in the "Ellenburger Group of Central Texas," published by the Bureau in 1948. The report is a classic. It includes a serious effort in paleoecology, preceding most of the work in that discipline by two decades. Another monographic work headed by Barnes was the "Stratigraphy of Pre-Simpson Paleozoic Subsurface Rocks of Texas and Southeast New Mexico," published in 1959. A comprehensive report by Barnes and W. C. Bell on "The Moore Hollow Group of Central Texas," is now being processed for publication by the Bureau.

While putting Central Texas geology on the map and making that region a part of the common parlance of geology, Virgil Barnes was also to become well-known for another quite different subject — black glassy specimens called tektites. These first came to Dr. Barnes' attention in 1936 when the Bureau was sponsoring a Works Progress Administration (WPA) mineral resource survey in Grimes County. At that time tektites were commonly thought to be meteorites. The Grimes County find piqued Barnes' interest; subsequent research and a review of the world tektite literature resulted in the 1940 publication of his now classic "North American Tektites." That work has been a basic reference during the Space Age. Although some researchers had the mistaken belief that tektites came from the moon, Dr. Barnes had earlier concluded that tektites had an earthbound origin. In fact, students commonly jested that the origin of tektites was "out behind the Barnes." Investigations during the great surge in tektite research in the sixties simply verified the conclusion earlier reached by Barnes.

Since 1960, the National Science Foundation has supported Barnes' research on tektites. During the ensuing years, Dr. Barnes has visited all the tektite-strewn fields: australites in Australia, javanites in Indonesia, indochinites in Thailand, Cambodia, South Viet Nam, and Malaysia, philippinites in the Philippines, Libyan Desert glass in Egypt, Ivory Coast tektites in the Ivory Coast, West Africa, and moldavites in Czechoslovakia. He also visited and studied certain impact structures, including among others, the spectacular Vredefort structure in South Africa, the Bosumtwi crater in Ghana, the Sudbury structure in Canada, and the nearly inaccessible craters of Wabar in the Rub' al Khali of Saudi Arabia and of the Oasis structure in Libya. In his worldwide travel and study of tektites, Dr. Barnes was accompanied by his wife, Milla, a remarkable person also. In many of the French-speaking countries, her facility in French was a true asset. Mildred Barnes has become a student of tektites in her own right, serving with Virgil as co-editor of "Tektites," a book in the Benchmark series of Dowden, Hutchinson & Ross, and co-authoring popular articles on tektites.

Although tektites are earthly objects, Dr. Barnes has published several papers on meteorites. He also participated in the study of lunar samples brought back by the first two Apollo landings.

Since 1961, Virgil Barnes has directed the Bureau's extensive Geologic Atlas of Texas project, which is designed to produce a modern geologic map of the state in 37 sheets and at a scale of 1:250,000. This long-term project is in its final stages, with four or five atlas sheets now being published each year.

From 1961 to 1968, Dr. Barnes served as Associate Director of the Bureau of Economic Geology, and since 1960 he has been a professor of geology in the Department of Geological Sciences as well as a research scientist at the Bureau.

Virgil Barnes has had arthritis most of his adult life, an ailment that hit him before the days of cortisone. A lesser man might have been deterred. Not Virgil Barnes. His dedication to geology and to the rigors of field work needed only some aid from aspirins to put him up and down the hills of Central Texas or through the jungles of Southeast Asia.

In his 40 years at the Bureau, Virgil Barnes has published well over 200 articles and reports, some of them major and comprehensive monographs. He adds to that list each year with no sign of slowing.

Few members of the geologic profession throughout the world do not know the name of Virgil Barnes. He has achieved an international reputation of the highest order and has brought worldwide attention to the Bureau of Economic Geology and The University of Texas at Austin. His is a record of which he can be justly proud and which we all envy.

The Bureau of Economic Geology salutes this senior member of its staff.

BUREAU OF ECONOMIC GEOLOGY
Annual Report for 1974

MARY JOSEPHINE CASEY

Born July 2, 1907, in Austin, Texas, while parents (Mr. and Mrs. Joseph W. Casey) and sister (Bertha F. Casey) were living there for a short time.

Mr. Casey's work with Southwestern Bell Telephone Company necessitated frequent moves, and the family called several South Texas towns "home" for varying periods of time. However, Austin became home base in 1913, at which time the residence at 305 West 38th Street was built. The two sisters still live in that house. Their father died in 1927; their mother, 1956.

Josephine attended the Austin Public Schools from the then "ward" schools through junior and senior high schools, graduating in 1923. She attended the University of Texas at Austin and was awarded a Bachelor of Arts degree with honors in 1928.

Preceding her long association with the Bureau of Economic Geology, she was employed from June to September 1926 at the business office of Southwestern Bell Telephone Company in Austin. It was there she received a telephone call from Miss Florence Stullken, in the University's School of Business Administration, setting up an appointment for an interview with Dr. E. H. Sellards, long-time Director of the Bureau of Economic Geology of the University of Texas. The position to be filled was that of stenographer, part-time, and on September 13, 1926, Josephine Casey became a part of the research family of the University. This part-time work continued until June 1927 when she, her mother, and sister spent three months with relatives in Missouri following the death of her father earlier that year.

Fortunately for Josephine, Dr. Sellards had another vacancy, and the part-time work was resumed in the fall of 1927 to continue through June 1928 when she went on full-time (and this was Mondays through Saturday mornings with occasional Saturday afternoons!).

In 1926 the Bureau was housed in old "B Hall," now only a memory. Some of the stored publications of the Bureau bore witness to the water fights staged (often on March 2) by would-be invaders of this men's dormitory!

The move in 1927(?) to roomier quarters on The Little Campus gave more adequate and comfortable space for staff offices, laboratories, drafting, library, and storage. And before the official "coffee break time" was decreed, lab beakers served as tea cups from time to time in the J Hall first floor southeast corner room.

The Bureau staff in 1926 consisted of Dr. J. A. Udden, Director (but not "officing"); Dr. E. H. Sellards, Associate Director; Dr. John T. Lonsdale (who returned as director in 1945) and W. S. Adkins, Geologists; and Miss Lelia Lambert (Mrs. George Howard Clark), Secretary. E. Manuel Hawtof held an assistantship.

Credit must be given and appreciation expressed to, first, Dr. Sellards for his tutelage on editorial work, to which he early introduced Josephine, and to his unfailing patience and understanding of problems as they arose in those early years. Second, to her predecessor, Miss Lambert, who as secretary showed by example and explanation how to cope with the duties of overseeing the daily routine. Such help was especially appreciated when Josephine became secretary in 1929 upon the resignation of Miss Lambert, who had become Mrs. Clark.

Josephine Casey's duties increased, and her job titles ranged from stenographer to secretary and through administrative secretary, administrative assistant (and informally to "assistant to the director"), and finally to Editor of Publications. She served under four directors — E. H. Sellards, J. T. Lonsdale, P. T. Flawn, and W. L. Fisher.

Her retirement on May 31, 1972 ended a 45-year, 9-month association with the Bureau of Economic Geology of The University of Texas.

JOSEPHINE CASEY

December 1977

ROSS ALAN MAXWELL

Ross A. Maxwell is noted for his studies of the geology of the Big Bend country of Brewster County, and for his investigations of State and National parks. His director at the Bureau of Economic Geology, Dr. Peter T. Flawn, stated:

> Dr. Maxwell has made fundamental contributions to the understanding of Texas geology in his studies of the Trans-Pecos region, particularly the Big Bend area. He has made a great effort to make geology meaningful to the public through his guidebooks to National and State parks.

Dr. Maxwell was born on a farm near Sparks, Oklahoma, on June 9, 1904. He grew up in this state, and attended the University of Oklahoma, receiving his Bachelor and Master of Science in Geology in 1929 and 1931. As a young doctoral graduate of Northwestern University, he went to the Big Bend for the first time in 1936 as a geologist for the National Park Service. He had purchased the cap and gown, doctoral "hood," and all the trimmings of the degree ceremony at Northwestern when the job offer came. The dean told him to take the job as they were hard to come by. On the day and the hour of the graduation ceremony in Evanston, Illinois, Ross Maxwell was sitting under a mesquite tree in the Big Bend, eating his lunch.

During the early years most of his work involved geologic mapping and searching for mineral deposits (the government did not want to create the Big Bend Park if it had any important economic minerals). In 1940 he was appointed Regional Geologist, directing geological research in National Park Service areas in Arizona, Arkansas, Colorado, New Mexico, Nevada, Oklahoma, Texas, and Utah. In 1944, Dr. Maxwell became the first Superintendent of the Big Bend National Park, a year after the park was officially established. He held this position until joining the staff of the Bureau of Economic Geology in 1952.

From 1952-1960, Dr. Maxwell and Dr. John T. Lonsdale worked on the geological information of Big Bend National Park. During this time, Dr. Maxwell also wrote Report of Investigations No. 43, *Mineral Resources of South Texas: Region Served Through the Port of Corpus Christi*, which was published in 1962. After Dr. Lonsdale's death in 1960, Dr. Maxwell continued the work on the Big Bend report and was senior author of University of Texas Publication 6711, *Geology of Big Bend National Park, Brewster County, Texas*. He was joined by R. T. Hazzard and J. A. Wilson in the completion of this work. In 1968, there followed Bureau of Economic Geology Guidebook 7, *The Big Bend of the Rio Grande, A Guide to the Rocks, Landscape, Geologic History, and Settlers of the Area of Big*

Bend National Park. In 1970, there followed Guidebook 10, *Geologic and Historic Guide to the State Parks of Texas*, in which contributions were made by L. F. Brown, Jr., G. K. Eifler, Jr., and L. E. Garner. Both of these guidebooks have gone through several editions. His final work for the Bureau, written in collaboration with John F. Dietrich and published in 1971, was Report of Investigations No. 70, *Correlation of Tertiary Rock Units, West Texas*.

Dr. Maxwell retired from the Bureau in 1969. During several summers, beginning in 1971, he was a guest Professor of Geology for the summer sessions at Sul Ross State University, where he taught a very popular graduate and upper undergraduate course entitled "Big Bend Geology." He is now working on a manuscript on the history of the Big Bend National Park.

W. KEENE FERGUSON

APPENDIX II

Memorials

WALTER SCOTT ADKINS
(1890-1956)

Walter Scott Adkins, internationally known Texas geologist and foremost authority of his generation on the paleontology and stratigraphy of Gulf Coast Mesozoic strata, died of a heart attack, September 22, 1956, at his home, "Cerro de Tecolote," Austin, Texas. Graveside services and interment in Austin Memorial Park were conducted the following day. Mary Grace [Muse] Adkins, his cherished wife of 24 years, survives. Ensign Jack Scott Adkins (1922-1944), son of a former marriage, was lost in the Anzio landing operation in the Italian campaign.

Born, December 24, 1890, in the community of Coal Creek, now Lake City, Tennessee, Adkins was the only child of Elizabeth [Thomas] Adkins (1869-1893) and Richard Byron Adkins (1861-1902), physician at Coal Creek. Deprived of his mother at age two and orphaned by the loss of his father ten years later, Adkins was reared through adolescence by well intentioned if puzzled relatives. Endowed with intellectual traits that in later years were to make him a legendary figure among those who knew him, his youthful questioning of philosophic and religious dogmas was considered "wrong" by some and the oppressive consequences left few happy childhood memories.

The formal education of Adkins was largely in zoology. At the University of Tennessee, 1907-1911, he received his B.S. degree in 1910 and served as graduate assistant in the zoology department the following year. Graduate study in zoology was continued at Columbia University during the academic years 1912-1913 and 1915-1916 with summer study at marine research laboratories (Cold Spring Harbor, Long Island, 1911; Woods Hole, Massachusetts, 1912-1913, 1915; an appointment to the Helgoland Marine Station in 1914 was foregone due to the war). Although Adkins completed all course requirements for the doctorate and had a zoological thesis well advanced, his interest during this period was transferred to geology through association with W. M. Winton. Even so, his early training was reflected in tenures as assistant professor in biology, Texas Christian University (Fort Worth), 1913-1915; instructor in anatomy, University of Illinois Medical School (Chicago), 1916-1918; and assistant professor in anatomy, Baylor University Medical School (Dallas), 1918-1929.

Brought to Texas by Winton, long-time head of the combined Biology and Geology Department at T.C.U., the rapport of these two, similarly trained in natural sciences, was immediate. Week-end and holiday excursions together during the years 1913-1915, especially over the fossiliferous Cretaceous terranes adjacent to the campus and in the surrounding north Texas counties, developed into a program of joint geologic research that was to rejuvenate interest in the area dormant since the days of R. T. Hill. These two years were perhaps the happiest of Adkins' early life as attested by a number of circumstances, namely: his continued and unfailing respect for Winton despite the episode which resulted in Adkins' departure from T.C.U. (a barn in which some students and Adkins had sought convivial refuge during a field trip burned under circumstances indefensible even by Winton before the college brethren); the fact that he celebrated his return to Texas in 1918 ever after as an important date in his life (even though it was to teach in Baylor Medical School); and perhaps most important, he became acquainted with Mary Grace Muse, one of two embarrassed coeds in his zoology class, later his wife (and the one responsible for his nick-name "Si," a contraction for "The Scientist" as she referred to him at this time).

Crucial to later events was Adkins' reorganization of his vocational activities during the period 1916 into 1919. Although utilizing his zoological training as an instructor in medical schools, his desire for a life-time career had changed to geology. Fortunate indeed was his meeting with Emil Böse in the field in north Texas in early 1919. Böse, then famous for his work with the Mexican Geological Survey and an authority on Cretaceous ammonites, in particular, was "astonished by the sober, sane judgment this still young man had developed" and "surprised to find that Professor Adkins had . . . studied the north Texas Cretaceous more thoroughly than anyone else has tried to do." With these phrases, and others even more complimentary, Böse recommended Adkins to J. A. Udden, director of the Bureau of Economic Geology, University of Texas, for employment. Thus, with this incalculable element of a lucky break at the right time, Adkins shortly thereafter began his professional career, initially as curator of the museum at the Bureau. The profound influence of Böse at this time was reflected ever after in Adkins' veneration of Böse.

During the year 1919, joint papers with Winton based on studies of the preceding six years were submitted to the Bureau for publication as Univ. Texas Bull. 1931 [Geology of Tarrant County] and Univ. Texas Bull. 1945 [Paleontological Correlation . . . Fredericksburg and Washita . . . north Texas]. Noteworthy is the statement in both publications to the effect that the order (see bibliography) of author names "does not in any way indicate seniority." A third paper by Adkins, also published in 1920 [Univ. Texas Bull. 1856 on the Weno and Pawpaw formations], acknowledged the joint and preliminary work with Winton on this study. This was the first of a planned series of which Adkins was to monograph the argillaceous formations of the Comanche series in north Texas; similarly, Winton was to treat the predominantly limestone formations. Later parts of the series were not realized.

The quality and quantity of Adkins' first efforts, plus his distaste for a curator's tasks, were apparent to Udden, and Adkins was promoted to associate geologist in 1920. A memorable event was his first west Texas field trip, alone, to famous Cretaceous localities at Cerro de Muleros, Sierra Blanca, Kent, Fort Stockton, and the Malone Hills — plus the pleasant recollection of a starlit-night freight train ride into Sierra Blanca atop the box cars next to the engine. Prior to departure for Mexico and service with the Mexican Eagle Oil Company out of Tampico, Adkins completed the field work for the McLennan County geological report [published in 1924 as Univ. Texas Bull. 2340] and did much of the field work on the Fort Stockton Quadrangle in western Texas.

From mid-December, 1921, through August, 1925, Adkins was in Mexico with the Compañia Mexicana de Petroleo "El Aguila" working under Paul Weaver and D. R. Semmes, chief geologist and assistant chief geologist, respectively. Engaged primarily as a paleontologist in Tampico, he made numerous collecting trips to the Sierra Madre mountain front; other trips, largely into the younger Tertiary in Vera Cruz, Chiapas, and Tabasco, were made in conjunction with the vast reconnaissance of the Tampico Embayment. For a summary report on the Isthmus of Tehuantepec, compiled by H. J. Tschopp in 1924, Adkins did most of the paleontology of the larger fossils. In 1925 Adkins prepared a summary report "On the Paleontology of the Formations of the Tampico Embayment" in which he first proposed the names Guayabal (later restricted to Claiborne and published in 1927 by W. S. Cole) and Coatzintla (regarded as uppermost Oligocene but could be in part Miocene). Adkins' faunal identifications and the Tampico report were extensively used by his close friend J. M. Muir in the latter's *Geology of the Tampico Region, Mexico* (Amer. Assoc. Petrol. Geologists, Tulsa, 1936). In addition to the many contacts with North American geologists then engaged in Mexican oil work, Adkins valued the acquaintances gained at this time with Carlos Burckhardt, Hans Jenny, Arnold Heim, W. L. F. Nuttall, and other European geologists. He also met and became friendly with an Aguila land-man, Franz Blom, later to become famous as an authority on Central American archeology. This friendship undoubtedly played a part in Adkins' subsequent accumulation of a valuable archeology library and his mastering of Nahuatl, the language of the Aztecs (to be added to previous command of German, Spanish, French, and Sanskrit). Nourished from these days forward was Adkins' enthusiasm for Mexico and its culture. His high regard for the Mexican people and their customs was reflected as much in the many vacations he and his wife spent there later as in the momentos of Mexico incorporated in their home. Natives and long-time residents south of their Rio Bravo were as amazed as *turistas* by his comprehensive knowledge of Mexico.

Prior to returning to the Bureau for a second period of intensive scientific work on the Texas Cretaceous, Adkins spent parts of 1925 and 1926 in further graduate study under E. Haug at the Sorbonne, Université de Paris. During this residence he added to his expanding private library on Mesozoic paleontology (a book-dealer in Paris, this past year, had vivid recollections of pleasant encounters with Adkins thirty years previously) and made special efforts to visit depositories in Europe of Texas Cretaceous holotype fossils. Notations, photographs, and plaster casts were made at the universities of Bonn and Breslau of Texas types described by Roemer, Schluter, Boehm, and Lasswitz. Marcou's southwest United States types at the British Museum of Natural History, South Kensington, the Lambert collection of echinoids in Paris, and others were similarly studied. These data were to be utilized in bulletins by Adkins issued by the University of Texas in later years [Bulletins 2738 and 2838].

The second period of Bureau service, from the summer of 1926 into spring of 1934, covered the most active and productive years of Adkins' career. In the miscellaneous chores of editing and proofreading (e.g., the reprinting of Deussen's classic Coastal Plain study), replying to queries routed to his desk (Mr. Adkins: Please handle this request."), special economic assignments (Rainbow Mine and Terlingua area), descriptions of well samples (Pecos and Bell counties), teaching the Mesozoic portion of Sellards' graduate course in stratigraphy at the University of Texas, he was diligent and cooperative. Memorable breaks in the routine were periods of field work concentrated largely in the Trans-Pecos area of western Texas, with side studies in northern Mexico, and a return to the central Texas Cretaceous section south of McLennan County previously reported. Data obtained were presented in bulletins on Bell County and the Fort Stockton Quadrangle, in descriptions of Taylor and Eagle Ford ammonites and their stratigraphic zonation, and in taxonomic paleontologic studies on rudistids and echinoids. Of particular significance to the increasing number of petroleum geologists were the issue of the "Handbook of Cretaceous Fossils" [Univ. Texas Bull. 2838] and "The Mesozoic Systems in Texas" [Part 2 of Bull. 3232], the stratigraphy volume on the geology of Texas. Although now out-of-date, the foregoing two syntheses are still the best available sources of Texas Cretaceous summarized information.

In addition, and in contrast to later years, Adkins was active in professional society affairs, particularly those of the Society of Economic Paleontologists and Mineralogists. A charter member (1927), he was appointed chairman of the initial S.E.P.M. Research Committee (1929-1932) and elected to the vice-presidency (1931-1932). He presented papers at all annual meetings from 1929 through 1933 (except the 6th when he was out of the country) and led a 1929 field trip to the Rio Grande valley where the party "listened to lectures on Mexican art and Cretaceous fossils by Adkins and afterwards enjoyed meals of venison, quail, wild turkey and sparkling liquids across the Border in the land of frijoles and large ammonites." He participated in the reactivation of the Texas Academy of Science (Fellow, 1931) with papers at the 1932 and 1933 annual meetings and served later on the Handbook of Science Committee (1938-1940). Older and long-continued affiliations were with the American Association of Petroleum Geologists (1920), Geological Society of America (Fellow, 1921), Paleontological Society (1922), Société Géologique de France (1920), and Sociedad Geológica Mexicana.

A signal honor in 1931 was Adkins' appointment as a John Simon Guggenheim Memorial Fellow, providing opportunity for further study abroad. A contemporary notice stated "This is the first Guggenheim fellowship for paleontologic research and it may almost be said the first to a geologist." Most of the six months tenure was spent at the British Museum of Natural History, in work on Lower Cretaceous fossils, especially types, and in the close company of L. F. Spath. From these days forward, Adkins treasured the friendship, cooperation, and opinions of Spath possibly more than those of any other paleontologist. The respect was reciprocal.

The insecurity fostered by lack of tenure for Bureau staff members, salary cuts in the depression period of the early thirties, and periodic legislative activities to curtail (even abolish) University extramural activities, of which the Bureau of Economic Geology was a part, led to Adkins' excursion of 1934-1950 into oil-company work with Shell. Of the change, he stated it "must be regarded as an unwelcome, but totally necessary diversion of my life, undertaken under the sheer necessity of solidifying my financial future." Setting aside his personal research and foregoing essentially all outside professional activity, he "bulldozed" into a second major career.

Not knowing at the time of his application to the Shell that Nuttall, in The Hague office, was submitting his name as a prospective company research paleontologist, Adkins was hired at a most modest salary for the specific purpose of doing paleontologic research work. With the advent of R. R. Morse into the company's Texas-Gulf area, the outstanding qualifications of Adkins were recognized and inequities adjusted. Progressive advancements to chief stratigrapher and head of the special problems research group were marked by such recommendations as ". . . there are few, and perhaps none, with a wider knowledge of Texas stratigraphy . . . a man of national reputation that we are fortunate to have . . . his enthusiasm for investigative work is peculiarly valuable for cold-blooded analysis of a wide range of stratigraphic and structural problems . . . both as a research worker and as director of our paleontologic and stratigraphic activities, his mental alertness, resourcefulness, and constant supply of constructive ideas have contributed much in the past and promise much for the future. . . ."

In December, 1945, the Shell Oil Company established its exploration and production fundamental research division under the direction of H. Gershinowitz. With Adkins as the nucleus of stratigraphic research and as an advisor to the director on many interrelated problems, the geologic section weathered the difficult early years of organization. In retrospect it is clear that the close liaison and mutual respect of capabilities between Gershinowitz and Adkins were significant factors in the organization's present favorable position in geologic research. Adkins' synthesis of the principles of age of folding and related oil accumulations represented the economic fruits of a lifetime of deductive reasoning and critical disciplined imagination. This final company report was proof positive of the practical value of strictly scientific reasoning.

To younger and less experienced staff associates during his Shell career, Adkins' ability to phrase a problem in the minimum of words, to isolate fact from speculation, to expose the relative strength of alternative solutions, and to specify the crucial area of investigation were objective demonstrations of a scientific mind at its best. Those fortunate to have received the "needle" through his blunt query, and often demoralizing rebuttal, as to the significance of a particular approach or result soon realized the difference between random "nitpicking" and aimed "nut-cracking." Such lessons are seldom forgotten.

At the end of 1950, Adkins retired to "Cerro de Tecolote," his combined home and personal research laboratory, located about 5 miles west of Austin on a hilltop overlooking the city. With his unexcelled private library and fossil collections at hand, he resumed non-commercial research with three large projects of public service in mind: (1) an illustrated manual of Cretaceous guide fossils, primarily for students and oil company trainees; (2) a complete revision of the Mesozoic stratigraphy of Texas, designed "to separate fact from hypothesis, extricate stratigraphic geology from its morass of speculation, and make an attempt to guide students and trainees into known facts and principles, after which they may be enabled to face new situations with judgement and maturity"? and (3), if time permitted, a comprehensive paleontology of the Texas Mesozoic. Excepting a six-month period (late 1951-early 1952) when he was called out of retirement to consult with Compañia Shell de Venezuela, he was occupied with these projects till his death. In tribute to the serious intent and the recognized value of these planned volumes, several oil companies contributed a research fund to expedite the work. That none was completed, due in large part to intervals of ill health, is a loss to knowledge probably unmatched in the geologic literature of Texas.

The final months of his life, despite arduous medical treatment, were occupied with concern as to the success of the Mesozoic and Cretaceous symposia at the XX International Geological Congress, Mexico City. A member of the symposia panels, his suggestions were praised by chairman L. W. Kellum. In what was to be Adkins' valedictory address, he presented an analysis of correlation by ammonites that was attentively received and warmly applauded. The effort of the Congress trip and delivery was physically taxing but typified his life. A year or so previously, in what he referred to as the Last Round-Up, he had written:

> I always believed in *living* while I was going through life, in travel, seeing various countries, civilizations, history, music, restaurants, etc. This distracts from the path of the bulldozer but it is worth it.

By way of an epilogue, here was an uncommon man sustained by the courage of his non-conformist convictions. A genius whose actions were often contrary to those of the main current, he was intolerant only of intolerance. A Texan prouder than most native-born Texans of their birthright, his services always were directed to the ultimate benefit of all, but especially to those of his adopted state. A scientist dedicated to fundamentals, he was most critical of his own deductions. No effort to get at the truth was personally too great, no serious request for assistance was ever more generously met. Though he can never be replaced in his field of accomplishment, his memory will be perpetuated through the vigorous precepts of scientific work he passes on.

FRANK E. LOZO[1]

Houston, Texas
January 22, 1957

---

[1]Senior geologist, Shell Development Company, Houston, Texas.

Acknowledgments. — Data of the "El Aguila" days in Mexico were supplied by R. W. Barker, Paul Weaver, and W. A. Baker, Jr. Recollections by Mr. and Mrs. W. M. Winton and their early correspondence with Adkins were of material aid. J. T. Lonsdale, preparing a similar memorial for the Geological Society of America, and H. B. Stenzel clarified the record in many ways as did pertinent articles by Mary Smith in the *Shell Pecten.* Special acknowledgment is due Mrs. W. S. Adkins for her diligent and devoted assistance in every manner. To those others who also do not gauge a man's service to science by his allegiance to the mores of his tribe, the writer is equally grateful.

GEOLOGIC PUBLICATIONS OF W. S. ADKINS

1920a. [With W. M. Winton]. "The Geology of Tarrant County [Texas]," *Texas Univ. Bull. 1931*, 123 pp., 6 pls., 6 figs., 2 maps.

1920b. [And W. M. Winton]. "Paleontological Correlation of the Fredericksburg and Washita Formations in North Texas," *Texas Univ. Bull. 1945*, 128 pp., 22 pls., 6 figs.

1920c. "The Weno and Pawpaw Formations of the Texas Comanchean," *Texas Univ. Bull. 1856*, pp. 1-172, 13 figs., 11 pls.

1924. "Geology and Mineral Resources of McLennan County [Texas]," *Texas Univ. Bull. 2340*, 202 pp., 10 figs., 4 pls. (incl. map).

1927a. [With J. T. Lonsdale]. "Euhedral Orthoclase Crystals from Sierra Blanca, Texas," *Amer. Mineralogist*, Vol. 12, No. 6, pp. 256-59; 1 fig.

1927b. "The Geology and Mineral Resources of the Fort Stockton Quadrangle [Texas]," *Texas Univ. Bull. 2738*, 166 p., 8 figs., 6 pls. (incl. map).

1928. "Handbook of Cretaceous Fossils," *Texas Univ. Bull. 2838*, 385 pp., 37 pls.

1929. "Some Upper Cretaceous Taylor Ammonites from Texas," *Texas Univ. Bull. 2901*, pp. 203-22; 2 pls.

1930a. "New Rudistids from the Texas and Mexican Cretaceous," *Texas Univ. Bull. 3001*, pp. 77-100; 1 fig., 6 pls.

1930b. "Texas Comanchean Echinoids of the genus *Macraster*," *Texas Univ. Bull. 3001*, pp. 101-20; 2 pls.

1930c. "Some Recent Literature on the Western Mesozoic," *Jour. Paleontology*, Vol. 4, No. 1, pp. 73-87.

1930d. In "Notes on New Publications," *Jour. Paleontology*, Vol. 4, No. 2, pp. 206-10. [Notes on 13 German paleontological articles or journals, mostly in 1928-29.]

1930e. [And M. B. Arick]. "Geology of Bell County, Texas," *Texas Univ. Bull. 3016*, 92 pp., 2 figs., map.

1931a. [With E. H. Sellards]. "Section from Marathon to the Rio Grande," p. 93a in "Geologic Descriptions of States Included in Central United States Cross-Section," *Kansas Geol. Soc., Guidebook 5th Ann. Field Conf.*, 96 pp.

1931b. "Some Upper Cretaceous Ammonites in Western Texas," *Texas Univ. Bull. 3101*, pp. 35-72; 2 figs., 4 pls.

1933a. "New Ammonite Fauna from Mexican Turonian [abstract]," *Pan-Amer. Geologist*, Vol. 59, No. 3, p. 238. [*Cf.* Kummel and Decker, *Jour. Paleontology*, Vol. 28, No. 3, pp. 310-19, Pls. 30-33, May, 1954.]

1933b. "The Geology of Texas, Vol. I, Part 2: The Mesozoic Systems in Texas," *Texas Univ. Bull. 3232*, pp. 239-518; 15 figs.

1936. "Upper Cretaceous Unconformities in Texas," *Texas Univ. Bull. 3501*, pp. 141-49.

1939. "Megafossils from Smackover Limestone, Ouachita Co., Ark.," pp. 166-69 in "Notes on the Comanche and Pre-Comanche? Mesozoic Formations of the Ark-La-Tex Area...," by R. T. Hazzard. *Shreveport Geol. Soc., Guidebook 14th Ann. Field Trip*, pp. 155-78.

1941. Review ["Lower Cretaceous and Jurassic Formations of Southern Arkansas," by R. W. Imlay]. *Bull. Amer. Assoc. Petrol. Geol.*, Vol. 25, No. 2, pp. 329-33.

1944a. "*Mortoniceras vespertinum* (Morton) Lower Cretaceous Ammonite Genotype [abstract]," *Dallas Digest* [Joint Ann. Mtg. AAPG-SEPM-SEG], p. 102, Dallas, Texas.

1944b. "New Texas Lower Cretaceous *Gryphaea* (Ostreidae) [abstract]," *Bull. Geol. Soc. America*, Vol. 55, No. 12, p. 1465.

1945. [With T. L. Bailey and F. G. Evans]. "Revision of Stratigraphy of Part of Cretaceous in Tyler Basin, Northeast Texas," *Bull. Amer. Assoc. Petrol. Geol.*, Vol. 29, No. 2, pp. 170-86; 1 pl., 3 figs.

1946. [With R. E. King]. "Geology of a Part of the Lower Conchos Valley, Chihuahua, Mexico," *Bull. Geol. Soc. America*, Vol. 57, pp. 275-94; 4 pls., 2 figs.

1949. "Eagle Ford Condensed Zone in Travis County, Texas," *Shreveport Geol. Soc., Guidebook 17th Ann. Field Trip* (Cretaceous of Austin, Texas, area), pp. 95-96.

1951. [And F. E. Lozo]. "Stratigraphy of the Woodbine and Eagle Ford, Waco Area, Texas," *Southern Methodist Univ.*, Fondren Sci. Ser., No. 4, pp. 101-75; 26 figs., 6 pls.

1956a. "Cretaceous Section, Medina County, Texas to Val Verde County, Texas," *San Angelo Geol. Soc., Guidebook 2nd Field Trip (Four Provinces)*, pp. 105-08.

1956b. "Means of Correlation: Ammonites [abstract]," *XX Congreso Geologico Internacional [Mexico], Resumenes*, p. 23.

1956c. "Age of Folding in East Texas Salt Domes [abstract]," *op. cit.*, p. 24 [Erroneously listed as joint-authored with F. E. Lozo.]

1956d. [And F. E. Lozo]. "The Cretaceous of Texas: General Features [abstract]," *op. cit.*, pp. 24-25.

Reprinted from American Association of Petroleum Geologists Bulletin, v. 41, no. 4, April 1957, p. 783-789.

ESTHER RICHARDS APPLIN
(1895-1972)

"In recognition of her performance as a trail blazer to the science of micropaleontology, leader in the organization of and charter member of the Society, and pioneer in the subsurface stratigraphy of the eastern Gulf Coast" — so reads the certificate of honorary membership in the Society of Economic Paleontologists and Mineralogists presented to Esther Applin at the awards luncheon in Houston, March 30, 1971. As a part of the presentation, Mark Hanna said, "It is impossible in a few short paragraphs to give due recognition to Esther Richards Applin, a pioneer and trail blazer in an almost new field of scientific endeavor — micropaleontology (dominantly Foraminiferida)." Now we have lost Esther — she died July 23, 1972 in Jackson, Mississippi — and even many long paragraphs cannot adequately express what this loss means to her many friends, colleagues, and the geological profession.

Esther was born November 24, 1895, at Newark, Ohio, of the marriage of Gary F. Richards and Jennie DeVore. During her early childhood, the family lived in Newark and other cities in Ohio before moving to Fort Des Moines, Iowa, where her father was employed as a civil engineer by the Quartermaster Department of the U.S. Army. When she was 12 years old, a year after the San Francisco earthquake of 1906, her father was transferred to California to construct the now famous prison on Alcatraz Island. From that time until 1920 when she went to Houston to work for the Rio Bravo Oil Company, her home was on the island from which she travelled by ferry to attend school in San Francisco and Berkeley.

In 1919, Esther Richards graduated from the University of California at Berkeley with an A.B. degree in paleontology, geology, and physiography. During that summer, she was employed as a paleontologist by the Rio Bravo Oil Company working under E. T. Dumble, vice-president and general manager. Esther soon realized that the megafossils to which she had devoted much study at Berkeley would be of little value in subsurface geology because the fossil fragments in the drill cuttings could not be specifically identified. She concluded that microfossils in drill cuttings offered real hope in correlating subsurface formations, and returned to the University of California to earn her Master's degree with research in micropaleontology. Thereafter she resumed work with the Rio Bravo Oil Company in Houston and began a trail-blazing career in applied micropaleontology by attempting to show the profession and the oil industry that microfossils could be used for stratigraphic correlation.

R. Dana Russell in writing the history of SEPM from 1926 to 1946, took especial note of Esther's effort to convince the profession of the value of microfossils.

> As usual, it took a while to convince people. In December of 1921, Esther Applin read a paper, authored by Dumble, at the Amherst meeting of the Paleontological Society, demonstrating the use of Foraminifera in establishing the age of several subsurface formations around the salt domes of south Texas. Professor J. J. Galloway rose and said: "Gentlemen, here is this chit of a girl, right out of college, telling us that we can use Foraminifera to determine the age of a formation. Gentlemen, you know that it can't be done!" But a year later he was advertising his ability to do it.

Esther Richards and Paul L. Applin, a young geologist who had come to the Gulf Coast after studying geology at Dartmouth and Yale, were married in 1923. Esther remained with the Rio Bravo Oil Company until 1927; during those years she was retained also by East Coast Oil Company, Gulf Oil Corporation, and by several other companies. From 1927 to 1942, she was a consulting paleontologist and subsurface geologist in Fort Worth, Texas and, from 1942 to 1944, an assistant professor of geology at The University of Texas and a consulting paleontologist for Sun Oil Company. During 1943-1944, Esther and Paul joined the U.S. Geological Survey, and the Survey gained one of the most effective research teams in its history.

The broad scope of Mrs. Applin's interests is evidenced by her early work in California, Mexico, and South Dakota, but most of her early studies were related to petroleum exploration in Mesozoic and Tertiary rocks of the western Gulf Coast. Later her studies extended eastward around the Gulf to include all of the southeastern United States, and after she joined the U.S. Geological Survey most of her work was done in that region.

Esther produced many definitive papers on micropaleontology during her career with the Survey. Some were co-authored with J. A. Cushman, Louise Jordan, and W. S. Cole. However, the more comprehensive reports on regional stratigraphy and structure based on micropaleontology were produced by the team of Applin and Applin. One of these was a pioneering report on the regional stratigraphy and structure of Florida and Georgia published in 1944 in the AAPG Bulletin. Much of later work by others, as well as the Applins, was built around this framework. Their summary Professional Papers on the Comanche Series and the Gulf Series of the Cretaceous, published in 1965 and 1967, are classics of three-dimensional geology in the southeastern states.

Wide recognition and many honors came to Esther Applin in her later years. In 1960, she received the Gulf Coast Association of Geological Societies award for outstanding contributions to Gulf Coast geology. E. H. Rainwater made the presentation and said in part:

> Mrs. Applin is a dedicated scientist who has contributed greatly to our understanding of Gulf Coast geology. She pioneered in the field of applied micropaleontology, and is still pioneering in the subsurface geology of the lesser known areas of the Gulf Coast province. Her publications are many, and her conclusions have been tried and proved correct. The geological ages and regional correlations of many subsurface formations, which Gulf Coast geologists accept without question, were first established by this eminent paleontologist.

In addition to her professional duties Mrs. Applin has devoted much time to unselfishly helping many geologists better understand the correlations and depositional history of this important area. It is no exaggeration to state that Mrs. Applin, world renowned, has contributed through publication, teaching, lectures, and stimulating discussion more to the knowledge of Gulf Coast stratigraphy than any other person.

Esther Applin retired from the U.S. Geological Survey, December 31, 1962, and shortly thereafter she received the U.S. Department of the Interior Citation for Meritorous Service. Secretary of the Interior Stewart L. Udall stated as a part of this citation:

Mrs. Applin has become the foremost authority on the smaller foraminifera of the coastal region of the Gulf of Mexico. Her sound scientific studies of these fossils and the rocks in which they are found have led to an authoritative knowledge of the stratigraphy and environment of deposition of a wide variety of sediments in a large region. Mrs. Applin made her scholarly knowledge available to other geologists through many publications and equally important personal consultation and correspondence. The data obtained and the conclusions reached in her studies have been applied extensively by many geologists in their search for oil and gas in the Gulf Coast region. Publication of significant regional geologic studies conducted jointly with her husband, Paul L. Applin, immediately preceded the period of maximum successful exploration for oil and gas in the region. Mrs. Applin's success at cooperative research is attested by the large number of papers published jointly with outstanding members of her special field of research.

After her retirement Esther continued to work for the U. S. Geological Survey on a part-time basis with me on the regional framework of the Atlantic coast and the eastern Gulf of Mexico. This work resulted in joint authorship of AAPG Cross Section Publication 6 on the eastern Gulf Coast, and a chapter on stratigraphy in USGS Professional Paper 659 on the geologic framework and petroleum potential of the Atlantic coastal plain and continental shelf. Neither could have been prepared without Esther.

Esther Applin was a charter and honorary member of the Society of Economic Paleontologists and Mineralogists, an honorary member of the Gulf Coast Section of SEPM, and of the Mississippi Geological Society, a member of the American Association of Petroleum Geologists, and a Fellow in the Geological Society of America. She is cited in *Who's Who of American Women*, and *American Men and Women in Science*.

Esther is survived by her husband, Paul, a son, Col. Paul. L. Applin, Jr., of Newport News, Virginia, a daughter, Mrs. Emmett Lawless, Jr., of Owensboro, Kentucky, twelve grandchildren, two great grandsons, and a sister, Miss Helen Richards of Washington, D. C. All can be very proud of the love, honor, and respect felt by so many for Esther, and of the knowledge that her scientific and professional contributions are locked in the minds and literature of geologists all over the world.

JOHN C. MAHER

Menlo Park, California
October 19, 1972

Reprinted from American Association of Petroleum Geologists Bulletin, v. 57, no. 3, March 1973, p. 596-597.

CHARLES LAURENCE BAKER
(1887-1979)

Charles Laurence Baker was one of the premier reconnaissance geologists to work in Texas. He was very close to Dr. Udden, and like Udden, was deeply interested in every facet of geology. He was the first geologist hired by Udden. Although he was with the Bureau only from 1914 until 1917, and then again for a year or so in the early 1930's, as a leading geologist with the Rio Bravo Oil Company, he continued to conduct marvelous geological studies in the state and always worked closely with the Bureau geologists. He was especially active in the Trans-Pecos and the High Plains.

Baker's exceptional career spans most of the years of this history. He himself is exceptionally knowledgeable about the story of geological explorations in Texas from the 1880's up to the 1940's. His career is a story in itself. He wrote the following biographical notes while in his seventies. The note that accompanied them, which was forwarded by his daughter, Virginia Baker Cook, reads:

Box 125, Cordova, Ill. 61242 Charles
19 Dec. 1972 Laurence
Baker

Dear Ferguson:

I much regret that I'm now so old (85) and my memory so poor that writing of 50 or 60 years ago have escaped me. You speak of Udden, Böse et al with whom I worked who were all kind and capable, who proudly I recall. If you do not know their previous careers I can help. But of personal idiosyncrasies, I forget. The term "economic" was given to the Bureau by its founder, the mining engineer Wm Battle Phillips.

Time makes idiots of us all.

Sincerely,
Baker

CHARLES LAURENCE BAKER

Record of Charles Laurence Baker

A. Scholastic:
- Student, 2 yrs. Monmouth College, Asst. in biology, 1 yr.
- Student, 2 yrs. Oberlin College, Asst. in vertebrate zoology, 2 yrs.
- Graduate student, University of Chicago, Asst. in physiography, 1 yr. Fellow in vertebrate paleontology, 2 yr.
- Graduate student, University of California, Fellow in vertebrate paleontology, 1 yr.

B. Positions held:
- Field asst., U. S. Geological Survey, summers 1908 and 1910
- Instructor in geology, Univ. of California, 1911
- Geologist, Southern Pacific Railroad, 1912-13, 1917-18
- Geologist, British syndicate, 1914
- Geologist, Univ. of Texas Bureau of Economic Geology, 1914-17, 1922, 1931-35

Geologist, various private interests, 1917-21, 1923
Geologist, Standard Oil Co. of California, 1921-22
Geologist, Indian Territory Illuminating Oil Co., 1923
Chief Geologist, East Coast Oil Co. of Mexico and Southern Pacific Railroad of Mexico, 1924-26
Chief Geologist, Rio Bravo Oil Co. and Southern Pacific Railroad, 1926-31
Head, Dept. of Geology, A.&M. College of Texas, 1935-44
Geologist, Tidewater Associated Oil Co., 1944-45
Geologist, State Geological Survey of South Dakota, 1945-53
Sometimes professorial lecturer in geology, Universities of Texas and South Dakota and Northwestern University.

* * * *

Cordova, Ill.
14 May, 1959

Dear children:

There was an episode in my life which perhaps I had best explain since otherwise sometime you may be perplexed.

Graduating from Port Byron Academy, where I had my best teachers, I had completed both scientific and classical courses and so upon marticulating [sic] at Cornell University I had sophomore standing, consequently was never a freshman. When I left Monmouth College in 1905 I had 102½ semester hours credit — 15 A's, 6 B's and 2 C's. That summer I earned 9 more hours at Minnesota Seaside Station, Port Renfrew, B. C.

Consequently upon entering Oberlin College, autumn, 1904, I already had 111½ semester hours out of a total of 120 required for a bachelor's degree. The first lemon they handed me was arbitrarily to reduce my credits to 75 though teachers at Monmouth were fully as good as the Oberlin ones. Oberlin had one first-class man, the chemist F. F. Jewett, who had been Willard Gibbs's assistant at Yale and so was one of the first physical chemists. Modern language teachers at both schools were lazy, hence poor.

I took no math at Monmouth and was rusty in it when resuming its study at Oberlin. The department head there was a Swiss, the poorest teacher I ever have had the misfortune to know. I dropped his course and resumed it under an underling, also very poor. When final exam came I was ill and asked additional time to finish it; this was refused and I walked out with the exam paper. However, I got a passing grade in the course. There remained 3 required hours of math but Branson, the geology instructor, told me that faculty action waived that requirement. Presumably I had to be punished for non-conformity and insubordination because a month before commencement they served notice I couldn't graduate. Over half my Oberlin work was A grade; I left there with 170½ semester hours college credits — only 9½ hours short of 6 full years of college work. I must have been a glutton for punishment for I sat in the gallery of the old First Church and saw my class graduate. The following June University of Chicago gave me the S.B. degree — and without any more mathematics. Likely it was my Scot (that is, Scandinavian) heritage which gave me the stubborness requisite to accomplish more than that entire Oberlin faculty of 1907!

The experience fed me up with self-righteous religiosity and stuffed shirts. Also it left me with little respect for academic degrees and unalterable rules. Mark Twain said that if he could not go from New York to Chicago without passing through Oberlin he would descend and go by way of Hades. His "The man that corrupted Hadleyberg" is about Oberlin.

/s/ Dad

* * * *

AS I HAVE KNOWN IT

I have little memory of earlier days and of languages. I recall 1893, year of the panic and Chicago world's fair. Our parents went there, leaving us with mother's uncle and aunt, my brother's being ill one night. The rest of the nineteenth century was slaving and saving on the farm but keeping out of debt. I had a moronic and sadistic schoolmaster in 1897-98. He whipped me every day so I was taken to Chicago in spring 1898 where I saw troops marching to Spanish-American war. I remember Dewey's May first Manilla Bay [sic] victory and the naval victory of Santiago. I had a very good teacher in 1898-99, Ruth Phillips, who, besides teaching everything from kindergarten to eight [sic] grade, managed to teach me Latin, universal history and physical geography.

The best school I ever knew was Port Byron Academy, entered in autumn, 1899. We had five teachers; of the four principals three became college professors and the other was a competent linguist who taught me Greek. All the economics, physics and chemistry I learned was there. Tuition was $26 per year and very good room and board $2.75 per week. I graduate [sic] in 1903 in both scientific and classical courses with sophomore standing in the best colleges and universities.

At Monmouth College, 1903-5, I note work in biology, European history and English. Had my first geology course in spring, 1904 — last year there was assistant in biology. In summer, 1905, went to Port Renfrew, west coast Vancouver Island and studied marine biology with Univ. of Minnesota faculty. A stopover on return in the Canadian Rockies made me a geologist. Theretofore I used the microscope, thereafter I could see nothing smaller than a mountain.

In Sept., 1905, at age 18, had enough credits for the bachelor's degree. Then went to Oberlin College as assistant in vertebrate geology. It was then in the doldroms [sic], coasting on past glories. Found my three poorest teachers there. Its real handicap was dominance by the theological seminary, too goody-goody for this sad world. However, I had good work in geology, philosophy and ornithology.

1907-10, as assistant and fellow was at University of Chicago where my best teachers were Williston, Leith, Willis and Iddings. Two summers were on U.S. Geological Survey mapping and valuing Wyoming coal land, summer 1909 was in northwest Wind River Mountains — in Sept. I found the Craddock Ranch bone bed, Baylor County, Texas, greatest assemblage of Permian amphibian and reptilian fossils yet known. The last year in Chicago was a fight against tuberculosis and I was indeed fortunate to be able to get West again. Completing all graduate work except thesis for the doctorate I went on another fellowship to the Univ. of California where my colleagues and teachers were a most notable group: Lawson, Louderback, Eakle, Wood and Merriam. (also Holway in geography). I visited Lawson's seminar, benefiting greatly by his analytically critical methods. When he returned to Canada to resume work in the early nineties he insisted I teach his classes. Work incidental to fossil mammal collections in the desert of California and Nevada was published in the Univ. of Calif. publications. Holway offered me an instructorship in meteorology which I declined because ignorant of the physics of gases and work in geology. At various times in California I mapped considerable of the Coast Ranges and worked in the Sierra Nevada. First and last perhaps no one else knew better western mountains and the other country between Missouri River, Gulf of Mexico and Pacific Ocean. The bane of the Western Hemisphere is the north wind, shut out by mountain from the Pacific Coast and Mediterranean, therefore dominant.

In mid-1912 began an association with Southern Pacific Railroad geologists lasting till Dumble's death and the panic. Besides him were Taft, Cummins, Drake, and Kennedy of the Texas Geological Survey of the early nineties — all were capable and great gentlemen. Suman's (my student) and my work in east Texas coastal plain was published by Dumble in "Geology of East Texas" in Univ. of Texas publications; my most important finds there were fossil animals at Burkeville and Cold Springs and much volcanic dust having a long history. My work subsequent is treated in more than fifty publications — the most important I think is mountain geology. Sixty years retrospect shows great fortune, health, finest of friends and associates and most work being what I desired, consequently with little drudgery.

Charles Laurence Baker

December 9, 1967

* * * *

A ROCK HUNTER

This will be mostly an account of friendships and associations in which I have been exceptionally fortunate. Also I have been favored by living in the latter days of the golden days of pioneer exploration, being privileged to contribute to it — in western United States, Mexico and Brazil.

A few words are in order about my mother's family. The Drennens lived originally in the southwest corner of Dundrennen Abbey, built for the Cistercian monks in the Twelfth Century, the word dun meaning castle. They were taken by James the First to Ulster in north Ireland. The one who came to America had been Lord Mayor of Dublin and got a crown grant near Parkersburg in what is now West Virginia. His son, my great-grandfather, was a strong and ardent anti-slavery man, freeing his own slaves. Becoming unpopular he moved to Missouri; having here the same experience he moved to Princeton, Iowa, where my mother was born in 1854. He and my grandfather married Wolf sisters. Grandfather married Phebe Wolf; all that I know of that family is that one was a Revolutionary general, dying in the battle of the Brandywine. My grandfather serving in the Civil War was away when his wife died in Princeton of measles and exposure, leaving my mother, still quite young, to do the household work for her father and three brothers. Afterward she served as nurse and housemaid for father's invalid and partly insane first wife. The brains of the family, James F., first Federal Judge of Dakota Territory, persuaded her to go to California to get some education. She worked there as seamstress and housemaid, finally becoming a country schoolteacher, and so she taught me to read, write and cipher before I entered the country school at the northeast corner of our farm. There Rodney and I did janitor work for two dollars per month. The teacher there was a great moronic bully who whipped me every day. I was taken away, visiting Chicago in the spring of the Spanish-American War. My teacher next year was a natural born one, Ruth Phillips, who, besides teaching from kindergarten to eighth grade, found time to teach me such preparatory subjects as advanced arithmetic, physical geography, universal history and beginning Latin. In autumn, 1899, I entered Port Byron Academy, an excellent school; we had five good teachers and about 25 or 30 pupils. Among them were May Rowe, teaching botany, physics, elementary chemistry and qualitative analysis. Grace Wood, who taught Latin and Greek, and Ora Maxwell, history and English, and the last of four principals, G. H. Bretnall, teaching zoology. Being made professor of biology in Monmouth College I accompanied him there, having completed four years in both classical and scientific courses, entering college with sophomore rating. I took all his biology courses and his elementary geology. Other good teachers there were Robinson in English, Florabel Patterson in modern European history, Glass in advanced Greek, Swan in chemistry; Mrs. Bretnall taught me three courses in German in the academy. After two years there I spent a summer studying marine biology at the University of Minnesota Seaside Station, west coast of Vancouver Island. There I was much influenced by C. W. Hall, who had taught all sciences at Minnesota and was the aged State Geologist, training a number of eminent geologists. On the return we made stopovers in the Selkirts and Canadian Rockies, which determined my future. Theretofore I was using the compound microscope; thereafter I could see nothing smaller than a mountain. Upon return I already had enough credits for a bachelor's degree.

I went to Oberlin College in autumn, 1905. There Branson taught me all his geology and paleontology. The great men there, F. F. Jewett, one of the early physical chemists, mainly responsible for cheap aluminum, taught me thoroughly his hobby, mineralogy. The able philosophy professor, MacLennan, taught some of us seniors the evolution of religion. The two poorest teachers I ever knew were in the mathematics department. Having 65 credit hours more than necessary for the bachelor's degree I petitioned for relief from a required analytic geometry course. According to Branson—on the course committee—it was granted. A month before commencement I was told I couldn't graduate. I had been assistant in vertebrate zoology and histology; my boss, Lynds Jones, taught me birds. Together we made the record for finding the greatest number of birds in one day. I studied advanced French and economics, elocution and physics.

In autumn, 1907, I began three years graduate study in geology at the University of Chicago. There were many good fellow graduate students and a strong faculty, especially Williston, Weller, Salisbury, Iddings, Emmons, Johannsen, Leith, Sargent, Atwood, Barrows and Bailey Willis. Williston was my second father. He met me one day on the museum steps the first spring, and offered me a fellowship in paleontology. He was an extraordinary man, an old-time naturalist, one of the best of teachers. He was an orphan. Before he was 21 he built the Fort Smith and Little Rock railroad but getting malaria was forced to quit engineering and construction. It was years later before he regained an equal salary. He wrote the manual of two-winged flies. He became a physician and was dean of Yale Medical School. Afterwards he was one of those who made the University of Kansas where he became a vertebrate paleontologist. When he came to Austin to inaugurate a chapter of Sigma Xi I found he was a very sick man. I obtained relief from his Chicago work but he died in middle age, virtually from overwork. Once when I was discouraged he wrote me "You will succeed but you must work like the devil"—best advice I ever had.

In summer, 1909, I went to Wind River Mtns., Wyoming, finding there westward thrusting and peneplanation. Then I joined Williston and Paul Miller in fossil collecting in the Permian delta of Baylor County, Texas, where we got malaria and chronic dysentery. My search had been a failure till the last half day when I wandered into higher rocks and found the greatest assemblage of Permian amphibians and reptiles yet known.

My next year in Chicago was misery. Weakened by Texas I unknowingly contracted tuberculosis, being well one week and ill the next. Williston saved me by having J. C. Merriam offer me a fellowship in the University of California whence I was to collect fossil mammals in the Mohave and Nevada deserts where I recovered my health. The University published my first two papers, the subject being desert geology. I visited the seminar of the department's head, Andrew C. Lawson, who made me instructor to teach his courses during his absence in Canada. Holway of the geography department then wished me to teach meteorology; I had to decline because I was a geologist and lacked the mathematics to teach physics of gases. Colleagues at Berkeley were Merriam, Clark, Louderback, Eakle and Wood. Merriam recommended me to E. T. Dumble who wished me to take charge of geologic work in Nevada. Disappointed in this he set me to work for the Southern Pacific Railroad in the Cenozoic of southeast Texas with one of my California students, John R. Suman, as assistant; I found three fossil mammal localities, Dumble publishing the work in his "Geology of East Texas." Thereupon began a long association with some able men, gentlemen all: Dumble, Kennedy, Cummins, Drake and Taff of the original Geological Survey of Texas.

I first left the Southern Pacific when two West India hurricanes, close together, washed out two million dollars worth of tracks in South Texas. Frank M. Anderson, a California acquaintance, engaged me to hunt oil for a British Syndicate in Texas and Oklahoma, Walter Stalder and Henry Truman Beckwish being associates. The assignment lasted till the First World War made the Britishers withdraw. I will narrate only briefly events of the next forty years, largely recorded in my published works.

Next Wm. Battle Phillips asked me to come to the University of Texas for a first term of three years. When I asked him who recommended me he said it was Dr. J. A. Udden whose son John had been a fellow student in Chicago. Udden had been for years geology professor at Augustana College, only 20 miles from my Cordova home but we had never met. My life of him will be found in the Bulletin of the Geological Society of America. Phillips belonged to two influential North Carolina families. He made a success of the Chisos mine, best producer in the Terlingua quicksilver district. His wife died there by a fall from a horse. Then he came to Austin as professor of mining engineering, founding the University Mineral Survey and later Bureau of Economic Geology. He was red-haired and a great hater of politicians. Among early colleagues in Austin were Edwin Leight Porch Jr., very admirable and lovable, clean-minded and clean-living, mining engineer graduate of Michigan College of Mines, Emil Boese [Böse], noted paleontologist, fellow student of Merriam under the noted Karl A. von Zittel at University of Munich and J. N. Beede, Permian paleontologist from University of Kansas. My most important work was exploration of the highly deformed old mountains of the Marathon Basin, Trans-Pecos Texas; also noteworthy is the "Review of the Geology of Texas" which had three editions totaling 50,000 copies. This geologic "best seller" sustained the bureau for two years when the governor vetoed the appropriation; I wrote two-thirds of it. Later associates were Adkins, Winton and the two Plummers, man and wife. Next I returned to the Southern Pacific with work in California, afterwards serving four years as consultant for various clients, notably L. E. Hanchett. For him I chanced to find the most productive of United States oil fields, finding late-date gravels dipping away from each other in Signal Hill Calif., flowing water to the east at Compton and a late uplifted arch, formerly sea bottom. When I told Hanchett that the city of Long Beach owned a waterworks site on the structure top he declared that he, after sad experience with the union depot site in Los Angeles, would have nothing more to do with politicians. His decision lost him about a hundred million.

Ralph D. Reed and I found that in Reef Ridge rocks thickened eastwards towards San Joaquin Valley. The first sentence of my report to E. W. Marland read: "When you are willing and able to drill between six and seven thousand feet deep you can test Kettleman Hills oil possibilities." When they could drill that deep that great oil field was found.

I went to Brazil for Standard Oil Co. of Calif. in 1921-22. There, in the Parana basin, I mapped the world's greatest lava field. It and another great one in South Africa had source beneath the South Atlantic Ocean, at least extended and deepened by removal of once molten rock from beneath. I returned to Austin in 1922 and the following year worked with Beckwith for the Indian Territory Illuminating Oil Co., finding an oil field beneath the city of Winfield, Kansas, which the vice-president, penurious "Birdy" Blue, wouldn't lease. Then for six years I was chief geologist for Southern Pacific in Mexico and Texas—in the former finding eastward thrusting in the mountains of reef rocks over eastern deeper-water limestones; main facts are published in my Panuco oil field paper though more complete mountain structural work may be published by Geological Society of America. Then I returned for four years to University of Texas, afterwards for nine years was head of the geology department at Texas A&M with the country's largest student enrollment and a staff of Turner, Cline, Goldich, Burt and Bacon.

My final work was eight years on the South Dakota Geological Survey with some teaching in the University of South Dakota; perhaps my most notable work being discovery that the White River Oligocene, world's largest fossil graveyard, was volcanic dust. I found also two other South Dakota formations were of volcanic origin ("Geology of Harding County") and worked out the subsurface rocks by microscopic study of well cuttings and in the Black Hills found that the Black River Ordovician only North American potash bentonite formation had been overlooked, being covered largely by debris from overlying rocks.

I have taught or lectured in five universities. Reed, Longnecker and I found Pliocene fossil mammals in the northern Texas Panhandle for which the Hemphillian stage has been named. My most competent assistants were J. P. Buwalda in the California Coast Ranges and Nevada desert and Ralph D. Reef helping me to find depth of Kettleman Hills oil. My lifelong friend, John M. Hamilton, and I learned photography in Monmouth, a very great aid in future work. In summer, 1907, I was mine examiner in the West for northern Ohio would-be investors though I dissuaded them from wasting money on worthless promotions. I had to go to Payson, northeast Arizona. Old polygamous Mormons involved in the Mountain Meadows massacre would neither sell nor rent me horses though I set out across the desert on foot and for the first and only time became lost. Finally I found water though suffering greatly from thirst. I wore out the soles of my feet and couldn't walk for two ensuing weeks. At end of Mohave Desert field work I made the first winter ascent of Mt. Whitney, freezing my feet in melting snow. It took two days crawling on hands and feet to descend to Owens Valley. The San Francisco papers wrote it up, one with the headline "Excelsior in real Life". My staunch friend, J. P. Smith, great Stanford paleontologist, told me: "when I heard of your stunt I had to say that the fools are not all dead yet". I was spared to have had published more than fifty original contributions to knowledge but much was done with no vacations, no holidays nor weekends taken.

Charles Laurence Baker

Sept. 26, 1968

JOSHUA WILLIAM BEEDE
(1871-1940)

Professor Joshua William Beede died at his home in Tulsa, Oklahoma, Tuesday, February 27, from shock resulting from a fracture of the thigh bone. Professor Beede had been in poor health for some time. He is survived by his wife, Mrs. Clara McKee Beede and four daughters, Genevieve (Mrs. G. G. Henderson), Lydia May (Mrs. L. O. Todd), Lucile Prosser Beede, and Clara Frances Beede.

Professor Joshua William Beede was born at Raymond, New Hampshire, September 14, 1871, son of Hiram Pratt and Lydia Marie (Brown) Beede. During his early boyhood the family moved to Kansas to make a home on the wide "Prairie" which Dr. Beede so greatly loved throughout his life. He attended the local schools and Washburn College, at Topeka, graduating with the B.S. degree in 1896. He received the Ph.D. degree from the University of Kansas in 1899, studying under the distinguished paleontologist, S. W. Williston. Also at Washburn College he came under the inspiring influence of Professor C. S. Prosser, at that time at Washburn, and later at Union College, Schenectady, New York, and still later head of the department of geology at Ohio State University. Dr. Prosser had the determining influence on Professor Beede's career. Through him Dr. Beede became interested in the Pennsylvanian and Permian rocks of Kansas, Oklahoma, and Texas, and made the study of these rocks and their fossils his life work. Dr. Beede was assistant in paleontology at the University of Kansas 1897-99, and teacher of science at the Atchison County High School, Effingham,

Kansas, 1899-1901. In 1901 he was appointed instructor in geology at Indiana University, assistant professor in 1906, and associate professor in 1909. He served in the latter capacity until 1917, when he accepted a position as geologist with the Bureau of Economic Geology and Technology of the State of Texas. Professor Beede was to have succeeded Dr. J. A. Udden as State geologist, but resigned in 1922 to become geologist of the Empire Gas and Fuel Company. During his connection with the Texas Survey and the Empire Company he published numerous additional papers on the geology and paleontology of the upper Paleozoic formations of Texas, including several that have become classics in this field. In 1928 he was invited to return to Indiana University in a temporary capacity as professor of geology during Professor Cumings' incumbency of the Waterman Research Professorship. This connection lasted till 1931. A few years later, his health and especially his eye-sight having failed considerably, he and Mrs. Beede returned to Oklahoma in order to be near their daughters. Professor Beede was a member of the Geological Survey of Kansas in 1896, 1898-99, 1903-10; of the Oklahoma Survey in 1911-17; and of the United States Geological Survey in 1901-02.

Dr. Beede was one of that small group of pioneers — Cragin, Haworth, Prosser, Girty, Gould — who in the last decade of the 19th century began the detailed study of the upper Paleozoic rocks of Kansas and adjacent states; and demonstrated the presence of the Permian system in America. Dr. Beede's studies, especially, fixed the lower boundary of this important succession of rocks, and determined the faunal succession and the remarkable evolutionary changes that swept over the faunas of these late Paleozoic times. He accumulated extensive collections of fossils, now housed mostly in the museums of the universities of Kansas and Texas. He was an assiduous and tireless collector. His detailed knowledge of the geological formations of the Mid-Continent region was the envy and despair of the younger geologists; and his name was a household word in Kansas and Texas geology.

Besides his many important papers on the upper Paleozoic formations of Kansas, Oklahoma, and Texas, Professor Beede published several papers on Indiana geology, including one on the cycle of subterranean drainage, which was the first statement anywhere of this important drainage phenomenon, and also of the phenomenon of underground stream piracy, so well illustrated in this region.

Professor Beede was a Fellow of the Geological Society of America; and a member of the Paleontological Society; the Society of Economic Paleontologists and Mineralogists; the American Association for the Advancement of Science; the American Association of Petroleum Geologists; the Indiana Academy of Science; and of the Kansas Academy of Science; Phi Beta Kappa; Sigma Gamma Epsilon; and Sigma Xi. He was one of the founding members of the Indiana chapter of the latter society.

Dr. Beede was a man of extreme modesty and unbounded generosity. So long as he was able to help others no one ever asked him for help in vain — whether it was for financial assistance, or for advice and comfort in stress and adversity. Though his manner and methods in teaching were somewhat difficult to follow, he was greatly beloved by his students because of his sympathetic helpfulness and his vast enthusiasm for scientific research. He was at his best in the field, in the out-of-doors, in his tireless investigation of the rocks and their fossils at first hand. His life and his work will not be forgotten.

E. R. CUMINGS

Indiana University
Bloomington, Indiana
March, 1940

Reprinted from American Association of Petroleum Geologists Bulletin, v. 24, Part II, 1940, p. 1855-1857.

EMIL BÖSE
(1868-1927)

Geology has sustained another great loss in the death of Emil Böse, leading paleontologist and stratigrapher of Mexico, who was injured in an automobile accident near Sabinal, Texas, on September 9, and died from the effects of it on November 8. Born at Hamburg, Germany, he was graduated as a paleontologist under Von Zittel, taking the doctor's degree in 1893. Most of his work was done for the Geological Survey of Mexico from 1898 to 1915, and then, due to the Mexican revolution, he migrated to the United States and gave his services to the Geological Survey of Texas for the following two years. After that time he was a consulting geologist, mainly for the Standard Oil Company of California. He was author of upward of 80 publications, dealing mainly with the Mesozoic of Mexico and the Permian of western Texas.

CHARLES SCHUCHERT

Reprinted from American Journal of Science, v. 215, 1928, p. 88.

HALBERT PLEASANT BYBEE
(1888-1957)

Halbert Pleasant Bybee died at his home in Austin, Texas, on March 30, 1957, after several months of prolonged illness. His passing leaves us with a sense of personal loss relieved only by many pleasant memories about him. He is survived by his wife, Ruth Woolery Bybee, three sons, and one daughter. Halbert H. Bybee, Robert W. Bybee, and Martha Ellen Bybee have followed their father in the profession of geology, and are rapidly advancing in the Texas petroleum industry. Wilbur C. Bybee is assistant manager of the Woolery Stone Company in Bloomington, Indiana.

"Hal" or "Doc" as he was usually called, was the son of William Lawson and Martha Kessler Bybee. He was born on January 7, 1888, on their farm near Rochester, Indiana. His early life was similar to that of other boys raised on farms in the Middle West, but by the time he finished high school he had developed two of the prime requisites of a geologist — the power of observation and intellectual curiosity. In later years, he said the reason he chose geology as a field of work was to find the explanation for the gravels and topography he had noted in the (glaciated) region around his home.

Except for weekends and some summers on the farm, the eight years following graduation from high school were spent in college or teaching. Hal received his

B.S. degree from Rochester College in 1908, and the following year became principal of a rural high school. The trustees hesitated to employ him because of his youth and warned him that previous principals had been unable to control the students. Bybee told the trustees, "You leave them to me and we will get along all right." Evidently this occurred, as almost fifty years later, when the 1909 graduating class heard he was seriously ill, he received a "get well" card signed by all thirteen members of the class. This faculty of inspiring the respect and devotion of young people characterized his entire teaching career.

In the fall of 1909, Hal entered Indiana University, but the next fall he left school to become an instructor in science at Clinton College, Clinton, Kentucky. Returning to Indiana University in 1911, he was awarded a B.A. degree in 1912, an M.A. in 1913, and a Ph.D. in 1915. While attending the University, he was a fellow in geology and spent summers in geologic field work.

While working on his doctorate, Hal became an instructor in geology at The University of Texas in January, 1914. At that time, the Department of Geology had a faculty of one professor and three instructors, an enrollment of about two hundred twenty and only about one dozen advanced students. In the fall of 1914, Hal had the choice of remaining at Texas with an increase in salary, but chose to accept an offer from Washington University, because St. Louis was closer to his home and the home of his recent bride Ruth Woolery. Ruth was a Phi Beta Kappa graduate at Indiana, receiving her B.A. in 1912 and an M.A. in 1913. She chose to be married on September 1 in preference to an offer to teach at Wellesley. This was a fortunate choice for both of them, and their married life was to be an inspiring model to their many friends. Her calm strong character and fondness for young people, equal to his, contributed greatly to his success as a teacher and geologist.

Fortunately for The University of Texas, a material increase in enrollment caused the head of the Department of Geology to wish to have Bybee back, and he was induced to return to Austin in November, 1914. During the next eleven years, the limited faculty of six or less carried an increasing load as the number of students in beginning Geology I doubled and the number of advanced students increased rapidly. Doc Bybee realized and stressed the importance of field work and he was responsible for several improvements and additions to the curriculum. He was promoted to Adjunct Professor in 1916 and Associate Professor in 1920. Bybee worked for the Bureau of Economic Geology two summers and for the Oklahoma Geological Survey two summers.

Overcoming some opposition, Bybee inaugurated the University's first summer camp in geology in 1917. Eleven boys, principally sophomores, attended the six weeks course in Llano County. Doc directed five of the first seven camps and he conducted the camps for the benefit of the students and not as an opportunity for the faculty to engage in research. This policy has been followed consistently so that only occasionally have the results been published by faculty members and these publications give the students credit for the major part of the field work. Mature for his years, Doc was at his best on the summer camps, planning the work and instructing and assisting the students. His method of instruction was between the two extremes of simply exposing the students to geology and leading them by hand. On Sundays and when the long days were over, Doc was just one of the group, and the students of his summer camps and field courses developed the greatest admiration and affection for him. Many of them came to feel almost a family relationship with him, and in later years, when these students returned to Austin with their children and grandchildren, they were welcomed in the Bybees' home.

Although Bybee was very fond of young people and of teaching, University salaries were low, graduates accepted jobs in the petroleum industry in preference to graduate study, and following the drowning of his oldest son Henry, Doc accepted a geological position with the Dixie Oil Company in West Texas in August, 1925. During his stay in San Angelo he helped organize the West Texas Geological Society and was its president in 1927.

The increasing oil activity in West Texas caused the Legislature to create the Board for Lease of University Lands. The University also established its Lands Division to handle its two million acres in its Permanent Fund. Bybee was persuaded to become Geologist in Charge in June, 1929. From time to time the policies of the Board for Lease of University Lands and the Board of Regents were altered. Many of these changes came as a result of his recommendations. He faithfully implemented and carried out all policies of the Boards. From 1929 to 1957 more than $200,000,000 was paid into the University's Permanent Fund from oil and gas rentals, bonuses, and royalties. Many people contributed to this achievement, but none more than Bybee. In addition to fulfilling his own responsibilities, Bybee worked in close cooperation with the other divisions of University Lands. His ability to work with people and to induce them to work with him was a valuable asset over a twenty-eight year period during which five different men were President of the University and the Board of Regents changed membership every two years. Bybee's character, ability, and good judgment help explain the fact that during the administration of eight governors and dozens of legislative sessions, the Permanent Fund was never involved in politics.

At the University, the enrollment in the Department of Geology continued to grow, and in September, 1936, the President, realizing the need for strengthening and reorganizing the department asked Bybee to return to the faculty as Professor in Geology, while continuing as Geologist in Charge of University Lands. Bybee was reluctant to undertake two full-time jobs, but duty seemed to require that he accept the assignment. He relinquished some of the Lands work to other members of the staff and naturally did not assume a full teaching load. He was Chairman of the Department from 1937 to 1941, and thereafter had an important role in all departmental affairs. He never lost sight of the fact that fundamentals rank first in training geologists, but his active mind continually worked to improve and add to the curriculum. He inaugurated and taught many of the new courses for a few years and then turned them over to younger men. As the years passed, the value of geology became well recognized and the Department's stature was enhanced, due in no small part to Doc Bybee's contributions. At the same time, his objectiveness and unselfishness enabled him to recognize the needs and possibilities of other departments; accordingly he also worked for the general welfare of the University as a whole.

As in the previous period as a teacher, a large part of Doc's time was given to the problems of students and associates. He always took time to listen to them, and the more worried they were the longer he would listen. His knowledge of young people and his observing nature often enabled him to know what was troubling the men before they came for help. Doc seldom lectured or "preached" to these men, and his comments, answers, and advice were usually brief, practical, friendly, and effective.

Doc Bybee's standards of integrity and honesty were exceptionally high and so firm and ingrained, he was not even tempted to depart from them. The Association's original detailed Code of Ethics was a good outline of his conduct. Though not without ambition, he was so selfless that his friends took his friendship, interest, and assistance for granted. He gave much to others and asked nothing in return. Doc was so straightforward and frank that occasionally he seemed blunt, but he could differ with people without offending or being offended, especially in scientific and business matters. He met problems openly and usually moved directly to their solution. Doc had a keen sense of humor, enjoyed "kidding" with young people and enjoyed what he received in return. He was tolerant of human frailties, but when a student got completely "out of line" or was not "doing right," as Doc usually phrased misdeeds, he could "straighten him out" and leave the young fellow a better friend than ever. His associates and other friends also stress his common sense, acumen, loyalty, patience, kindness, encouragement, understanding, and humanism.

Hundreds of Doc's students recall that he recommended them for a job. Doc had the happy faculty of recognizing each student's good qualities. He was wiser than many professors and employers in realizing that the students with the best grades would not always be the most successful and that there were more positions to be held below chief geologist level than above it. His evaluations of students were quite frank and dependable, and employers learned from experience they could accept his judgment as conclusive.

One man recently wrote that the period 1914 to 1925 was the greatest period of Bybee's life. No doubt the University Regents, faculty, and alumni think his work in connection with the University Lands was his greatest contribution. During the 1936-1957 period, Doc played the major role in enlarging and strengthening the faculty of the Department of Geology. The staff was doubled, only one of the present nineteen has more than twelve years' service and two of the top professors came from the oil industry, a reversal of the past process. The total enrollment is the largest in the country; there are over one hundred graduate students and more than twenty graduate degrees awarded annually. Contrasted with a single camp of eleven boys in 1917 were the three camps of eighty-two seniors and graduate students and a three-week camp of one hundred ten juniors. The results of the early camps seldom were published, but as advanced or graduate standing were required, bulletins and other articles became common. The greater number of students that came under Doc's influence the past twenty-one years, together with the staff of relatively young, extremely capable and enthusiastic geology teachers he left, should prove that this period was his greatest.

Evaluations of Bybee as a geologist ranged from a good "practical" geologist by those who only had limited contacts with him, to an outstanding geologist by those who knew him best. Even the former were frequently surprised by his reasonable interpretations and sound conclusions of geologic problems. He would have enjoyed engaging in considerable research, but he placed people ahead of geology, and he thought his most important duty lay in the development of students. Much of the time he might have spent in geological research was devoted to listening, advising, and helping students, associates, friends and even strangers. His greatest contributions to geology have and will come from the hundreds of geologists he trained and other geologists who came under his influence.

Bybee had been an active member of the Association since 1919, attended most of the annual meetings, served on several committees, and delivered a number of papers. In November, 1952, at the Geology Homecoming Day at Austin, a dinner was held in his honor and he was presented with a Life Membership. Friends in the oil industry also gave him a citation for "Distinguished Service to the Oil Industry and the State of Texas." On December 20, 1956,* Bybee was elected an Honorary Member of the Association, one of only sixteen geologists who have devoted most of their lives to teaching to have received this honor. Bybee was a member of many other scientific societies, including the American Association for the Advancement of Science, Geological Society of America, Sigma Xi, Sigma Gamma Epsilon, American Geophysical Union, American Geographical Society, and Texas Academy of Science (Honorary Life Member). Unlike most scientists, he was active in civic affairs for over thirty years in the Chambers of Commerce, Kiwanis Clubs (ex-president), and Boy Scout Councils in Austin and San Angelo.

Bybee was a very religious man and a "practical Christian" who may have had as much influence as many ministers. He was a deacon in Baptist churches for over forty years. In a beautiful funeral eulogy his minister spoke of his profound but simple faith, of his dependability and faithfulness, and of the minister's custom of arranging for young people who were worried over what they thought was a conflict between science and religion to have a talk with Bybee, with almost universally favorable results.

In November, 1954, Bybee and his associates in the Department of Geology, in conjunction with the Development Board of the University, established the Geology Foundation, the primary aim of which is to develop superior training for geology students. At the first meeting of the Advisory Committee after Doc's death, plans were made to establish the "Hal P. Bybee Memorial Fund" to be administered as an integral part of the Foundation.

A letter to Mrs. Bybee from the first Chairman of the Regents' Committee on University Lands, who had become a great admirer of Bybee, quoted a poem by William Allingham that expressed a part of Doc's philosophy:

> "No funeral gloom, my dears, when I am gone,
> Corpse-gazings, tears, black raiment, graveyard grimness;
> Think of me as withdrawn into the dimness.
> Yours still, you mine, remember all the best
> Of our past moments, and forget the rest,
> And so, to where I wait, come gently on."

**Editorial Note.* — The date of March 29, 1957, inadvertently published in the list of honorary members in the March, 1957, membership list, was the date on which the executive committee confirmed the mailed ballot election of December 20, 1956.

The memories of his family and friends are "all of the best," and they are so real it is difficult to believe he has "withdrawn into the dimness." We appreciate Bybee for many things, and especially for the pleasure of remembering.

L. T. BARROW

Houston, Texas
August 20, 1957

BIBLIOGRAPHY OF H. P. BYBEE

1914. (With Clyde A. Malott). "The Flood of 1913 in the Lower White River Region of Indiana," *Indiana Univ. Bull. 12*, No. 11, pp. 105-223; 1 pl., 50 figs., 8 maps.

1916. (With J. A. Udden). "The Thrall Oil Field," *Texas Univ. Bull. 66*, 93 pp., 7 pls., 7 figs., 1 map.

1921. "Some Recent Notes on the Thrall Oil Field of Williamson County, Texas," *Bull. Amer. Assoc. Petrol. Geol.*, Vol. 5, No. 6, pp. 657-60.

1925. (With R. T. Short). "The Lytton Springs Oil Field," *Texas Univ. Bull. 2539*, 69 pp., 9 pls., 3 figs., 2 maps.

1927. (With Fred M. Bullard). "The Geology of Cooke County (Texas)," *ibid., Bull. 2710*, 170 pp., 11 pls., 6 figs., 1 map.

1930. (With E. H. Sellards and H. A. Hemphill). "Producing Horizons in the Big Lake Oil Field, Reagan County, Texas," *ibid., Bull. 3001*, pp. 149-203; 6 figs.

1931a. "Some Major Structural Features of West Texas," *ibid., Bull. 3101*, pp. 19-26; 1 fig.

1931b. (With E. F. Boehms, Cary P. Butcher, H. A. Hemphill and G. E. Green). "Detailed Cross Section from Yates Area, Pecos County, Texas, into Southeastern New Mexico," *Bull. Amer. Assoc. Petrol. Geol.*, Vol. 15, No. 9, pp. 1087-93; 2 pls.

1937. (With Berte R. Haigh). "Developments in West Texas and Southeastern New Mexico in 1936," *ibid.*, Vol. 21, No. 8, pp. 1034-41; 1 fig.

1938a. (With Berte R. Haigh and Taylor Cole). "Developments in West Texas and Southeastern New Mexico in 1937," *ibid.*, Vol. 22, No. 6, pp. 694-700; 1 fig.

1938b. "Possible Nature of Limestone Reservoirs in the Permian Basin," *ibid.*, Vol. 22, No. 7, pp. 915-18; 1 fig.

1939. (With Berte R. Haigh and Surce John Taylor). "Developments in West Texas and Southeastern New Mexico During 1938," *ibid.*, Vol. 23, No. 6, pp. 836-43; 1 fig.

1952. "The Balcones Fault Zone — An Influence on Human Economy," *Texas Jour. Sci.*, Vol. 4, No. 3, pp. 387-92.

Reprinted from American Association of Petroleum Geologists Bulletin, v. 41, no. 10, October 1957, p. 2394-2398.

ALEXANDER DEUSSEN
(1882-1959)

During the early morning hours of Saturday, September 5, 1959, in Flow Hospital, Denton, Texas, near his ranch-home at the little town of Ponder, Texas, death came to Alexander Deussen. The sudden end had not been expected, but it came peacefully. Those of his immediate family who survive him are his wife, Sue Campbell Deussen, their daughter Lucile Deussen McRae and his grandson, Alexander Deussen McRae.

Perhaps no other individual has contributed more to the welfare of this Association than Alexander Deussen. He served as its second president during the fateful period of its infancy when its survival was by no means assured. At the end of his own administration, with consummate judgment and tact, he selected and (all but single-handedly) secured the election of his successor, William Embry Wrather. Once he relinquished the office of president, his continuing deep concern for the success of the Association gained for him the unofficial, but generally recognized, title of "elder statesman." That title he retained throughout the rest of his life.

At its first meeting after his death, the Executive Committee adopted a special resolution of regret and sympathy worded as follows:

WHEREAS, Alexander Deussen, an honorary member of The American Association of Petroleum Geologists, passed away on September 5, 1959; and

WHEREAS, Mr. Deussen had an outstanding career as a petroleum geologist, and enjoyed the affection, admiration and respect of the geological profession; and

WHEREAS, Mr. Deussen was one of the organizers and second president of the A.A.P.G. and saved it from collapse after a period of stagnation; and

WHEREAS, we have lost our mentor, our guide and beloved friend, who will endure in our hearts;

NOW THEREFORE, be it resolved that our high esteem for our friend, Alexander Deussen, be published in the A.A.P.G. bulletin, and copies be delivered to his family as an expression of sympathy from the members of the A.A.P.G.

Repeatedly during his lifetime the Association honored Alexander Deussen. In 1947 it bestowed upon him, in recognition of his outstanding contribution to petroleum geology, the Sidney Powers Memorial Award. In 1953 he was elected to Honorary Membership.

From the biographical data recorded in the citation for the Sidney Powers Award[1] the following quotations are excerpted.

Alexander Deussen was born in San Antonio, Texas, January 19, 1882. His parents were likewise natives of San Antonio. Of his grandparents, three were emigrants from Germany who landed at the Port of Galveston and thence continued by oxcart to San Antonio, reaching their destination, one after another, during the years 1851 and 1852. The fourth grandparent was born in South Carolina, of English descent, and came to San Antonio in 1848. Deussen's father, at the time of his death two years ago, was vice-president of the Groos National Bank of San Antonio, the oldest financial institution (established in 1854) now active in the state. He had gone to work for the bank in 1880 and continued in its service for more than sixty years.

As a child, Deussen attended the German-English Grammar School, which gave instruction in both the German and English languages. Later, he entered San Antonio High School and upon graduation was admitted to the University of Texas. From this institution he obtained his Bachelor's Degree in 1903 and his Master's Degree in 1904. During his undergraduate work Deussen distinguished himself, and helped defray in part the cost of his schooling by winning two successive scholarships in political science. In his post-graduate year he instructed in zoology, and after he obtained his Master's Degree, he became a member of the faculty of the university, where he taught geology. He continued in his position for ten years. From 1907 to 1915, Deussen also held the position of assistant geologist, United States Geological Survey. In 1915 he was employed as geologist by the Guffey Petroleum Company, the forerunner of the present Gulf Oil Corporation, to work out of the company's headquarters at Houston, Texas. He resigned his position after one year and opened his own office as a consulting geologist in Houston in 1916. He has maintained his own practice continuously ever since, except for the years 1924 to 1930, inclusive, during which

[1]Wallace E. Pratt, 1947, "Alexander Deussen-Sidney Powers Memorial Medalist," *Bull. Amer. Assoc. Petrol. Geol.*, Vol. 32, No. 5 (May), pp. 829-34.

period he devoted all his time to the Marland Oil Company, as vice-president in charge of operations in the Gulf Coast region.

In 1924 Deussen initiated and directed the operations of the first seismograph party that undertook to explore the Gulf Coastal Plain. As an executive on the original board of directors of the first service corporation to make electric logging available to American oil producers, he pioneered in the development of this now indispensable technique. Again, in 1924, he established one of the early micropaleontological laboratories for research on Gulf Coast sediments. Much earlier he recommended and directed the drilling of the well which discovered the Blue Ridge oil field in Fort Bend County, Texas. He made similarly direct contribution to the early development of numerous other Texas oil fields, including White's Point, South Dayton, and Thompson.

In 1907 and 1908 Deussen did the field work for Water-Supply Paper No. 335, "Geology and Underground Water Resources of the Southeastern Part of the Texas Coastal Plain," which the United States Geological Survey published in 1914. In this exhaustive study on an area of more than 36,000 square miles, which the text defines as about equal in size to the state of Indiana, Deussen seized his first opportunity to examine closely oil and gas occurrences. In the report he remarks: "It is probable that the gas in the Mexia District is struck in these discontinuous and locally warped sands on the east side of the fault." In 1921 the discovery well in the first of the great fault-controlled fields of East Texas was drilled near the axis of "the Mexia anticline," as determined by other geologists. A few months later as development proceeded amidst intense excitement, everyone suddenly realized that the oil field had "skidded over," as the local wags expressed it, to come to rest again nearly one-half mile west of the "anticline" but squarely against the fault Deussen had previously mapped and described. Thus, many of us belatedly came to comprehend the obscurely displayed structure which Deussen had unerringly deciphered, in his field work fifteen years earlier.

Deussen's most ambitious study is Professional Paper No. 126, "Geology of the Coastal Plain of Texas West of the Brazos River," which the United States Geological Survey published in 1924 and the University of Texas re-published in 1930. In this report Deussen says: "The descriptions and the conclusions here presented are largely the results of field examinations made by the writer from 1909 to 1913." These examinations of an area which covers nearly 39,000 square miles revealed to Deussen the Darst Creek fault, the position of which is accurately shown on his map, and led him to predict that the oil and gas accumulations already in evidence in the Corpus Christi district would prove to be controlled by the system of faults he had detected and mapped there. Yet it was not until 1929 that other geologists recognized the conditions which Deussen had perceived in the field work he had begun twenty years earlier, and finally persuaded themselves to recommend the exploratory wells which discovered the major oil field which subsequently developed at Darst Creek.

The distinctive quality of Deussen's mind is indelibly stamped on the pages of these two geological studies. The wealth of detail; the patient, accurate observation; the sure dedication; the restraint of all impulse toward speculation; all these stand out. His is not only the infinite capacity for taking pains. There is also his almost intuitive detection and interpretation of obscure surface evidences of underlying geological structure in the expressionless outcrops of the sediments he maps and describes. Even more remarkable is the clear vision which enabled him to anticipate, many years before fault-controlled oil fields became a reality, that this type of geologic structure would prove to be of major importance to the oil industry in Texas.

Many years ago, Alexander Deussen and Sue Campbell Deussen, his wife, established their commodious ranch-home, Deussendale, in the broad valley of Elm Creek, where that creek meanders across the maturely dissected outcrop of the Lower Cretaceous, in northeastern Texas. Significant of the character of the man is the fact that in the intimacy of their home, Mrs. Deussen who, as a girl was a student in his classes, still refers to her husband with half-mischievous, half-deferential affection as "the Professor." By a singular (but appropriate) coincidence, the neighboring village is called Ponder, and it was inevitable, in view of this circumstance, that among his friends Deussen should come to bear the sobriquet, Sage of Ponder.

Throughout his adult life Alexander Deussen kept a diary. It is a document which brings into relief the character of its author — painstaking, precise, detailed. It recorded fully and conscientiously the day's activities; the time he arose; the weather; the business of the day; the time he went to lunch; where he lunched and with whom; the time he returned to the office and the time he left for the day. On more than one occasion this exact record accompanied him when he went to court as an expert witness. Always it constituted unimpeachable evidence. However unusual the incidents recorded, the methodical style of the diary went unruffled:

"August 23, 1905, this day I married Sue Burnett Campbell at Denton, Texas. Left for Austin, Texas, for our future home this night via M.K.&T. Railway."

Of medium physique and far from erect in posture, Alexander Deussen achieved, nevertheless, a singular dignity of bearing and a forceful presence. His was a diffident nature and the austere, all but gruff, manner he often displayed toward strangers doubtless reflected this trait. Recently, an intimate friend, Morgan Davis, has portrayed him graphically:

> It was in 1928 when the train carrying a party of Houston geologists to the Los Angeles AAPG convention stopped for an hour in Albuquerque.... His reserved manner which was somewhat in contrast to his highly intelligent, alert brown eyes; his kindly but searching questions regarding the geology of the area; his old-fashioned, high Belmont collar and conservative clothes, again somewhat in contrast with a twinkling sense of humor which seemed about to break out momentarily; these are still vivid impressions of my first meeting with Alexander Deussen.

Throughout his long professional career Deussen maintained an office as consulting geologist in Houston, Texas, dividing his time between that office and his ranch home. As the years passed he constantly restricted his clientele, devoting himself more and more exclusively to his own independent search for oil. Nevertheless, his prestige continuously expanded and he came to be known very widely as "dean of Gulf Coast petroleum geologists."

Although he took no conspicuous part in public affairs, he nourished a keen sense of civic responsibility. Among other extensive benefactions he conveyed to Harris County, Texas, in 1954, a beautiful 200-acre site on Lake Houston and dedicated it to the public as a park and play grounds. This popular recreation center came to be known as Alexander Deussen Park.

Alexander Deussen was a member of the honorary, professional, and scientific societies listed here:

Phi Beta Kappa
Sigma Xi
American Association of Petroleum Geologists (Past-President) (Awarded honorary membership, 1953)
Geological Society of America (fellow)
American Institute of Mining and Metallurgical Engineers
American Association for the Advancement of Science
Houston Geological Society (honorary member)
Houston Philosophical Society
Society of Economic Geologists
Texas Geographical Society
Society of Exploration Geophysicists

Society of Economic Paleontologists and Mineralogists
American Petroleum Institute
Texas State Historical Association
West Texas Historical Association

A list of Alexander Deussen's publications is appended.

LIST OF PUBLICATIONS BY ALEXANDER DEUSSEN

"Has the Railroad Commission of Texas Succeeded in Preventing Discrimination?" (Thesis which won the Hancock Fellowship in Political Science for the years 1902-1903). *Univ. Texas Record*, Vol. IV (December, 1902), pp. 422-62, Austin, Texas.
"The Geology and Soils of Texas," *Texas Almanac* (1904), pp. 73-76, Dallas, Texas.
"The Mineral Resources of Texas," *ibid.*, pp. 147-53.
"Cement Resources and Industry of Texas," *The Tradesman* (November 15, 1906).
"The Beginnings of the Texas Railroad Systems," *Trans. Texas Acad. Sci. for 1906* (1907), Vol. IX, pp. 42-74.
"Lee Hager (1874-1944)," *Bull. Amer. Assoc. Petrol. Geol.*, Vol. 28, No. 11 (November, 1944), pp. 1666-68.
"Physiography and Geology of Texas," *Texas Almanac* (1910), pp. 59-64.
"Mineral Resources of Texas," *ibid.*, pp. 65-72.
"The University's Contribution to the Economic and Material Development of the State," *Jour. Proc.*, 22d Annual Meeting, Southern Educational Association, Houston, Texas (1911), pp. 411-14.
"Notes on Some Clays from Texas," *U. S. Geol. Survey Bull. 470* (1911), pp. 302-51.
"Natural and Geographic Conditions of Texas," *Texas Almanac* (1911), pp. 139-61.
"Topography and Geology of Texas," *Texas Almanac* (1912), pp. 154-60.
"The Mineral Resources of Texas," *ibid.*, pp. 161-68.
"The Survey of the Artesian Resources of Southwest Texas," *Irrigationist*, No. 1 (1913), pp. 9-11. San Antonio, Texas.
"The Geographical Units of Texas," *Texas Almanac* (1914), pp. 152-55.
"Texas Mineral Production, 1914," *ibid.*, pp. 156-63.
"Geology and Underground Waters of Southeastern Texas Coastal Plain," *U.S. Geol. Survey Water-Supply Paper 335* (1914).
"Ground Water in LaSalle and McMullen Counties, Texas," *ibid.*, No. 375 (1916), pp. 141-77.
"Humble, Texas, Oil Field," *Bull. Southwestern Assoc. Petrol. Geol.*, Vol. 1 (1917), pp. 60-84.
"Review of Development in the Gulf Coast Country in 1917," *Bull. Amer. Assoc. Petrol. Geol.*, Vol. 2 (1918), pp. 16-37.
"Captain Anthony F. Lucas," *ibid.*, Vol. 5 (1921), pp. 688-89.
"Salt Domes of Texas and Louisiana," *Oil Weekly*, Vol. 24, No. 4 (January 21, 1922), pp. 11, 16, and 18.
"Oil Development in Gulf Coast Fields during 1922," *Trans. Amer. Inst. Min. Met. Eng.* (1923).
"Geology of the Coastal Plain of Texas, West of Brazos River," *U. S. Geol. Survey Prof. Paper 126* (1924). Reprint, *Univ. Texas Bur. Econ. Geol.* (1930).
"Hockley Salt Dome, Harris County, Texas" (with Laura Lee Lane), *Bull. Amer. Assoc. Petrol. Geol.*, Vol. 9, No. 7 (1925), pp. 570-99.
"Oil Reserves of the Gulf Coast," *Oil Weekly* (March 20, 1933), p. 16.
"Oil Producing Horizons of the Gulf Coast in Texas and Louisiana," *Bull. Amer. Assoc. Petrol. Geol.*, Vol. 18, No. 4 (April, 1934), pp. 500-18.
"Two Decades of Progress in the Art of Oil Finding," *ibid.*, Vol. 18, No. 7 (July, 1934), pp. 942-44.
"Use of Electric Logs for Correlation in the Gulf Coast of Texas and Louisiana" (with Eugene Gilbert Leonardon), *Amer. Petrol. Inst. Bull. 216* (1935), pp. 13-18.

"Electrical Exploration of Drill Holes" (with Eugene Gilbert Leonardon), *Drilling and Production Practice 1935, Amer. Petr. Inst.* (1936), pp. 47-50.
"Thirty-five Years of Progress in the Knowledge of the Geology of Texas," *Univ. Texas Bull. 3501* (January 1, 1935), pp. 37-57.
"Texas-Louisiana Gulf Coast Oil Reserves, as of January 1, 1936," *Oil Weekly*, Vol. 81, No. 1 (March 16, 1936), pp. 47-48.
"Acre-Foot Yields of Texas Gulf Coast Oil Fields," *Petroleum Development and Technology, Amer. Inst. Min. Met. Eng.* (1936), p. 53.
"Orange (Texas) Oil Field" (with E. W. K. Andrau), *Bull. Amer. Assoc. Petrol. Geol.*, Vol. 20, No. 5 (May, 1936), pp. 531-59. Abstract, *World Petroleum*, Vol. 7, No. 8 (August, 1936), p. 404. Reprinted in *Gulf Coast Oil Fields, Amer. Assoc. Petrol. Geol.* (1936), pp. 880-908.
"Oil Producing Horizons of Gulf Coast," *Pan-Amer. Geologist*, Vol. 67, No. 3 (April, 1937), pp. 193-214.
"Use of Temperature Measurements for Cementation Control and Correlations in Drill Holes" (with Hubert Guyod), *Bull. Amer. Assoc. Petrol. Geol.*, Vol. 21, No. 6 (June, 1937), pp. 789-805.
"Discoveries (of Oil)," *Geophysics*, Vol. III, No. 3 (1938), pp. 177-97.
"Correlation of Surface and Subsurface Formations in Two Typical Sections of the Gulf Coast," *Bull. Amer. Assoc. Petrol. Geol.*, Vol. 23, No. 11 (November, 1939), pp. 603-34; *Oil and Gas Jour.*, Vol. 36, No. 44 (March 17, 1938), pp. 71-74; Vol. 37, No. 24 (October 27, 1938), p. 47.
"Laura Lee (Lane) Weinzierl (1900-1929)," *Bull. Amer. Assoc. Petrol. Geol.*, Vol. 13, No. 1 (January, 1929), pp. 94-96.
"Royalties," chapter in *The Elements of the Petroleum Industry, Amer. Inst. Min. Met. Eng.*, Seeley W. Mudd Ser. (1940), pp. 105-15.
"Memorial to Frederick William Simonds," *Proc. Geol. Soc. America for 1941* (1942), pp. 192-99.
"New Oil Reserves Must Be Found," *Oil Weekly*, Vol. 109 (April 5, 1943), pp. 47-48.

WALLACE E. PRATT

Carlsbad, New Mexico
January 13, 1960

Reprinted from American Association of Petroleum Geologists Bulletin, v. 44, no. 3, March 1960, p. 371-374.

ALVA CHRISTINE ELLISOR
(1892-1964)

On the night of September 8, 1900, a little black-haired girl of eight clung desperately to her mother at the top-most part of the roof of her Galveston home. Her mother clutched two baby brothers lest they be swept away by the waves of the hurricane which drowned thousands and destroyed so much property on that fateful night. A timber, tossed by the great seas, swept the babies from their mother's arms, but the little girl was saved to become one of the first, and one of the most distinguished, woman stratigraphers this country has produced.

Alva Christine Ellisor, for such indeed was this marooned child, was born April 26, 1892, at Galveston, Texas, the only daughter, and the oldest, among five children. After a life devoted to stratigraphic research, and marked by numerous publications on most phases of Cretaceous and Tertiary stratigraphy and paleontology of the American Gulf Coast, she passed away of a heart ailment September 22, 1964, at Galveston. Miss Ellisor was survived by a brother, William L. Ellisor, Jr., but he has since passed away.

Her father, William Lee Ellisor, Sr., was the grandson of one of five brothers, was a sea captain and lived at Cedar Bayou, near Houston, where his grandfather had settled. Her mother, Emma Osterman Ellisor, was born on Gotland, a large island in the Baltic Sea, off the southeast coast of Sweden.

Educated in Galveston's public schools and at Ball High School, where she was valedictorian of her class, Alva entered the class of 1915 at the University of Texas and soon became absorbed in geology, from which she was graduated with high honors. Her keen mentality and her ability to think a problem through were early recognized by her professors, Whitney, Bybee, and others, so that she easily gained a post teaching science at Ball High. The next year she returned to Texas University for two years where she taught, studied, but never gained her master's degree. Her first published work, "*Turritella* in the Buda and Georgetown Limestones," appeared in 1918 (University Texas *Bull.* 1840) and was outstanding. Fourteen of her species were included in W. S. Adkins' "Handbook of Cretaceous Fossils," the standard reference on Cretaceous megafossils.

As early as the summer of 1918 she worked briefly in the Ranger area of Texas, for the Humble Company, examining well cuttings of Pennsylvanian and Permian age. Thus she was probably the first woman to do work of this kind in the oil industry. This was at a time when there were even few men employed in commercial geology. During 1918 and 1919 Miss Ellisor was instructor in geology at the University of Kansas and later was briefly with Kansas Geological Survey. In late November, 1920, at the insistence of Wallace Pratt, she returned to the Humble Company to organize its paleontological laboratory in Houston, for the examination of megascopic fossils; Foraminifera had not yet been recognized on the Gulf Coast. In 1927, she became Humble's first research stratigrapher and paleontologist. In this work she remained until her retirement in April, 1947, contributing numerous published and unpublished articles on the stratigraphy of this important American Gulf Cost province.

Although she had never seen a foram until she went to work for Humble, she is believed to have been the first person to have noted Foraminifera in wells along the American Gulf Coast (in 1920, in a Humble well at Goose Creek). She early recognized the necessity for field work to correlate the subsurface and surface sections, and in 1924 began a series of field trips over a period of 10 years or more, with the writer and many other Humble geologists, to sample and study the Tertiary and Cretaceous sections.

In 1925, with Applin and Kniker, she co-authored an outstanding pioneer paper, "Subsurface Stratigraphy of the Coastal Plain of Texas and Louisiana" (A.A.P.G. *Bull.*, Vol. 9, No. 1, p. 79-122). Miss Ellisor wrote the part on the Oligocene, and helped with other portions, in which the three-fold division of the Middle Oligocene — *Discorbis*, *Heterostegina*, and *Marginulina* zones — was first described and is still accepted, although whether the age was Oligocene, as recognized at the type locality on the island of Antigua in the West Indies, or Miocene, later became a controversial matter. At any rate, these zones have been most valuable for many years as markers for regional structural maps and in positioning valuable producing zones. Such maps, in the early thirties, contributed to the discovery of a number of excellent coastal oil fields. Miss Ellisor was the first to recognize reworked Cretaceous forams in younger Miocene strata where they occurred in beds beautifully preserved, in their usual assemblages and at times even in their normal sequences. In 1926 her "Coral Reefs in the Oligocene of Texas" appeared in the *Bulletin* (Vol. 10, No. 10). In this paper she was the first to establish that the development of limestone in the *Heterostegina* zone, on the surface at Damon Mound, and around many other salt domes underground, was actually of coral reef origin.

Between 1931 and 1945 she collaborated with J. A. Cushman in a series of joint papers in which many well-known foraminiferal index fossils, later of great value in Gulf Coast correlation, were described and named. These papers appeared at various times in Cushman's Laboratory Foraminiferal Research Contributions. The last paper, "The Foraminiferal Fauna of the Anahuac Formation," appeared in the *Journal of Paleontology*, November, 1945.

Miss Ellisor's field work to correlate the surface section with that found in wells began in 1924 with trips to the chalk areas of northeast Texas and southwest Arkansas and showed, as she had for some time suspected, that the Annona (White Cliffs) Chalk of Arkansas was the same as the Pecan Gap (Taylor) of Texas, thereby correcting an earlier view that it was of Austin age: "The Age and Correlation of Chalk at White Cliffs, Arkansas, with Notes on Subsurface Correlations of Northeast Texas" (A.A.P.G. *Bull.*, Vol. 9, No. 8, 1925). This study also showed for the first time the sands (Blossom) producing below the Austin Chalk and the Brownstown Marl in southwest Arkansas and northwest Louisiana were really Veatch's sub-Clarksville sands and upper Eagle Ford in age. The true Blossom sands were thus the equivalent of part of the Austin Chalk. Later, as the result of field trips with L. F. McCollum and John Teagle, the Pecan Gap Chalk of northeast Texas was studied and a paper, "Correlation of Pecan Gap Chalk in Texas" (A.A.P.G. *Bull.*, Vol. 18, No. 11, 1934) was published.

Additional field trips continued with Knebel, Wendlandt, and others to study the Claiborne (Eocene) of northeast Texas. The results of these studies appeared in A.A.P.G. *Bull.*, Vol. 23, No. 10 (1929), "Correlation of Claiborne of East Texas with the Claiborne of Louisiana." Her work here further established the Crockett, Weches, and Reklaw Formations, proposed and defined by Knebel and Wendlandt. She is credited with reviving the name Cockfield, for the upper Claiborne, which had been replaced by the term Yegua.

Later Miss Ellisor tackled the Jackson and her results appeared in 1933 in Vol. 17, No. 11, of the *Bulletin*, "Jackson Group of Formations in Texas with Notes on the Frio and Vicksburg," in which the Jackson was divided into three main divisions: the Whitsett, McElroy, and Caddell. Many Humble geologists assisted in the field work because this study extended from the Rio Grande to the Sabine River and eastward into Louisiana and Mississippi.

In 1936 she introduced the *Potamides matsoni* zone as a valuable guide and mappable unit, A.A.P.G. *Bull.*, Vol. 20, No. 4, "'*Potamides matsoni*' Zone of Texas (Burkeville Beds)"; and in 1940 she published in the *Bulletin*, Vol. 24, No. 3, "Subsurface Miocene of Southern Louisiana." In the latter paper a number of species, now widely used in correlation of the Miocene in coastal Texas and Louisiana, were described and named. In 1944 she was asked by the Houston Geological Society, as a tribute to her pioneer work, to write on the Anahuac Formation which had come to include the three original divisions of the middle Oligocene — *Discorbis*, *Heterostegina*, and *Marginulina* zones — and described by her in an early paper (A.A.P.G. *Bull.*, Vol. 9, No. 8, 1925) of which she was co-author with Applin and Kniker. In this later paper, she expressed her original belief that these zones were of Oligocene age.

Alva's final publication (1947), prior to retirement, was a small volume (99 pages) entitled "Rockhounds of Houston," an informal history of the Houston Geological Society. It is well documented and constitutes a thorough account of this large Society and its many accomplishments since its official founding in 1923.

Miss Ellisor's published contributions, in addition to those mentioned, are numerous and they include at least six papers co-authored with Professor J. A. Cushman, in which many new and diagnostic foraminiferal species were named and described, many of them after Humble Company geologists and others, who had cooperated with her in her stratigraphic field work.

Alva was twice elected vice-president of the Houston Geological Society, in 1924 and again in 1930. Sponsored by Sidney Powers, of which she was very proud, she had been a Fellow since 1929 of the Geological Society of America; a member of the American Association of Petroleum Geologists; the Society of Economic Paleontologists and Mineralogists (a vice-president in 1941); the Paleontological Society; the honorary societies of Sigma Xi and Chi Upsilon; and since April, 1948, an honorary member of the Houston Geological Society. In 1953 she received honorable mention by the Desk & Derrick Clubs of North America as the outstanding oil woman of the year. In February, 1962, she was presented by L. T. Barrow, at Houston, with the Distinguished Geology Alumni Award of the Department of Geology of the University of Texas, although alumni such as Deussen, Taff, and others were also considered in making the selection. In the presentation address to Miss Ellisor, Barrow, in part, said:

> In recent years it has seemed to be the custom of most schools to honor or make awards to alumni who occupy, or who have occupied, high positions in industry or government, or who have become wealthy — especially if they have made large money contributions to the schools. I am particularly pleased our Department of Geology has adopted a different criterion in the present case, and a graduate is being honored because of her contributions to geology. . . .
>
> During the first few years as the work being carried on in the Houston laboratories gradually became known, prejudice and opposition to the validity of this work appeared, not only by petroleum geologists, but by quite a few prominent geologists and paleontologists in the academic world. Some warned the three young ladies they were making a terrible mistake in relying on microfossils. Besides the necessity of proving their work to their own companies, it became desirable to convince the rest of the geologic profession. Another difficulty Miss Ellisor faced, and this may be true of the other two, was that she had to work many well samples as "unknown" — not being given the name and location of the wells. This was not just to test her during the proving stage, but some Humble geologists wanted to be certain the determinations were based entirely on what was found in the samples, and not by any information as to what the age should be. . . .
>
> Although the usefulness and reliability of microfossils for both short and long range correlations rapidly became recognized, the prejudice and opposition continued to some extent for many years, as too many of us geologists are reluctant to accept and use new methods which we do not thoroughly understand. I get the impression that as these early workers look back, they think of the prejudice and opposition as being amusing. However, unless they were far more confident than most young geologists are, I doubt much of it was amusing at the time.

Following her retirement Alva had many years left in which to travel; attend meetings of professional societies and other groups throughout the world; and enjoy life in general. Beginning in 1959, after her mother's death, she began a series of trips to various parts of Europe and elsewhere, including her mother's native Sweden, Hawaii, and Panama. She attended the World Geological Congress at Copenhagen in 1960, and she had planned another European trip in 1965. On these trips she met many people in her profession. She kept a diary from the early thirties which it is hoped may form the basis for a more extensive review of the life and accomplishments of this remarkable woman. She had six scrap-books with professional references to herself and friends.

Two years prior to her death she gave her valuable and remarkably complete geologic library, consisting of more than 3,000 volumes and many separates, to the Geology Library at the University of Texas. She was also financially generous to the various University funds, in fact much more so than many who were far better able to contribute. A past-president and always active in the Altrusa organization at Houston, she was instrumental in the formation of an Altrusa Chapter at Galveston, of which she had also been president.

When Miss Ellisor entered upon the scene Gulf Coast subsurface geology was almost a hopeless muddle. Drillers' logs were inaccurate and unsatisfactory. Her discovery of forams and their significance on the Gulf Coast, came just in time to bridge the gap of seven or eight years before the introduction of electric logging. Although forams have lost some of their value because of electric logging methods, they are still indispensable in many Tertiary areas of the world, particularly the Louisiana coastal region where lithologic changes in many areas are frequent and abrupt.

Hers was surely an inquiring spirit of research. When she had an idea she never gave it up until it had been settled one way or another to her satisfaction. She was supremely confident and persistent in her pursuit of the solution of a problem. Her industry, alertness, and tenacity kept her hammering at an idea until a geologist took her to see what she wanted to see. Her clear thinking and analytical mind at times seemed to out-run that of geologists who had spent much more time on the same field problems with which she worked. She possessed an unusual acumen in working with forams in that she seemed able to recognize, almost at a glance, the proper stratigraphic position of a fauna. Surely we geologists have much for which to be grateful to this brilliant woman.

L. P. TEAS

Houston, Texas
January 7, 1965

Reprinted from American Association of Petroleum Geologists Bulletin, v. 49, Part I, 1965, p. 467-471.

JULIA ANNA GARDNER
(1882-1960)

Julia Anna Gardner died on November 15, 1960, at her home in Bethesda, Maryland, at the age of 78. She left as her heritage contributions to the science of geology which have been and will continue to be of inestimable value to all geologists interested in the development of the mineral and human resources of the

Atlantic and Gulf Coastal Plains; a host of friends and admirers from all walks of life; and an example of perseverance and courage, tempered by kindliness, that is rarely equalled.

Julia Gardner was born in Chamberlain, South Dakota, on January 26, 1882, the only child of Dr. Charles Henry Gardner and Julia Brackett Gardner. Her father, a physician, died within four months after she was born, leaving the rearing of their child to her mother. During most of Julia's childhood, the mother and daughter remained in South Dakota, and Julia's early education was by private tutoring. In 1898 they moved to North Adams, Massachusetts, where Julia graduated from Drury Academy. She then enrolled at Bryn Mawr, graduating in 1905. She taught, for a short period, in the grammar school in Chamberlain, but, having become interested in geology, she soon returned to Bryn Mawr, where she received her Master's degree in 1907.

In the same year, Miss Gardner entered Johns Hopkins University as a graduate student and assistant in paleontology. She was awarded the Ph.D. degree in 1911, but remained at Johns Hopkins as assistant in paleontology and part-time geologist with the Maryland Geological Survey until 1915. Her work with the Maryland Survey dealt chiefly with the Molluscan faunas of the Upper Cretaceous of that state. It resulted in two comprehensive papers that were published by the Maryland Survey in two volumes which appeared in 1916. Despite the short time that was available, the work was so carefully and accurately done that these papers remain essential reference works for the many others who have since done detailed work on the Cretaceous of Maryland and other areas.

Julia was fascinated by the rich faunas of the Upper Tertiary sediments of the Coastal Plain of Virginia and North Carolina, which had been the subject of her doctor's dissertation, and was eager to continue her studies in this field. So when the opportunity came in 1915 to renew these studies, she moved to Washington and began work at the U.S. National Museum under contract with the U.S. Geological Survey. In the meantime, the first World War had begun and she was horrified by the thought of the terrible waste of life and the destruction which seemed to her so useless. She felt that in some way she must make a personal sacrifice to help alleviate the suffering of the millions of people who were caught up in this disaster which had engulfed all of Europe. In 1917, shortly after the death of her mother, she joined the Red Cross and was in France before the end of the year. At first she worked with the American Canteen Service and later with the ambulance service and in the hospitals as an auxiliary nurse. After the armistice, she worked with the American Friends Service in the devastated areas of France until 1919, when she was hospitalized in Rheims as the result of an accident incurred in the line of duty.

Julia Gardner loved people and with rare tact and skill was able to make them talk about themselves, their families, and their activities, but she rarely volunteered any information about herself or her achievements. This was particularly true of her war experience. Almost the only information on this part of her life is taken from her diary in which she wrote of her day-to-day activities. However, it is clear that she was no rear-area non-combatant, but was often in or near the front lines where the going was rough. This is also clear from the fact that her small unit was decorated by the French Government for its service with distinction and valor.

Miss Gardner, after her return to the States, joined the U.S. Geological Survey and resumed her studies of the Coastal Plain, which she continued until after her retirement with comparatively few interruptions. One of these was during the second World War when she joined the newly formed Military Geology Unit of the Geological Survey, which provided information essential to the military forces by describing terrain conditions and geologic factors involved in planning, establishing, and maintaining beachheads in many parts of the world. She brought to the task knowledge obtained in her travels overseas and the ability to read fluently in several languages. Although she was now nearing retirement, she worked as hard as any of this hard-working unit, putting in long hours, sometimes around the clock, and was always ready to lend a helping hand to her younger fellow workers with translations, ideas, and encouragement. Her maps were beautifully prepared, and her texts were superior. Although it was the custom among this group to use first names only, she inspired such respect and devotion that she was known to all as "Miss Julia."

Again in 1946-1947 Miss Julia recessed her studies of the Coastal Plain for a tour of duty with the Natural Resources Section of the Headquarters, Supreme Allied Command in Japan, where she worked with Japanese geologists and biologists. Although her headquarters were in Tokyo, she traveled widely, visiting universities and marine biological laboratories and making friends with Japanese scientists and their families. In this way she did much to restore normal relations between our two countries by winning respect and admiration as a scientist, and affection as a most charming lady.

After World War I, Miss Gardner made at least three trips to Europe. She was Delegate to the International Geological Congresses in Madrid in 1926 and in Moscow in 1937, taking part in several of the excursions sponsored by the Congresses. In 1929 she collected fossils from the type localities of the Danian in Denmark, and visited several other type localities and a number of the museums of Europe, including the British Museum.

Miss Gardner's principal work lay in the Coastal Plain extending from Maryland southward into Mexico and, except for her pioneering work in the Cretaceous in Maryland, all of it was concerned with the Tertiary beds. Her doctor's thesis, which was concerned with the Upper Tertiary of Virginia and North Carolina, was completed and published many years later as U.S. Geological Survey Professional Paper 199-B. In the meantime, she had extended her field work southward into Florida and the Alum Bluff group, which contained more than 800 species of mollusca. These were carefully described and profusely illustrated in several chapters of Professional Paper 142, which were published over a period of 25 years. All except one of these chapters have been reprinted. From Florida, Miss Gardner's work took her westward to the Mississippi Embayment, where profound changes in facies occur, and thence westward into Texas, where she made extensive studies of the Midway group. Although the fauna was much more limited, she covered it with her usual thoroughness. Her studies were aided by the loan of fossil collections and maps from practically all the oil companies in Texas. She worked out the stratigraphy and showed the correlations of the Midway with rocks of similar age, both in this country and abroad. Finally, Miss Gardner's work took her across the Rio Grande Embayment into Mexico, where, with the aid of hundreds of collections from oil geologists and a grant from the Geological Society of America, she was able to correlate the Tertiary faunas of northern Mexico with other Gulf Coast faunas, despite the major facies change caused by the Rio Grande Embayment. These fossils, in addition to those she collected, were fully described and interpreted in Memoir 11 of the Geological Society of America.

In addition to the five major contributions mentioned Miss Gardner published 40 or more papers and brief notes. Only nine of her papers were prepared with one or more co-authors. Space does not permit the listing of her impressive bibliography here. When she retired from the Survey, Miss Gardner was engaged in the study of collections of the Cenozoic fossils made by Survey teams in the islands of the Western Pacific, and in the preparation of a chapter of the Treatise on Invertebrate Paleontology. Unfortunately, her unexpected illness brought these two projects to an untimely end.

Julia Gardner was a member of the Paleontological Society and was its president in 1952. She was vice-president of the Geological Society of America in 1953, the third woman to be so honored. She was also a member or fellow of the following societies: American Association for the Advancement of Science, American Association of Petroleum Geologists, Washington Academy of Sciences, Geological Society of Washington, Biological Society of Washington, Phi Beta Kappa, and Sigma Xi. When she was retired she was awarded the highest honor of the Department of the Interior — the Distinguished Service Medal.

Any memorial to Julia Gardner, the scientist, would be incomplete without a word about Julia Gardner, the woman. She was no "ivory-tower" scientist, although her standards were high and exacting. She achieved a remarkable balance between her work and her social life. For many years she lived at the Washington Arts Club, of which she was a charter member, and often entertained her friends in the attractive rooms whose walls were hung with the latest works of Washington artists. Her guests included scientists, writers, painters, teachers, businessmen, and many others, but, despite the wide differences in the interests of her guests, her parties were always delightful. When in the Museum, Miss Julia took a coffee break in the middle of the afternoon. Guests and colleagues alike were most welcome. It was doubtless the strongest, blackest coffee in Washington. On one occasion, when she was busy with something else, I undertook to make the coffee according to my usual recipe. When she had time, she took one look at it, doubled the amount of coffee, and poured out half the water. It was big medicine, and it awakened and refreshed her guests, dispelled any drowsiness, and restored the staff to their morning vigor. For relaxation, Miss Julia read omnivorously and remembered what she read, seeming to have almost total recall of everything from philosophy to the latest fiction.

In the field, she usually traveled alone. If there were no hotel accommodations she stayed at the nearest farmhouse, and more than once when she was out in the boondocks at nightfall she slept in her car, which she drove with more daring than skill, although she always seemed to get through. She seemed to be almost without the sense of fear. During the ten years when I took many trips with her, the only time she showed fright (perhaps it was really surprise rather than fright) was when she climbed up the Bordas escarpment to find herself staring at eye-level at a coiled rattlesnake. Wherever she went she made friends with young and old, the prosperous, and the "not so prosperous." In South Texas, the Mexicans affectionately called her "Tia Julia" (Aunt Julia). Perhaps the outstanding characteristic of this remarkable woman was her deep sense of humanity. She was forever unobtrusively, almost stealthily, performing acts of kindness — some small, but many of them quite large. For example, when T. Wayland Vaughan became ill and finally blind, she spent several evenings each week reading to him aloud from his favorite books. Her kindness did not stem from lack of understanding of the frailties of her colleagues. Indeed, her views on professional and personal behavior and also on political matters were strong and firm, and she expressed them well and generally wittily. If the situation demanded, she could criticize concisely and succinctly, but never unkindly. Those of us who knew her will always be grateful for this privilege which has meant so much to us each in accordance with his needs.

In preparing this memorial I have leaned heavily on the memorial prepared for publication in the Bulletin of the Geological Society of America by Harry S. Ladd, who, in turn, received much helpful information from Mrs. Eleanor B. Knopf and Wendell P. Woodring.

A. NELSON SAYRE

Washington, D.C.
March 15, 1961

Reprinted from American Association of Petroleum Geologists Bulletin, v. 45, Part II, 1961, p. 1418-1421.

RALPH ALEXANDER LIDDLE
(1896-1963)

Ralph Alexander Liddle died of a failing heart on January 16, 1963 at his home in Fort Worth, Texas. Great is the sorrow of his bereaved wife and daughter, and of his two brothers and two sisters. His friends lost a loyal comrade and a colleague always ready to help.

My own acquaintance with Ralph dates from April 1922, when I worked with Apex Oilfields in Trinidad. A little Indian office boy announced a "Mester Ledal" who wished to see me, and soon I saw before me a slender, slightly younger man with a friendly smile in his eyes. He introduced himself as Ralph Liddle, Geologist of Standard Oil Company of New Jersey, and explained that during 2 years' field work he had studied the Cretaceous of Eastern Venezuela and now was keen to get to know its development in Trinidad. These were the days when many people suspected the Cretaceous bituminous shales of Guanoco and Naparima Hill to be the source rocks of all the oil in the region, including that of the productive Miocene sands. His visit to Trinidad lasted several months in which he was admired by his colleagues for the unfailing enthusiasm with which he tackled our mutual problems.

We met again during the International Geological Congress in 1933 when Ralph conducted a small group of petroleum geologists from 11 different countries through the Van Field in Texas. During the summer of 1960, Ralph visited Switzerland with his great friend Theron Wasson. From a high point in the Jura Mountain, we looked toward the Alps when Ralph spoke of his much beloved wife Pearle. He wished to bring her to Switzerland, together with their daughter Sandra, of whom he spoke with great fatherly affection and pride.

Ralph Liddle, of Scotch Presbyterian parentage, was born on a farm near the town of Argyle in eastern New York on November 18, 1896. He graduated from Argyle High School, spent 1914 at Monmouth College, Illinois, and then went with his brother Albert to Cornell University where he received a B.A. in February 1918. During his last year, he served as student assistant to his greatly admired teacher Professor G. D. Harris. He also was a member of Phi Beta Kappa and Sigma Xi. He worked as Associate Geologist with the Bureau of Economic Geology at the University of Texas, and in 1918 he served in the 472nd Regiment of U.S. Engineers, from which he was honorably discharged in January 1919.

In 1920 Ralph Liddle joined the Standard Oil Company of New Jersey to serve as Petroleum Geologist in Venezuela and Trinidad. During 6 years, he crossed Venezuela from one end to another, sometimes with a colleague, but often only with a few natives and some mules. Exploration work in those days meant rough living conditions, but Ralph's healthy constitution enabled him to cope with them.

When he was offered a transfer to the East Indies, Ralph decided to discontinue tropical exploration work. He accepted an offer from The Pure Oil Company as Chief Geologist of the Texas Producing Division at Fort Worth. He organized a department and as manager and administrator carried on investigations which led to important oil discoveries in East and West Texas. Among these was the Van Field, found in 1929, in Van Zandt County.

In 1934 Ralph married Miss Pearle Goolsby of Mexia, Texas. She helped him in his work, in the preparation of manuscripts, accompanied him to Ecuador, and kept a beautiful home open for their mutual friends. In 1936 Ralph left The Pure Oil Company to take up valuation work for the National City Bank in Chicago. He returned to Western Venezuela for a restudy of the Rio Cachirí section in Perijà. In 1940 he explored an oil concession near Cuenca of Eastern Ecuador, and in 1946 he entered into business for himself.

Theron Wasson, whose moving tribute to Ralph Liddle[1] is the source for much information used in this memorial, described his friend Ralph as a student of many interests during all his life. Ralph translated part of the Greek New Testament while in college and read Greek for pleasure. In his later years, he showed particular interest in the history of the first century A.D. and the events which preceded it. He collected literature on the Etruscans, for on his way to the Geological Congress in India 1964 he hoped to revisit Etruscan monuments in Italy. These few examples may illustrate the wide cultural outlook of Ralph Liddle, worthy of attention by all scientists whose horizons may be narrowed by ever increasing specialization.

In December 1921 Ralph Liddle was elected a Fellow of The Geological Society of America. He was also a member of the Société Géologique de France, the American Association of Petroleum Geologists, and the American Institute of Mining Engineers. He was one of the founders of the Paleontological Research Institution, Ithaca, New York; in 1933 he became a trustee and in 1946 was elected president. In 1961 he became a member of the Council of Cornell University. He was a member of the Board of Trustees of Monmouth College, Illinois.

[1]Wasson, Theron, Ralph A. Liddle 1896-1963: Memorial printed by the Paleontological Research Institution, Ithaca, N.Y., April, 1954.

Liddle's contributions to our science can be divided into those concerning the Gulf Coast region and those dealing with South America. His studies were primarily directed toward the search for oil, and accordingly the Cretaceous and Tertiary sediments received most of his attention.

His capacity to arrive at independent economic appraisals, based on the appreciation of regional as well as local problems, is already manifest in his first (1918) paper on the Marathon Fold. In those days, the subsurface structure and trend of the Marathon Fold were poorly known. Liddle stressed the importance of detailed studies of the exposed Comanchean sediments, for even their slight disturbances were considered reactivated features pointing to stronger folding in the underlying Carboniferous strata, particularly in Brewster County. Liddle, together with Prettyman, extended the survey into Crockett County mapping about 3000 square miles. The results were laid down in a comprehensive report dealing with the physiography, flora, stratigraphy, and structural geology, the emphasis being on the oil prospects. The differences between Triassic and Permian sediments were brought to light, and the stratigraphy of the Cretaceous along the Pecos River was closely investigated.

The excellent results of Liddle's investigation of Medina County were published in 1920 in Texas University Bulletin 1860. Soils and vegetation assisted Liddle in mapping about 20 formations ranging from Cretaceous to Recent, covering 2380 square miles in an area having a population of about 1700 people. Liddle introduced the terms Elstone and Squirrel Creek formations. Both were subsequently considered lithologic facies members of the Midway Formation, previously studied by R. T. Hill, T. W. Vaughan, L. W. Stephenson, and E. H. Sellards. Liddle's Buttsgin Formation of the Lower Wilcox Group is about 500 feet thick, and his Seco Formation on top of the Wilcox Group is about 100 feet thick. A test of the Elstone structure found commercial oil shortly after Liddle had mapped it.

After his return from Venezuela in 1926, Liddle again investigated the Balcones Fault Zone in East Texas. Detailed surface exploration suggested the town Van in Van Zandt County. This was verified by a seismic survey. In 1929 Liddle reported the discovery of commercial oil by The Pure Oil Company in the Woodbine Formation at a depth of 2710 feet which led to the development of the Van Field. The final report on the Van Field appeared in 1936. It represents an exemplary description of an oilfield developed by The Pure Oil Company with the participation of Shell Oil Company, The Texas Company, Sun Oil Company, Humble Oil and Refining Company, and Atlantic Refining Company. Twenty-seven plates, stratigraphic columns, and maps give a graphic picture of the structure believed to be underlain by an enormous salt mass which disturbed an area of 500 square miles. Silurian rocks were presumed to be the source of the salt, about 8000-9000 feet below the surface. The field with its ultimate production of a half billion barrels of oil of about 35 degrees API was operated largely as a unit. The history of the field's proration and its water disposal is presented in an admirable form.

Liddle's paper on the tectonics of the Maracaibo Basin, Venezuela, appeared in 1927. The *en échelon* arrangement of its structures, with their flanks inclined toward the axis of the flask-shaped basin, was explained by clockwise rotational movement, suggested by the northward plunge of structures west of the basin, and southward plunge east of the basin. The two major masses influencing the shape of the basin were thought to be the ancient Gondwana Land and Schuchert's Antillea. The Caribbean Range was thought to include Paleozoic sediments and to form a great anticlinorium with a flank steeply inclined to the north, toward the source of orogenic movements.

Liddle's major contribution toward our science was his "Geology of Venezuela and Trinidad" published in 1927. It was reviewed by Williston and Nichols[2], who added some pertinent information on the geology of Falcon. Liddle's book is a store of information which will maintain its value for as long as geologists need to unravel the geology of Venezuela. He produced the first useful geological map of Venezuela, accompanied by a correlation chart. The 169 half-tones are all originals, partly of historical value. The impact of Liddle's work is also evident in the "Stratigraphical Lexicon of Venezuela" (Caracas, 1956). Not only did he establish more than 20 formational terms, but he also defined many other groups, formations, and members locally used by petroleum geologists of various companies operating in Venezuela and Trinidad. His systematic treatment of each formation under headings such as name, stratigraphic position, physical character, extent, thickness, age, correlation, and topographic expression allow for a quick orientation and comparison.

In 1941 Liddle gave a detailed account of the geology of the Cuenca Basin in the headwaters of the Rio Paute, the main gateway to "El Oriente." The introduction illustrates his ability to give in a short description the essential details concerning orography, vegetation, and inhabitants. The map includes partly unpublished information of Theron Wasson, Axel Olsson, R. F. Dickerson, and A. L. Bowden.

The Rio Cachirí of the Sierra de Perijà, between Venezuela and Colombia, is well known for its fossiliferous Devonian section, about 8000 feet thick. The interval between this Devonian and the core of the Sierra de Perijà was not known. In 1942 Liddle undertook the arduous ascent of the river amid wild, rugged, forested peaks. The results of the expedition were published in 1943. G. D. Harris and J. W. Wells contributed the part on paleontology. Liddle divided his Devonian Cachirí Series into three formations. He established the pre-Devonian, probably Archaeozoic, Sierra de Perijà Series, and supplied additional information on the Permian-Carboniferous Palmarito Series on Cretaceous, Paleocene, and Eocene formations.

In 1946 Liddle sponsored the publication of the revised and considerably enlarged edition of "The Geology of Venezuela and Trinidad" which was reviewed by E. Mencher[3], who concluded his review by stating: "Taken as a whole, the new edition has value in bringing together a great mass of factual data on a poorly known part of the world. Its detailed information must be critically examined, however, before being accepted."

HANS G. KUGLER

Natural History Museum
Basel, Switzerland

[2]Am. Assoc. Petroleum Geologists, Bull., v. 12, p. 445-451.
[3]Am. Assoc. Petroleum Geologists, Bull., v. 32, p. 382-383.

BIBLIOGRAPHY OF RALPH A. LIDDLE

1920 The Marathon Fold and its Influence on Petroleum Accumulation: Univ. Texas Bull. no. 1847, p. 9-16

—— (with T. M. Prettyman), Geology and Mineral Resources of Crockett County with Notes on the Stratigraphy, Structure and Oil Prospects of the Central Pecos Valley: Univ. Texas Bull. no. 1857, p. 7-86

1921 The Geology and Mineral Resources of Medina County: Univ. Texas Bull. no. 1860, p. 7-157

1927 Tectonics of the Maracaibo Basin, Venezuela: Am. Assoc. Petroleum Geologists, Bull., v. 11, no. 2, p. 177-186

—— The Geology of Venezuela and Trinidad: J. P. Mac Gowan, Fort Worth, Texas, 552 p.

1928 Petroleum Development in East Texas and along the Balcones Fault Zone: Am. Inst. Min. and Met. Eng., Petroleum Dev. and Technology, p. 628-639

—— Venezuela the most rapidly growing Source of Supply: Oil Field Engineering, June 1, p. 14-22

1929 Van Field, Van Zandt County, Texas: Am. Assoc. Petroleum Geologists, Bull., v. 13, no. 12, p. 1557-1558

1930 Magnetometer Survey of Little Fry Pan Area, Uvalde and Kinney Counties, Texas: Am. Assoc. Petroleum Geologists, Bull., v. 14, no. 4, p. 509-516

1931 Discussion on "Limestone as Source Rock in Venezuela" by Hollis D. Hedberg: Am. Petroleum Geologists, Bull., v. 15, no. 3, p. 244-246

1932 Outlines of Venezuelan Stratigraphy: Pan-Am. Geologist, v. 57, p. 161-185

1936 The Van Oil Field, Van Zandt County, Texas: Univ. Texas Bull. no. 3601, p. 9-79

1941 The Geology of the Cuenca-Azógues-Biblian Region, Provinces of Canar and Azuay, Ecuador: Bull. Am. Paleontology, v. 26, no. 100, p. 5-30

1943 The Rio Cachirí Section in the Sierra de Perijà, Venezuela: Bull. Am. Paleontology, v. 27, no. 108, p. 273-321

1946 The Geology of Venezuela and Trinidad: 2d. ed., Paleontological Research Institution, Ithaca, New York, 890 p.

Reprinted from Geological Society of America, Proceedings for 1966, September 1968, p. 285-289.

JOHN TIPTON LONSDALE
(1895-1960)

John Tipton Lonsdale died suddenly of a heart attack at his home in Austin, Texas, October 5, 1960, at the age of sixty-four. He had been Director of the Bureau of Economic Geology and Professor of Geology at The University of Texas for fifteen years and was active on the campus up to two days before his death. He is survived by his wife, Edna Gertrude Lonsdale, Austin, Texas; a sister, Miss Bernice Lonsdale, Holly Hill, Florida; and a brother, Richard R. Lonsdale, Daytona Beach, Florida.

Born at Dale, Iowa, November 8, 1895, John Lonsdale was the oldest son of Eva Mary (Connor) and John Dye Lonsdale. He completed his elementary education at Dale, Iowa, and his high-school education at Guthrie Center High School. He was awarded the Bachelor of Arts degree in 1917, the Master of Science degree in 1921, both from the University of Iowa, and the Doctor of Philosophy degree in 1924 from the University of Virginia.

His distinguished professional career began in 1921 as Assistant Professor of Geology at the University of Virginia and geologist for the Virginia Geological Survey. He moved to Norman, Oklahoma, in 1924 to become Assistant Professor of Geology at the University of Oklahoma and geologist for the Oklahoma Geological Survey, and in 1925 he came to The University of Texas as geologist in the Bureau of Economic Geology. In 1928 he became Professor of Geology and Head of the Department of Geology, Agricultural and Mechanical College of Texas at College Station, Texas. He was the leader in developing a strong geology curriculum at College Station and was chiefly responsible for the planning and design of a new geology building on that campus. While at College Station he was geologist for the Missouri Pacific Railroad. From Texas he went to the Iowa State College at Ames where, except for military leave, he stayed from 1935 to 1945 as Professor of Geology and Head of the Department of Geology. He returned to Austin in 1945 to re-join The University of Texas as Director of the Bureau of Economic Geology and Professor of Geology with graduate rank.

Following service in the United States Army as Lieutenant from 1917 to 1919, wherein he was cited for bravery, Dr. Lonsdale developed a strong interest in military affairs and was active as a reserve officer between World Wars I and II. He served in the Army again from 1942 to 1945 as Lieutenant Colonel and Colonel and continued his interest in reserve activities even while holding the rank of a retired officer. In recognition of his services to his country, his ashes were interred at Arlington National Cemetery.

In addition to his professional and military careers, he was active in the Austin Rotary Club. He was also an outdoors man and a gun collector and participated with friends of similar inclination in numerous deer- and antelope-hunting expeditions. Among his closest friends of long-standing acquaintance was a group of well known University colleagues and Austin professional men who together have held a deer lease for more than fifteen years. Dr. Lonsdale was well known as a delightful companion in camp, a stimulating but modest conversationalist, and, as some have put it, "in every way a swell guy." Those who knew him well recognized him as a genuinely humble man.

Dr. Lonsdale was a member of a long list of scientific and honorary societies including Sigma Xi, American Association of Petroleum Geologists, American Institute of Mining, Metallurgical, and Petroleum Engineers, American Geophysical Union, Society of Economic Paleontologists. He was a Fellow of The Geological Society of America, the Mineralogical Society of America, and the American Association for the Advancement of Science. He served his professional societies faithfully by work on various committees such as Committee for Stratigraphic Nomenclature of the American State Geologists, the Research, Business, and Well Sample committees of the American Association of Petroleum Geologists, the Committee for Nomination of Fellows of the Mineralogical Society of America, and the Policy and Administrative Committee of The Geological Society of America. He also served on various official committees including the Texas Topographic Mapping Advisory Committee, Mapping Committee of the Texas Water Conservation Association, Geology and Minerals Committee of the Arkansas and Red River Inter-agency Committee, Texas Advisory Committee on Conservation Education, and the Natural Resources Committee of the Houston Chamber of Commerce.

His contributions to geologic research ranged from detailed analyses of complex rock and mineral systems to the economic geology of many different industrial mineral products. Among his specialties were investigations of ground-water resources and the general geology of the Big Bend National Park. His geological investigations of that Park were nearly complete at the time of his death and will be published as part of the comprehensive report on the geology of the Big Bend Park. His work also had a strong practical slant toward economic geology. During his years as Director of the Bureau of Economic Geology, he began a program of investigation of the State's mineral resources, concentrating on minerals other than oil and gas. The full importance of this work to the development of Texas and the diversification of its industrial base has yet to be realized, but in the future it will be ranked as one of his greatest contributions.

Dr. Lonsdale had a way of developing a strong bond with his students that endured long after they left the classroom. Many of these students have risen to high places in the scientific and business world. Because of their deep affection and esteem for him, they continued to visit and correspond with Dr. Lonsdale through the years and up to the time of his death. Many came hundreds of miles to pay their last respects and many have written sincere tributes to his influence through the years.

Among the important tasks assigned to him by University officials shortly after returning to The University of Texas in 1945 was to aid in the recruiting and building of an outstanding Department of Geology. He served on the Department's Budget Council during those difficult years, and his counsel and guidance had a marked effect on the Department's growth.

Not only was Dr. Lonsdale a distinguished man in the geological profession, but also he was admired and trusted by all who worked with him. To his associates, the quality which set him apart was complete and unswerving integrity. He was devoted to his wife; they were married more than 40 years. The loss of this man can not be measured. His memory will remain as an example of how a man should live his life.

SAMUEL P. ELLISON, *Chairman*
H. H. POWER
JAMES R. D. EDDY
P. T. FLAWN

University of Texas
Faculty Committee
Austin, December 1960

PUBLICATIONS OF JOHN TIPTON LONSDALE

Some effects of heat on the properties of minerals: Am. Mineralogist, v. 8, no. 8, p. 242-247, August 1923.

Post-Cincinnatian granites of northeastern Piedmont Virginia: Jour. Geology, v. 34, no. 2, p. 159-166, 3 figs. (maps), February-March 1926.

A Piedmont Virginia magmatic complex: Am. Jour. Sci., 5th ser., v. 11, p. 505-513, 2 figs., June 1926.

Niter and soda niter from Brewster County, Texas: Am. Mineralogist, v. 22, no. 7, p. 189-190, July 1926.

The occurrence of caliche in Oklahoma (abstract): Iowa Acad. Sci., Proc., 1925, v. 32, p. 357, 1926.

The occurrence of caliche in Oklahoma: Oklahoma Acad. Sci., Proc., v. 5, 1925 (Oklahoma Univ. Bull., new ser., no. 330, Univ. Studies no. 22), p. 132-136, April 1, 1926.

Geology of the gold-pyrite belt of the northeastern Piedmont, Virginia: Virginia Geol. Survey, Bull. 30, 110 p., 11 figs., 9 pls., 1927.
(with Adkins, W. S.). Euhedral orthoclase crystals from Sierra Blanca, Texas: Am. Mineralogist, v. 12, no. 6, p. 256-259, 1 fig., June 1927.
The Florence meteorite of Williamson County, Texas: Am. Mineralogist, v. 12, no. 11, p. 398-404, 2 pls., November 1927.
Igneous rocks of the Balcones fault region of Texas: Texas Univ. Bull. no. 2744, 178 p., 20 figs., 9 pls. (incl. map), November 22, 1927.
(with Crawford, David J.). Pseudo-igneous rock and baked shale from the burning of lignite, Freestone County, Texas: Univ. Texas Bull. no. 2801, p. 147-158, 1 fig., 3 pls., 1928.
Analcite from Brewster County, Texas: Am. Mineralogist, v. 13, no. 8, p. 449-450, August 1928.
Diphrite and associated contact minerals from the Franklin Mountains of Texas: Am. Mineralogist, v. 14, no. 1, p. 26-32, January 1929.
An underground placer cinnabar deposit (Brewster County, Texas): Econ. Geology, v. 24, no. 6, p. 626-631, 1 fig., September-October 1929.
Euhedral magnesite crystals from Winkler County, Texas: Am. Mineralogist, v. 15, no. 6, p. 238-239, June 1930.
(with Metz, M. A., and Halbouty, Michel Thomas). The petrographic characters of some Eocene sands from southwest Texas: Jour. Sed. Petrology, v. 1, no. 2, p. 73-81, 2 figs., November 1931.
Underground water resources of Atascosa and Frio counties, Tex.: U. S. Dept. Interior Press Memo. 66110, 9 p., 2 maps, October 13, 1932.
(with Day, James R.). Ground-water resources of Atascosa and Frio Counties, Tex.: U. S. Geol. Survey Water-Supply Paper 676, 90 p., 8 pls., incl. geol. map, 4 figs. incl. index map, 1935.
Investigation of underground water resources (abstract): Pan-Am. Geologist, v. 65, no. 4, May 1936.
Investigation of underground water resources of Texas (abstract): Iowa Acad. Sci. Proc. 1936, p. 249 (1937?).
(with Day, James R.). Geology and ground-water resources of Webb County, Tex.: U. S. Geol. Survey Water-Supply Paper 778, v. 104, p. 13 pls. incl. geol. map, 6 figs. incl. index map, 1937.
The Plantersville meteorite, Grimes County, Tex.: Am. Mineralogist, v. 22, no. 8, p. 877-888, 13 figs., August 1937; abstracts, no. 3, p. 213, March 1937; Geol. Soc. America Proc. 1936, p. 87, June 1937; Pan-Am. Geologist, v. 68, no. 2, p. 157, September 1937; Iowa Acad. Sci. Proc., v. 44, p. 135-136, 1937.
Igneous rocks of the Terlingua-Solitario region, Texas: Geol. Soc. America Bull., v. 51, no. 10, p. 1539-1636, 7 pls. incl. geol. maps, 8 figs. incl. index map, October 1, 1940.
Origin of analcite in igneous rocks of the Terlingua district, Texas (abstract): Iowa Acad. Sci. Proc. 1939, v. 46, p. 252, June 1940.
The Peña Blanca Spring meteorite, Brewster County, Texas: Am. Mineralogist, v. 32, nos. 5-6, p. 354-364, 6 figs., May-June 1947; abstracts, nos. 3-4, p. 203, Mar.-Apr. 1947; Geol. Soc. America Bull., v. 57, no. 12, pt. 2, p. 1214, Dec. 1946.
(with Dickson, Kathryn O.). Analcime suite of igneous rocks, Brewster County, Texas (abstract): Geol. Soc. America Bull., v. 59, no. 23, pt. 2, p. 1337, December 1948.
(with Dickson, Kathryn O.). Analcime suite of igneous rocks, Brewster County, Texas (abstract): Am. Mineralogist, v. 34, nos. 3-4, p. 380, Mar.-Apr. 1949.
(with Maxwell, Ross Allan). Petrology of Big Bend National Park, Texas (abstract): Geol. Soc. America Bull., v. 60, no. 23, pt. 2, p. 1906, December 1949.
(with Maxwell, Ross Allan). Petrology of Big Bend National Park (Texas) (abstract): Am. Mineralogist, v. 35, nos. 3-4, p. 286, Mar.-Apr. 1950.

(with Barbeck, A. H.). Mineral and geological resources of Red and Arkansas basins in Texas: Minutes Fifth Meeting AWRB Inter-Agency Committee, Appendix H, p. 1-3, 1951.
Texas minerals: Manufacturers Record, p. 109-110, November 1951.
Preservation of well samples and cores: Shale Shaker, March 1952.
Brown iron ores of Texas: Preprint, American Inst. Mining Engs., Sept. 1952.
(with Dickson, Kathryn O.). Native lead, Presidio County, Texas (abstract): Geol. Soc. America Bull., v. 63, no. 12, pt. 2, p. 1276, Dec. 1952.

Reprinted from American Association of Petroleum Geologists Bulletin, v. 45, no. 4, April 1961, p. 561-564.

LEROY THOMPSON PATTON
(1880-1957)

On June 22, 1957, the Geological profession lost a great leader and scholar. Dr. Leroy T. Patton died of a heart attack while returning home from his farm near Abernathy, Texas. Until two years before, he had enjoyed good health; at that time a slight heart attack had acted as a warning, causing him to relax from his active career and settle down to the comparatively quiet life of a farmer.

Leroy Patton was born December 5, 1880, in Fairpoint, Ohio. He was the son of John and Lauretta (Thompson) Patton. The family moved to Arkansas City, Kansas, while he was a small boy. His early boyhood was spent on a Kansas farm, but his family returned to Ohio to the town of St. Clairsville when he was sixteen years of age. This remained his home for some time.

During the years that followed while obtaining a high-school and college education, Leroy Patton worked at many jobs, including employment in steel mills and teaching in country schools. He continued to pursue his education with diligence, receiving a B.A. degree from Muskingum College, B.S. degree from the University of Chicago, M.S. degree in geology from the State University of Iowa, followed by his Ph.D. in geology from the same school. After completing his college education, he became the high-school principal and later the superintendent in a public school in Ohio, to be followed by becoming a member of the faculty of Geneva College in Beaver Falls, Pennsylvania. It was during this time, 1915-1916, when he took a leave of absence to finish his doctorate degree at the University of Iowa. From 1918 to 1922, he was a member of the faculty of Muskingum College, New Concord, Ohio, after which he became an associate geologist with the Bureau of Economic Geology, University of Texas at Austin. Following this, he became technical advisor to the Texas State Board of Water Engineers.

In 1925 the State Board of Education was in search of a capable geologist to head the geological department of the newly organized Texas Technological College, Lubbock, Texas, and could have made no wiser choice than that which brought Dr. Patton to head the department. There he remained for the next twenty-three years. In these fruitful years the department grew from one professor and 91 students in 1926, to 12 professors and 842 students at the close of 1949. It was during these years of the mushrooming growth of young Texas Tech that the staff was very much overworked taking care of the dire necessity of teaching the ever increasing number of students interested in geology. Dr. Patton, with his knack for selecting capable teachers, met this challenge, and with the help of M. A. Stainbrook, W. I. Robinson, Raymond Sidwell, and a few student assistants, built the foundation for a department which has since been recognized as one of the outstanding geological departments in the south. Petroleum engineering was added to his manifold duties in 1949 when 199 students enrolled in that school.

His devotion to the supervision of his department, to teaching, and to scholarly research was surpassed only by his respect and love for his students. During the entire period from 1925 until his death, he maintained a card file on all the graduates in his department and corresponded with them most generously.

During the early days of college life on the Texas Tech campus when many duties crowded every minute of his day, Dr. Patton spent tireless evenings coaching fencing and boxing. His agile body and quick eye were the envy of his students, not only in the fencing class, but also on the mountain trails where he enjoyed their fellowship while leading the summer field trips. His close association with these men around the camp fires and on the trail made him to be remembered as a man's man and an everlasting friend.

As the head of the department of geology, Dr. Patton spent freely of his time, not only with other members of his staff but with the student body as well. His office was always open to any student desiring counsel on matters either personal or scientific. As a teacher he was dignified, scholarly, and efficient. His sincerity in purpose inspired his students, and I am sure many remember his oft quoted phrase, "Observe every phenomena as if you were the first to see it and describe it so that if you are the last, the scientific facts will be passed on to others."

It was after coming to Texas Tech in 1925, that Dr. Patton met Miss Bertha Eubanks, teacher of home economics at the college. Following a courtship of two years, they were married in August, 1927. Those who remember Dr. Patton during the following years can remember no one who was more proud of his wife and two sons, Joseph Thompson, born August 8, 1928, and Bruce Bransford, born February 15, 1930. Both sons are graduates of Texas Tech in engineering. Joseph is a senior engineer with the Texas State Highway Department, and Bruce is an industrial engineer with Chance-Vought Aircraft Company, Dallas, Texas. Mrs. Patton is a graduate of Texas State College for Women where she majored in home economics. She holds a Master's degree from Texas Tech in English and education. She is a member of the Daughters of American Revolution and Daughters of Texas Republic. The men who made up the first field parties for the Tech summer field course remember her with a great deal of gratitude. It was Mrs. Patton's completely planned menus that were followed for every meal during these first field camps. It was the rotating chore of the men attending the course to serve as cooks for three days in succession. Mrs. Patton's careful planning of the menu along with complete recipes for the preparation of each meal prevented mutiny, and preserved the health of the camp. It is not often that a college junior can make a rhubarb pie on an open fire that is edible, much less palatable.

The summer field course was a class that was thoroughly enjoyed by its leader. For many successful years after its origin, Dr. Patton led the junior and senior students in numbers from 12 to 40 over the mountains of Colorado and New Mexico. The course was well planned and executed with careful supervision, bringing out all phases of field geology.

Dr. Patton continued his teaching and administrative duties as head of the department from 1925 to 1948, after which he acted as professor of geology and director of the Seismological Observatory at Texas Tech until 1951, when he became professor emeritus.

After retirement, he diverted his boundless energy toward managing his 640-acre farm near Abernathy, Texas. It was during this period that he found time to finish his text book on geology, which was commenced as a project of research and careful tabulation while executing his other duties at the college. This text was being presented to publishers at the time of his untimely death. Another unpublished book was written on the "History of Texas Tech." This was written as a hobby and brought Dr. Patton much personal pleasure. Another of his works, "In Defense of Alumni" was purchased and published by a popular publishing company.

Since childhood, Dr. Patton was a deeply religious man. He was an active member of the Presbyterian Church. Those who knew him best can never remember hearing him speak an unkind word about his fellowman.

His name was rightly included in Who's Who in America, American Men of Science, Leaders in Education, Who's Who in Texas, Who's Who in Higher Education, and Internationale Kalender für Geologie and Mineralogie.

He had served as a member of the executive committee of the Ohio Athletic Conference, technical advisor to the Texas State Board of Water Engineers on geological questions concerning building of dams for flood control and conservation, and a member of the committee on water resources of the Texas State Planning Board.

Dr. Patton was a fellow of the Geological Society of America; fellow of the American Association for Advancement of Science; member of the American Association of Petroleum Geologists, the Society of Sigma Xi, the American Association of University Professors, and the Lubbock Geological Society.

G. C. CLARK

Tyler, Texas
November 25, 1957

BIBLIOGRAPHY OF LEROY T. PATTON

1. "In Support of Gardiner's Theory of the Origin of Certain Concretions," *Jour. Geol.*, Vol. 30, No. 8, pp. 700-01, 1 fig., Nov.-Dec., 1922.

2. "The Geology of Potter County, Texas," *Texas Univ. Bull. 2330*, 184 pp., 4 figs., 9 pls. (incl. map), Nov., 1923.

3. "Geology and the Location of Dams in West Texas," *Econ. Geol.*, Vol. 29, No. 8, pp. 756-61, Dec., 1924.

4. (and Hicks, Clifford), "Notes on the Occurrence of Glacial Material Beyond the Border of the Drift in Muskingum County, Ohio," *Ohio Jour. Sci.*, Vol. 25, No. 2, pp. 97-98, March, 1925.

5. "The Sandstone Dikes Around Rockwall, Texas," *Holland's Magazine*, Dallas, Texas, Vol. 44, No. 6, p. 586, 4 figs., June, 1925.

6. "Geology and the Location of Dams on the Canadian River, Texas," *Econ. Geol.*, Vol. 20, No. 5, pp. 464-69, August, 1925.

7. "The Stratigraphy of Western Oklahoma and Adjacent Parts of Texas," *Amer. Jour. Sci.*, 5th Ser., Vol. 12, pp. 193-97, 1 fig., Sept., 1926.

8. "The Geology of Stonewall County, Texas," *Texas Univ. Bull. 3027*, 77 p., 4 figs., 1 pl., map, Dec. 1930.

9. "Paleogeographic Wall Maps," *Science*, N.S., Vol. 76, pp. 474-75, Nov. 18, 1932.
10. "Ripple Marks of the Merkle Dolomite of Western Texas and Their Paleogeographic Interpretation," *Jour. Sed. Petrology*, Vol. 3, No. 2, pp. 77-82, 2 figs., Aug. 1933.
11. "Geological Field Courses in American Colleges," *Pan-Amer. Geol.*, Vol. 60, No. 1, pp. 25-33, Aug., 1933; abstract, No. 4, pp. 307-08, Nov., 1933.
12. "Crystalline Rocks of the Las Vegas, New Mexico Quadrangle," paper presented before the Southwestern Division of the American Association for Advancement of Science, April, 1934.
13. "A Holder for a Diamond Saw," *Econ. Geol.*, Vol. 29, No. 7, Nov., 1934.
14. "Earth Science and Military Education," *Military Engineer*, Vol. 27, No. 152, pp. 87-89, Mar.-April, 1935.
15. "Some observations on the So-Called Lakes of the Llano Estacado of Texas," paper read before the joint meeting of the Amer. Assoc. for Adv. Sci. and Geol. Soc. America, 1934; abstrat, Proc. Geol. Soc. America 1934, p. 451, June, 1935.

Reprinted from American Association of Petroleum Geologists Bulletin, v. 42, Part 1, 1958, p. 690-693.

WILLIAM BATTLE PHILLIPS
(1857-1918)

William Battle Phillips came from a family rich in traditions of scholarship. According to The University of Texas Record, Vol. 11, No. 4, Dec., 1900, his father, the Rev. Charles Phillips, occupied for ten years each the chairs of Civil Engineering and of Mathematics in the University of North Carolina; while his grandfather, the Rev. James Phillips, was for over thirty years Professor of Mathematics in the same university.

Dr. Philliips was born at Chapel Hill, North Carolina, July 4, 1857. He was graduated from the University of North Carolina in 1877 and received his doctorate there in 1883. He studied also at the School of Mines at Freiburg, in Saxony, then the most famous mining school in the world. Positions held by him were as follows: Chemist, North Carolina Experiment Station, 1877-82, and of the Navasso Guano Company, 1882-85; professor of agricultural chemistry and mineralogy, University of North Carolina, 1886-88; mining engineer, Birmingham, Alabama, 1888-92; professor of chemistry and metallurgy, University of Alabama, 1891-93; chemist, Tennessee Coal, Iron, and Railway Company, 1894-98; staff of Engineering and Mining Journal, American Manufacturer and Iron World, 1893 and 1897-98; instructor in economic geology, University of Texas, 1900-05; director, University of Texas Mineral Survey, 1901-05; mining engineer, 1905-08; director, Bureau of Economic Geology and Technology, University of Texas, 1909-15; president, Colorado School of Mines, 1915-16. He died at Houston, Texas, June 8, 1918.

Dr. Phillips was the author of about 300 articles and books. He was a proficient mining engineer, chemist, geologist, mineralogist, metallurgist, and technologist. Some of his more notable accomplishments were the development of the Chisos quicksilver mine, his work on iron making in Alabama, on the utilization of lignite, coal, petroleum, and natural gas in Texas, and directing the Mineral Survey and the Bureau of Economic Geology and Technology of The University of Texas.

Dr. Phillips was memorable for sincerity and integrity, fearlessness and independence in mind and character, a widely-interested, versatile, well-stocked, and broadly-liberal mentality, untiring industry, thoroughness and earnestness of purpose. The very efficient utilization of these qualities enabled him to accomplish numerous important works in a relatively short and crowded lifetime occupied in doing well very many exacting and laborious tasks. Rudyard Kipling's poem "If," one of the finest tributes ever made to the great Washington, brings back to his friends William Battle Phillips; so does likewise Hamlet's appreciation of his friend Horatio:

> "for thou has been
> As one, in suffering all, that suffers nothing;
> A man that fortune's buffets and rewards
> Hast ta'en with equal thanks; and blest are those
> Whose blood and judgement are so well commingled
> That they are not a pipe for fortune's finger
> To sound what stop she please."

Reprinted from University of Texas Bulletin 3501, *Quarter-Centennial Memorial Volume of the Division of Natural Resources*, 1935, p. 10.

FREDERICK BYRON PLUMMER
(1886-1947)

Frederick Byron Plummer, born August 31, 1886, Hanover, New Hampshire, died suddenly of a heart attack, near San Marcos, Texas, February 17, 1947. Of Scottish or English origin, the American Plummers are believed to be descendants of Francis Plummer, who came to Newbury, Massachusetts, about 1633. Frederick Byron was the son of Alvah A. and Laura E. Merrill Plummer and was of the tenth or eleventh generation of his family in New England. For more than a century the Plummers have lived in New Hampshire and for most of it in or near Hanover.

Plummer graduated, B.S., Dartmouth College, 1909. He was an assistant in the department of geology, University of Chicago, 1909-11; instructor in geology, Vassar College, 1911-13; and a fellow in geology, University of Chicago, 1913-15. In his graduate work at Chicago he completed practically all requirements except the thesis for his doctor's degree.

His earliest geological work was done in the White Mountains. During the summer of 1914 he was a field assistant in the iron-ore districts of north-central Wisconsin for the Wisconsin Geological and Natural History Survey.

While still at Chicago, Plummer was engaged by Richard A. Conkling, chief geologist for the Roxana Petroleum Company of Oklahoma, and beginning about the middle of 1915 he worked as a field geologist in that state. In early May, 1916, with Clarence E. Hyde, he was transferred to the Houston office of the Roxana and spent some months in investigations on the Gulf Coast and in East Texas. In the early winter of 1916 he moved to North Texas with headquarters at Mineral Wells and took charge of mapping the northern part of the Pennsylvanian outcrops. He gave a preliminary report on this work at the Dallas meeting of the Association in March, 1919. At that time Wallace Pratt said of the work, "most of us in North Texas believe the Roxana Petroleum Company is doing the most thorough, the most extensive geological work of any of us." Plummer was congratulated upon this paper by Charles Schuchert and David White. The final report, "Stratigraphy of the Pennsylvanian Formations of North-Central Texas," was published in 1921 under the joint authorship of Plummer and Raymond C. Moore. Plummer was in charge of geological work for Roxana in Texas and Louisiana from 1917 to 1920.

From late 1920 until late 1922, Plummer was a member of the headquarters staff of geologists for the Royal Dutch-Shell group at The Hague. Upon his return to the United States he continued with the Roxana until late October, 1923, when he joined the Rycade Oil Corporation, chief geologist, Donald C. Barton, as field geologist. He continued in this position, except for leaves during the spring quarter of each year, until the end of 1925. His first assignment was a reconnaissance examination of the petroleum prospects of Maverick County. From early December, 1923, until early November, following, he mapped in detail the fault-line area extending northward from Mexia and then spent several months in the Red River area. The writer had the pleasure of a field trip with Plummer upon the completion of this work, in order to be shown the criteria on which his zoning of the Midway was based. From July, 1925, until mid-September, Plummer studied West Texas, including the trans-Pecos region. The rest of the year was spent completing gaps in his fault-line work.

Plummer spent the entire year of 1926 as chief geologist in Texas and New Mexico to the Amerada Petroleum Corporation, reporting directly to John Lovejoy. In 1927 and 1928 he was consulting geologist to the Vacuum Oil Company, engaged in interpreting in geological terms, on a regional basis, and with valuable results, torsion-balance and magnetic surveys in the Permian basin.

He joined the Bureau of Economic Geology of the University of Texas in November, 1928, and continued to work with that Bureau until his death. As a geologist, his outstanding achievements were his masterful summary, including much original work, of the Cenozoic system of Texas which makes up a third of Volume I of *The Geology of Texas*. His study of the Pennsylvanian of Texas has been noticed. Of like importance are his studies of the Carboniferous formations of the Llano uplift, much of which is yet to be published.

Plummer was largely responsible for the establishment of a department of petroleum engineering at the University of Texas. Beginning with the 1928-29 session, the University offered a course leading to the Bachelor of Science degree in petroleum production engineering. The course was catalogued to be given by Professor E. H. Sellards but it is Dr. Sellards' recollection that he taught the course only for about a month before Plummer arrived. In 1930 a department was organized with Plummer as chairman and full professor. He offered two courses, one in petroleum production methods and another in oil-field management. Plummer continued this chairmanship until he was succeeded by Harry Power in the autumn of 1936 and he continued his professorship in the school, including membership in the graduate faculty, until his death.

As a petroleum engineer, Plummer was interested chiefly in research in the areas where the field is overlapped by the sciences of chemistry and geology. His chief accomplishments were his studies of the waters of the Woodbine sand and as a teacher. He published a series of fifty-odd articles on the fundamentals of petroleum engineering in the *Oil and Gas Journal* during 1944-45.

Plummer became a member of the American Institute of Mining and Metallurgical Engineers in 1917, of this Association in 1919, of the Institute of

Petroleum (London) in 1922, and a fellow of the Geological Society of America in 1925. He was a founder of the Society of Economic Paleontologists and Mineralogists and it was chiefly through his efforts that the five-thousand-dollar fund was raised in 1926 which made the publication of its journal possible. He was secretary of the Society in 1927-29 and secretary-treasurer in 1930.

He was a member and one-time president of Sigma Xi at the University of Texas, president of the Texas Academy of Science, 1945-46, a fellow of the American Geographical Society, some-time secretary and national president of Phi Eta, a member and founder of Pi Epsilon and a member of Sigma Gamma Epsilon and Phi Kappa Tau.

He was a lecturer in petroleum geology at the University of Chicago during the spring quarters of 1924, 1925, and 1926 and at Northwestern University, 1929-1932.

From August to September, 1938, Plummer did geological work in Santo Domingo for Harold C. Bishop. He spent most of the last half of 1946 in geological studies of the Maranhao basin, northern Brazil, for the Conselho Nacional do Petroleo.

In 1918 Plummer married Helen Jeanne Skewes, who had been an office geologist with the Illinois Geological Survey, 1914-17, and with the Roxana Petroleum Company, 1917-18. Mrs. Plummer is an able geologist and eminent paleontologist in her own right and one of the pioneers in the field of micropaleontology.

Fred Plummer, geologist, paleontologist and petroleum engineer, was a friendly man of wide interests and inquiring mind, a tireless worker and prolific writer. He ranks with those other New Englanders, Donald Barton and Sidney Powers, in his outstanding contributions to petroleum geology and to the science of geology. He is among the top half dozen men to whom we are indebted for our present comprehensive knowledge of the geology of the state of Texas.

E. DE GOLYER

Dallas, Texas
June 19, 1947

Reprinted from American Association of Petroleum Geologists Bulletin, v. 31, no. 9, Sept. 1947, p. 1698-1700.

HELEN JEANNE PLUMMER
(1891-1951)

On January 11, 1951, Helen Jeanne Plummer died in Austin, Texas, after a brief illness. She became suddenly ill on Saturday, January 6, and an emergency operation was performed from which, though her condition from the first was critical, she seemed to rally, particularly after blood transfusions were given on the following Wednesday. But the strain upon her weakened heart was too great, and she died Thursday morning about 7:30. According to explicit directions in her will, her body was taken immediately to San Antonio for cremation.

She was born Helen Jeanne Skewes on May 7, 1891, in Muskegon, Michigan. Sometime during her childhood her parents moved to Illinois, and in that state she attended both high school and college. After graduation from Evanston High School she attended Northwestern University, from September, 1909, to June,

1913, graduating with a B.A. degree and with Phi Beta Kappa honors. For most of the next school year she was a fellow at Northwestern. For a few years her formal study was interrupted by professional work and later by marriage, but was resumed in 1924 when she attended the University of Chicago during the spring quarter. In 1925 she received an M.A. degree from Northwestern University.

In April, 1914, she began her professional career as geologist with the Illinois Geological Survey, where she remained until April, 1917. In that month she became a geologist for the Roxana Petroleum Company at Tulsa, Oklahoma, and remained there one year. Her marriage to Frederick Byron Plummer on May 7, 1918, put an end to regular professional positions but not to her scientific interests, which she pursued wherever Mr. Plummer's work as a geologist led him. Undeniably their various residences — in several states of this country, in The Hague, and for a short time in the Dominican Republic — broadened the range of her knowledge and experience. In 1928 they moved to Austin, Texas, which was their home for the remainder of their lives. By 1925 Mrs. Plummer was well known as a consulting paleontologist — a semi-professional activity which she continued until 1948. In 1930 and 1932 she was a special lecturer on micropaleontology at Northwestern University. From 1933 until 1948 she was consulting geologist at the Bureau of Economic Geology in Austin; from 1948 until her death she was a regular member of the Bureau staff.

Her work in micropaleontology, especially on foraminifera, gave her an international reputation. By 1931, for example, she was known and respected in England as one of the best American micropaleontologists. Her publications, both solid and extensive, on various foraminiferal subjects attest her serious interest and her painstaking, accurate work. Such interest is further attested by the number of scientific societies to which she not only belonged but in which she participated actively: American Association of Petroleum Geologists, American Geophysical Union, Paleontological Society of America, Geological Society of America, American Association for the Advancement of Science, Sigma Xi, Texas Academy of Science.

Because of the bulk and quality of her own work many who knew of it were not also aware of her constant help to Mr. Plummer in his scientific work — editing his manuscripts, doing with rare skill the necessary difficult drawings, painstakingly checking results, accompanying him on short field trips and even on some consulting expeditions. How much he depended on her only their intimates knew.

This unwearying aid was due not only to her personal loyalty and devotion to Mr. Plummer but her devotion to science as well. She was the living embodiment of the scientific spirit at its best; intellectually she had all the highest scientific qualities except the bold but rigidly disciplined imagination that marks the scientist *par excellence*. She had an intense regard, almost a fierce reverence, for scientific work; and her own high intellectual integrity and objectivity led her to attribute to others engaged in it the same selfless motives as her own. She spared no effort to help others; she gave generously of her materials, her time, and her unflagging spirit. It is to be deplored that a few whom she helped so untiringly lacked steady serious purpose and hence, in making an unwarranted demand upon her strength, helped shorten her valuable life; but it was part of her character to make any sacrifice in the name of what she felt was, or even might be, scientific advancement. And deplore it as they may, because of the drain upon her own energy and scientific productivity, it was to her friends one of her most admirable and endearing traits.

Her scientific and objective approach she carried into almost all phases of her private life. She was incapable of petty gossip or malicious judgment, though articulate and forthright if occasion arose. She was tolerant of motives and actions that stemmed from honesty and sincerity; but hypocrisy and self-seeking, especially social contacts for the sake of professional advancement, she could not and would not tolerate. Phases of domestic life that Mrs. Plummer did not like she attacked with scientific thoroughness. She did not particularly like gardening and would have preferred to leave uncultivated a considerable part of the extensive grounds surrounding her home. But Mr. Plummer preferred that they be a garden spot. So Mrs. Plummer made a botanical study of the plants and shrubs in the yard and experimented in the greenhouse built a short distance from the house. She joined the local Garden Club and, for some members at least, revolutionized the concepts of that organization by furnishing solid botanical information and suggesting suitable scientific methods. She did not like to cook, though nobody familiar with the Plummer cuisine and not with what lay behind it would have suspected such dislike. She subordinated cooking to her professional activities but as nearly as possible reduced it to streamlined, unvarying efficiency. She devised techniques to eliminate failure, as in breadmaking, and to improve flavors; friends profited both in enjoyment of the Plummer hospitality and in the recipes she generously gave to those who wanted them. The beautiful and scientifically complete Plummer home in Austin was always hospitably open to scientists, individuals, or groups. Few visiting geologists, among those well known to the Plummers, were allowed to stay at a hotel unless they wished to. Many an impecunious student found that home a haven where he might study and work at little or no expense; almost countless young people have reason to revere the name of Helen Jeanne Plummer for the practical help she gave them and for the inspiration of her uncompromising scientific spirit.

Those who knew Helen Jeanne Plummer only as a scientist — a dedicated, perhaps even formidable scientist — or who did not penetrate beyond her reserved, somewhat repressed manner were probably not aware of her rich emotional nature, which she lavished upon her friends as unstintingly as she lavished her scientific knowledge upon science itself. To the numerous children of her neighborhood she was known as Auntie Plummer; they had the freedom of her house, particularly if she felt they could benefit from the rich treasures of books or of botanical and geological collections available there. She gave careful thought to their budding intellectual interests and also to the quirks in their developing personalities. Her Christmas lists were long, and behind each gift lay careful consideration of what the recipient might like and enjoy most. Like Mr. Plummer she was generous with money as with time, but not ostentatious; only the recipients of bonds or of other substantial aids for later education know how generous.

The sudden death of Mr. Plummer in 1947 was a blow from which she never fully recovered, but she faced her loss with characteristic determination and courage. By late 1950, in addition to carrying on her regular professional work, she had arranged and properly disposed of Mr. Plummer's books, professional papers and collections, had sold the large house and moved across the street to a small house she also owned — in short, by that time had successfully arranged her life on a more restricted scale that would make possible the maximum scientific progress and personal satisfaction. Friends who saw her last in the opening days of 1951 commented with joy on her apparently improved physical condition and her final successful adjustment to a solitary home life. She seemed to have put personal sorrow behind her and to be literally ready for a new life. It is one of the most tragic of the world's many ironies that within ten days she was to be snatched from the very threshold of that life.

It is undeniable that Mrs. Plummer's neglect of self while she devoted herself untiringly to her personal relationships and to whatever she conceived to be the cause of science was in large part responsible for her too early death. Her scientific achievements will live on in her published work, in her valuable collections (now at Cornell University), and anonymously in her contribution to the publications of others. As a person she will be an imperishable memory to those privileged to call themselves her friends and intimates; they will cherish their recollections of one of the rarest spirits among humankind.

MARY GRACE MUSE ADKINS

Austin, Texas
May, 1954

Reprinted from American Association of Petroleum Geologists Bulletin, v. 38, Part II, 1954, p. 1854-1857.

EDWIN LEIGHT PORCH, JR.
(1889-1948)

On June 9, 1948, Edwin Leight Porch, Jr., consulting geologist of San Antonio, Texas, died in the Nix Hospital in San Antonio.

Edwin, son of Edwin Leight, Sr., and Nina Dortch Porch, was born at Kisseems, Florida, on June 19, 1889. He graduated from Bell High School in Galveston, Texas, in 1906. The following year his parents moved to San Antonio and Ed attended the University of Texas during the term of 1906-1907. The following three years he attended the Michigan College of Mines at Houghton, Michigan, and received his B.S. and Engineer of Mines degrees from that school in 1911.

The following year Ed was employed as a mining engineer for the Espada Mines Company in the state of Jalisco, Mexico, and when the revolutionaries made it impossible to continue mining operations in that part of Mexico, he entered the University of Texas as a special student, taking courses in technology of fuels and in electrical engineering during the year 1912-1913. The next year was spent on various projects and from June, 1914, to July, 1917, he was employed by the Texas Bureau of Economic Geology.

World War I interrupted Ed's professional work and he was in the Air Corps from July, 1917, until March, 1919, as an instructor. He received his pilot's wings in 1918 and was discharged as a First Lieutenant.

Following army service, the next two years were spent in San Antonio as a consultant and from April, 1921, to December, 1923, Ed was employed by "El Aguila" and held the position of acting chief geologist in Tampico, Mexico.

After the termination of this second tour in Mexico, and until his death, Ed was engaged primarily as a consulting geologist with offices in San Antonio, making many trips to Mexico for various clients to examine mineral properties and prospects, both oil and mining. Most of his work, however, was in Texas, Louisiana, and New Mexico.

Ed was the author of the University of Texas *Bulletin 1722*, "The Rustler Springs Sulphur Deposits" (1917) and he compiled or assisted in compiling data used in the publication of University of Texas *Bulletin 35*, "The Fuels Used in Texas" (1915), *Bulletin 44*, "Review of the Geology of Texas" (1916), *Bulletin 1814*, "Chemical Analyses of Texas Rocks and Minerals" (1918), *Bulletin 1839*, "Road Building Materials in Texas" (1919), and *Bulletin 1932*, "The Geology and Mineral Resources of Bexar County" (1920).

He was a member of the American Association of Petroleum Geologists, American Institute of Mining and Metallurgical Engineers, South Texas Geological Society, and the Society of Economic Paleontologists and Mineralogists, and a fellow of the American Association for the Advancement of Science, and honorary member of Sigma Gamma Epsilon. He was a member of the Sons of The American Revolution. His primary interests were scientific as reflected by his extensive library and his many activities in the South Texas Geological Society, of which he was a charter member.

Ed was always willing to lend a helping hand, particularly to younger members of his profession. He had a cheerful disposition and his innate love for his fellow man was reciprocated by a wide circle of friends that loved and respected him in turn. Those who knew him best had the greatest respect for his scientific mind and ability.

EDWARD LEE JOHNSON

San Antonio, Texas
August 3, 1948

Reprinted from American Association of Petroleum Geologists Bulletin, v. 32, no. 9, Sept. 1948, p. 1849-1850.

THOMAS MANN PRETTYMAN
(1888-1940)

On November 2, 1940, Thomas Mann Prettyman died suddenly of a cerebral hemorrhage at his home, 4842 Crestline Road, Fort Worth, Texas.

He is survived by his widow, Carolina Moreland Prettyman, two brothers, Frank and Cannon Prettyman of Summerville, South Carolina; and a sister, Mrs. Virginia Rhett of Charleston, South Carolina.

Mr. Prettyman was born in Cartersville, Virginia, on July 14, 1888. He attended the Virginia Military Institute from September, 1905, to June, 1908. Having decided on a business career, he enrolled at Eastman Business College, Poughkeepsie, New York, and studied there from September, 1908, to May, 1909. For several years thereafter, he was associated with his family in the lumber business in South Carolina. In 1916, he answered the call of the West and came to Texas where he attended the Texas School of Mines at El Paso. He received his E.M. degree from that institution in June, 1918.

During the World War No. I, he served in the chemical and metallurgical branches of the Ordnance Department, spending most of that time in Pittsburgh and Washington.

After the war he returned to Texas and worked as assistant geologist with the Bureau of Economic Geology at Austin. He was assigned to work on the oil possibilities of University Lands in West Texas under Dr. Udden and was co-author, with R. A. Liddle, of University of Texas *Bulletin* on the "Geology of Crockett County, with Notes on the Stratigraphy, Structure and Oil Prospects of the Central Pecos Valley." It was this bulletin, published in 1918, which, in a large measure, was responsible for the development of what is now known as the West Texas Permian basin, one of the largest producing areas in the United States.

From February, 1921, to February, 1923, Mr. Prettyman was assistant State geologist of Georgia. His work on the Georgia Survey is summarized in *Bulletin 40*, "Petroleum and Natural Gas Possibilities of Georgia," by T. M. Prettyman and H. S. Cave, published in 1923.

He again returned to Texas, early in 1923, this time as geologist for The Texas Company. During the Mexia-Powell fault-line "play," his work in this region combined with his high technical skill, made him one of the best informed geologists on the "fault line." In 1924, after the discovery of the Big Lake field on University Lands in Reagan County, West Texas became very active and on account of Mr. Prettyman's knowledge of this vast area, he was transferred by The Texas Company to their West Texas division with headquarters at Cisco. He continued his work in West Texas with this company until the partnership of Renaud and Prettyman, consulting geologists, was formed in 1925. This firm maintained offices in Cisco and Abilene and was active until 1929.

In 1929, the firm of Renaud and Prettyman organized the Westexas Oil and Royalty Corporation and moved to Fort Worth. Mr. Prettyman was vice-president and a director of the company from the beginning and was very active in its affairs until the time of his death.

Prettyman joined the Association in 1924, being sponsored by R. F. Baker, Wallace E. Pratt, and Fred C. Sealey. He was a member of the American Institute of Mining and Metallurgical Engineers, the American Petroleum Institute, the Fort Worth Club, the River Crest Club, the Fort Worth Boat Club, and Sigma Gamma Epsilon.

"Tom" Prettyman's passing was untimely and sudden. He leaves a host of friends who, with affection, will long remember his considerate, friendly, and noble nature.

C. L. RENAUD

Fort Worth, Texas
December 9, 1940

Reprinted from American Association of Petroleum Geologists Bulletin, v. 25, no. 2, Feb. 1941, p. 346-347.

ELIAS HOWARD SELLARDS
(1875-1961)

At 1:15 on the morning of February 4, 1961, Elias Howard Sellards, one of the country's best known and most respected geologists, died in Austin, Texas, at the age of 85. Dr. Sellards was widely known for his capable performance in the important administrative positions which he held during the greater part of his long professional career, for his many notable contributions to the scientific literature, and for his service as Professor of Geology in the University of Texas and the University of Florida. He was a distinguished member of many professional organizations including The Geological Society of America, American Association of Petroleum Geologists, American Association for the Advancement of Science, Paleontological Society, Society of Economic Paleontologists and Mineralogists, and the Society of Vertebrate Paleontology. He was an active member of the University Presbyterian Church in Austin. He is survived by his two daughters, Helen (Mrs. H. A. Hemphill) and Daphne (Mrs. F. H. McGowan), both of Houston, Texas; five grandchildren; eight great-grandchildren; and one brother, E. W. Sellards of Topeka, Kansas. His wife, Mrs. Anna (Alford) Sellards, died in Austin in 1953.

Elias Sellards was born May 2, 1875, in Carter, Kentucky. He spent his boyhood days and began his formal schooling in Carter, but while he was still a youth his family moved to Kansas and he completed his high-school education in Scranton, Kansas, in 1894. He attended Washburn Academy for a time, then went to the University of Kansas at Lawrence where he received the B.A. degree in 1899 and the M.A. degree in 1900. From 1901 to 1903 he attended Yale University, where he held both a scholarship and a fellowship, and in 1903 he was awarded the Ph.D. degree.

After completing his formal education Dr. Sellards served for one year as Instructor of Geology and Mineralogy at Rutgers College and three years as Professor of Geology and Zoology at the University of Florida. In 1907 he became the first State Geologist of Florida. He held that position until 1918 when he moved to Austin, Texas, where he was to spend the remainder of his career.

In Austin, Dr. Sellards served as geologist on the staff of the Bureau of Economic Geology from 1918 to 1925, as Associate Director of the Bureau from 1925 to 1932, and as Director from 1932 to 1945. He went on modified service with the University of Texas in 1945, but remained on the Bureau staff as Director Emeritus. In addition to Bureau duties, Dr. Sellards was also Professor of Geology and a member of the Graduate Faculty from 1926 until retirement, and he also served as Director of the Texas Memorial Museum from 1938 to 1957.

Despite the heavy load of administrative and teaching duties he carried during the greater part of his career, Dr. Sellards was an exceptionally productive writer. His complete bibliography lists 179 separate papers, including several major works. The first of these publications appeared in 1900, and the last in 1960. At the time his final illness set in a few weeks before his death he was engaged in an extensive revision of his book, Early Man in America — A Study in Prehistory.

His published works covered a broad range of topics relating to geology and allied sciences. Early in his career he received favorable recognition for publications on the results of his researches on fossil plants and insects from the upper Paleozoic of Kansas. While in Florida he published numerous important papers on the basic geology, land features, mineral resources, and vertebrate fossils of that area. Also, he discovered artifacts near Vero Beach in geological associations indicating that the antiquity of man in America was much greater than was then generally supposed. His published accounts of these discoveries attracted wide attention and revived scientific interest in the search for evidence of ancient human cultures. Dr. Sellards retained a keen interest in the problems of early man throughout his life, and published more than 30 papers on that subject.

Throughout his long tenure in Texas Dr. Sellards continued to publish on the results of his investigations, and these contributions to the geological literature of Texas are of outstanding importance. His numerous reports deal with structural, stratigraphic, and economic aspects of geology, and with a variety of related subjects such as stream terraces, land subsidence, meteor craters, earthquakes, early man, and vertebrate fossils. He is best known to geologists, however, for his exhaustive study of the Red River, which he made in conjunction with the Texas-Oklahoma boundary dispute, and for his major role in producing a two-volume compendium, The Geology of Texas (University of Texas Bulletins 3232 and 3401), which has been a standard reference on the stratigraphy, structure, and economic geology of Texas for many years.

Dr. Sellards' devotion to his work, and his ability to carry an extraordinary load of work, were recognized by all who knew him. These qualities were never more evident than during the seven years preceding his tenure on modified service — the period from 1938 to 1945. He served capably during this entire period as director of both the Bureau of Economic Geology and the Texas Memorial Museum, and at this time both institutions were especially active. During a substantial part of that period he also directed a state-wide paleontological and mineralogical project and made many extended field trips that were required of him in that capacity. In addition to his regular duties, he also served during 1938 as president of the Society of Economic Paleontologists and Mineralogists; from 1938 to 1940 as councilor of the Geological Society of America; during 1942 as president of the Paleontological Society, and during 1943 as vice-president of the Geological Society of America. He was an active member on several important committees, including the University of Texas Research Council. It seems remarkable indeed that during that seven-year period he also found the time to lecture at numerous professional meetings and to author 30 scientific reports.

Those of us who for many years were closely associated with Dr. Sellards knew him not only as a dedicated and hard-working geologist, but also as a man of highest integrity and exemplary personal habits. He always met disappointment and adversity with uncommon fortitude. Even in his last years when he was struggling valiantly against increasing physical infirmities he did not become disheartened and did not dissipate his energy fretting over an unalterable situation. Instead, he did as he had always done — he applied himself cheerfully and vigorously to some self-imposed problem in the science he loved so much.

GLEN L. EVANS

Midland, Texas
July 15, 1961

SELECTED BIBLIOGRAPHY OF E. H. SELLARDS

A new genus of ferns from the Permian of Kansas: Kan. Univ. Quart., v. 9, p. 179-189, pls. 37-42, 1900.

Codonotheca — A new type of spore-bearing organ from the Coal Measures: Amer. Jour. Sci., 4th ser., v. 16, p. 87-95, 1 pl., 1903.

A study of the structure of Paleozoic cockroaches with description of new forms from the Coal Measures: Amer. Jour. Sci., 4th ser., p. 113-134 and 213-227.

Types of Permian insects, Part I, Odonata: Amer. Jour. Sci., 4th ser., v. 27, p. 249-258, 8 text figs., 1906. Part II, Plectoptera: ibid., v. 23, p. 345-355, 13 text figs. 1906. Part III, Megasecoptera, Orcytoblattinidae, and Protorthoptera: ibid., v. 24, p. 151-173, 28 text figs., 1907.

Sketch of the geology of Florida: Fla. State Geol. Survey, 1st Ann. Rept., p. 17-53, 1908.

Fossil plants of the Upper Paleozoic of Kansas: Univ. Geol. Survey Kan., v. 9, p. 386-467, pls. 44-59, 1908 (distributed 1909).

A preliminary paper on the Florida phosphate deposits: Fla. State Geol. Survey, 3d Ann. Rept., p. 17-41, 1910.

The artesian water supply of eastern Florida (with Herman Gunter): Fla. State Geol. Survey, 3d Ann. Rept., p. 77-195, 1910.

The underground water supply of west-central and west Florida (with Herman Gunter): Fla. State Geol. Survey.

Origin of the hard rock phosphate deposits of Florida: Fla. State Geol. Survey, 5th Ann. Rept., p. 28-80, 9 pls., 1913.

Some Florida lakes and lake basins: Fla. Geol. Survey, 6th Ann. Rept., p. 225-259, 1914.

The pebble phosphates of Florida: Fla. State Geol. Survey, 7th Ann. Rept., p. 25-116, 52 text figs., 1915 (includes description of vertebrate fossils).

Human remains and associated fossils from the Pleistocene of Florida: Fla. Geol. Survey, 8th Ann. Rept., p. 123-160, 1916.

Fossil vertebrates from Florida: A new Miocene fauna: New Pliocene species; the Pleistocene fauna: Fla. State Geol. Survey, 8th Ann. Rept., p. 77-119, 1916.

Geology between the Ocklocknee and Aucilla rivers in Florida: Fla. State Geol. Survey, 9th Ann. Rept., p. 85-143, 1917.

The skull of a Pleistocene tapir including description of a new species and a note on the associated fauna and flora: Fla. State Geol. Survey, 10th Ann. Rept., p. 57-70, 1918.

Geology between the Apalachicola and Ocklocknee rivers (with Herman Gunter): Fla. State Geol. Survey, 10th Ann. Rept., p. 9-55, 1918.

Geology between the Choctawatchee and Apalachicola rivers in Florida (with Herman Gunter): Fla. State Geol. Survey, 11th Ann. Rept., p. 77-110, 1918.

Review of the geology of Florida with special reference to structural conditions: Fla. State Geol. Survey, 12th Ann. Rept., p. 105-141, 1919.

Structural conditions in the oil fields of Bexar County: Amer. Assoc. Petroleum Geologists Bull., v. 3, p. 299-309, 1919.

On the underground position of the Ellenburger formation in north-central Texas with a preliminary contour map: Univ. Tex. Bull. 1849, 21 p., 1918 (1920). Also Amer. Assoc. Petroleum Geologists Bull., v. 4, p. 283-298, 1920.

Geology and natural resources of Bexar County, Texas: Univ. Tex. Bull. 1932, 202 p., 1919 (1920).

On the petroleum possibilities of Florida (with Herman Gunter): Fla. State Geol. Survey, 14th Ann. Rept., p. 33-236, 1922.

Notes on the oil and gas fields in Webb and Zapata counties: Univ. Tex. Bull. 2230, p. 5-29, 1922.

The producing horizon in the Rios well in Caldwell County: Univ. Texas Bull. 2239, 40 p., 1922.

Investigations on the Red River made in connection with the Oklahoma-Texas boundary suit (with B. C. Tharp and R. T. Hill): Univ. Tex. Bull. 2327, 174 p., 1923. Original record verbal testimony, U. S. Supreme Court Records no. 23, p. 3800-4328 and 4834-4856, 1923.

The Luling oil field, Caldwell County, Texas: Amer. Assoc. Petroleum Geologists Bull., v. 8, p. 775-788, 1924.

The subsurface geology of the Big Lake oil field (with L. T. Patton): Amer. Assoc. Petroleum Geologists Bull., v. 10, p. 365-381, 9 figs., 1926.
Subsidence in Gulf Coastal Plain salt domes: Univ. Tex. Bull. 3001, p. 9-36, 9 figs., 1 pl., 1930.
Producing horizons in the Big Lake oil field, Reagan County, Texas (with H. P. Bybee and H. A. Hemphill, Sr.): Univ. Tex. Bull. 3001, p. 149-203, 6 figs., 1930.
The Valentine, Texas, earthquake: Univ. Tex. Bull. 3201, p. 113-138, 1 fig., 1932.
Oil fields in igneous rocks in Coastal Plain of Texas: Amer. Assoc. Petroleum Geologists Bull., v. 16, p. 741-768, 8 figs., 1932.
The geology of Texas (with W. S. Adkins and F. B. Plummer), Vol. I, Stratigraphy: Univ. Tex. Bull. 3232 (Aug. 22, 1932), 1007 p., 54 figs., 11 pls., incl. geol. map, 1933.
The geology of Texas, Vol. II, Structural and economic geology: Univ. Tex. Bull. 3401 (Jan. 1, 1934), 884 p., 41 figs., 7 pls., incl. structural maps, 1935; Pt. I, Major structural features of Texas east of Pecos River, p. 11-136, 14 figs., incl. maps, 2 pls., incl. map; Pt. 3, Economic geology of Texas (exclusive of petroleum): Mineral Production in Tex. (with C. L. Baker and others), p. 215-220, 1 fig., 1935.
Artifacts associated with Bull., fossil elephant (Roberts County, Texas): Geol. Soc. America, v. 49, p. 999-1009, 6 figs., 3 pls., 1938. Abst., Geol. Soc. America Proc. 1937, p. 112, 1938.
Structural map of Texas, 2d ed., revised: Univ. Tex. Bur. Econ. Geol., 1939.
Mineral locality map of Texas showing occurrence of useful minerals, rocks, and other geologic substances (abst.): Tex. Acad. Sci. Proc. and Trans., v. 27, 1943, p. 138, 1944.
Occurrence of oil and gas in Texas (with Leo C. Hendricks): Univ. Tex. Publ. 4301 (Jan. 2, 1943), p. 179-190, 1946.
Index to Texas mineral resources (with G. L. Evans): Univ. Tex. Publ. 4301 (Jan. 1, 1943), p. 359-383, maps, 1946.
Early man in America — A study in prehistory: Univ. Tex. Press, Austin, XVI, 221 p., illus., 1952.

Reprinted from American Association of Petroleum Geologists Bulletin, v. 45, no. 11, 1961, p. 1904-1906.

JOHAN AUGUST UDDEN
(1859-1932)

Dr. Udden was born at Lekasa, Sweden, 19 March, 1859, and emigrated to Carver, Minnesota, in 1861. He studied at Augustana College, from which he received his bachelor's degree in 1881 and the master's degree in 1889. He taught natural science, civics, and a number of other subjects at Bethany College, Lindsborg, Kansas, from 1881 to 1888, studying part of the year 1885 at the University of Minnesota. He was Oscar II professor of geology and natural history at Augustana College from 1888 to 1911. He was special assistant of the Iowa Geological Survey from 1897 to 1903, assistant geologist, University of Texas Mineral Survey, 1903-4, geologist, Illinois Geological Survey, 1906-11, special agent, United States Geological Survey, 1908-14, and geologist of the Bureau of Economic Geology and Technology, University of Texas, 1911-15. From 1915 until his death, on 5 January, 1932, he was director of the Bureau of Economic Geology of the University of Texas. He was made a member of the graduate faculty of the University of Texas in 1929.

Other fruitful work was done by Udden as geologist for the Orient Railroad, the New York and Texas Land Company, the Chisos Mining Company, and the Illinois Board of World's Fair Commissioners.

Dr. Udden received the following honorary degrees: Doctor of Philosophy, 1900, and Doctor of Laws, 1929, Augustana College, Doctor of Science, Bethany

College, 1921, and Texas Christian University, 1923. He was an honorary member of the American Association of Petroleum Geologists and of the Society of Economic Paleontologists and Mineralogists. King Oscar II of Sweden knighted him with the Order of the North Star in 1911.

He was married to Johanna Kristina Davis in 1882, his wife surviving him exactly five months. A daughter passed away while still very young and two sons, Jon, the geologist, and Anton, the meteorologist and physicist, died before their father. One son, Swante, manages a large power and light corporation in Texas.

The testimony of his students and the honorary degrees conferred upon him by the educational institutions which he served demonstrate that Dr. Udden was a great teacher. He was the first teacher and one of the founders of Bethany College, where in his earlier years he taught many subjects and in addition edited the local paper. At Augustana he taught throughout the tenure of his professorship the various courses in Botany, Zoology, Astronomy, Physiology, Meteorology, and Geology, and at times helped out in remote fields, such as History. He founded there a series of monographs, Augustana Library Publications, contributing five of the fourteen numbers issued so far.

Not content with bearing with conspicuous success a teaching burden the mere thought of which would fill with horror the mind of almost any present-day college or university teacher he managed to accomplish while at Augustana a very noteworthy amount of scientific research of a high order. He published 46 papers during the 23 years of his professorship there. These included very detailed reports on the areal geology of five counties in Iowa, five quadrangles in Illinois and three large areas in Texas, the writing of the geologic masterpiece on the Chisos country of Texas, and the fundamental papers on the work of the atmosphere, the mechanical composition of clastic sediments, the Pennsylvanian sedimentary cycles and the examination of well cuttings. He had studied in meticulous detail — and before motor cars came into use — the geology of the territory within a radius of 25 to 30 miles of Rock Island. These investigations were not all confined to the surface geology. Professor Norton could write over 30 years ago that, thanks to Udden, there were few areas anywhere in which the subsurface geology and artesian conditions were as well known as in the vicinity of the three cities of Rock Island and Moline, Illinois, and Davenport, Iowa.

Udden's success in research can be attributed to several factors. One of these was doubtless his thorough knowledge of all the basic natural sciences, a mastery only attained so well as by those who have successfully taught so many of them. A second was his great ingenuity and inventiveness, shown by his invention of one of the earliest successful adding machines, and by his ability to accomplish important results with a few of the cheapest and simplest devices. We will note one instance of this. He was the pioneer in the study of well cuttings and sedimentary petrography yet it was not until some years after he moved to Texas that he was able to use a binocular microscope, an appliance which had then been in general use for over a decade. A third factor in Udden's success was that, because the small denominational college of the Middle West in the late seventies and early eighties could afford no professional scientific training, he had to rely upon his own powers of observation and reasoning as well as to devise his own technique. This made him self-taught, self-reliant and self-sufficient. Patience, in a preeminent degree, he appears to have received as a natural endowment. He early came to the conclusion that the best contribution he could make, with the limited opportunities of his environment, would be in fields overlooked by others. Thus he came to emphasize the importance of the minute. His geologic observations are characterized by their minuteness as well as their comprehensiveness. Together with this penchant for the most minute detail there went a power for philosophic reasoning, generally rare in one so painstaking in small things, which was to bear great fruit in his subsequent career in Texas. Coupled with all the above there was a tireless industry, an ability to take a great deal of self-inflicted punishment in relentlessly self-imposed tasks which in the end meant overwork and premature aging. The teaching regimen alone in the typical mid-west college of a generation ago was sufficient to wear out many a sturdy soul but here was a man making many fundamental contributions, some of which involved a vast amount of arduous labors of the most painstaking sort, in addition to carrying a teaching schedule in itself intolerably burdensome.

When Udden left Augustana in 1911 he could have said — although his rare modesty prevented him from saying or, indeed, thinking anything of the sort — that he had made just about the best possible use of the facilities afforded to a geologist on the plains of Kansas and the prairies of Illinois and Iowa. Greater things were to come with the greater opportunities of Texas.

As early as 1896 Udden had written in the 17th Annual Report of the United States Geological Survey "An account of the Paleozoic rocks explored by deep borings at Rock Island, Illinois, and vicinity" and in 1893 had given the geologic section of northern Illinois in the report of the World's Fair commissioners. Later he extended his subsurface studies to other parts of Illinois in the reports of the State Survey. His first assignment after he went permanently to Texas was the preparation of a report on the Wichita and Clay County oil and gas fields. In this he devoted a section to drillers' logs. Even the statistical method failed to give any real indication of what strata were penetrated. It is generally known that one cause in the lack of reliability of well logs is that the helper on a well rig automatically becomes a graduate driller by the process of crossing the Mississippi River in a westward direction but the main fault with the system is that the driller is not and cannot be expected to be a geologist and inasmuch as it has proved impossible to persuade the oil and gas industry to drill wells properly by continuous coring it is necessary for geology to be pragmatic and do the best it can under the circumstances. This is what Udden accomplished for his science. His Wichita and Clay County report demonstrated the practicability and utility of the laboratory examination of drill cuttings in the search for oil, gas and water. Udden had established the first subsurface laboratory and E. T. Dumble followed his lead in the various oil companies of the Southern Pacific. Today it is becoming the exception to find even the remotest wildcatter who does not preserve drill cuttings. Considering alone the State of Texas, if drillers' logs proved worthless under the comparatively uniform sedimentation conditions of the Paleozoic they were even worse in the very irregular territory of the Upper Cretaceous and the Cenozoic. The present deep drilling beneath structural unconformities in Texas and elsewhere would scarcely be feasible unless Udden had not pointed the way and the means by which accurate subsurface data could be procured from the results of ordinary drilling. It is beyond dispute that Udden's pioneering in this line has benefited the oil and gas industry to the extent of many millions of dollars.

In 1914 Udden published the paper on the deep boring at Spur, in which his use of his own methods in the laboratory study of drill cuttings demonstrated for the first time the occurrence of potash salts in the great Permian Basin of the Southwest. Near the end of the same year the writer procured for him the second set of samples which showed the presence of potash and from 1915 on Udden was convinced that Texas and New Mexico probably contained workable reserves. At the present time one mine in New Mexico is exploiting water-soluble potash salts and Dr. Fraps has shown recently that crude polyhalite, abundant in the Permian Basin, is available as a fertilizer. Udden was thus the undisputed pioneer in southwest potash.

It was more than 15 years ago that Udden advanced to the writer and others his view that geophysical research might lead to the discovery of new oil and gas fields. With characteristic caution he delayed the publication of his ideas until 1920. Since then the use of geophysical methods has led to the discovery of some scores of proved and prospective oil and gas fields in Texas and Louisiana alone.

In 1916, partly on theoretical grounds in advance of his colleagues, and partly as the result of field studies he had made years previously, he advised the Regents of The University of Texas of the probable occurrence of oil and gas on The University of Texas lands in Reagan County. This led to the discovery of the Big Lake oil field, now the world's deepest profitably-producing field. Big Lake was the forerunner of other west Texas and New Mexico fields, one of which, Yates, ranks with the first half dozen of the world's most prolific producers. The probabilities are that the Permian Basin of west Texas and eastern New Mexico still contains vast unexplored reserves. Revenue from its oil lands, since the discovery of Big Lake, has increased the endowment of The University of Texas by more than $16,000,000.

The impetus given to discovery of oil in west Texas, and to the development of potash and of geophysical methods, is an instance of Udden's possessing the courage of his convictions. He did not lack for boldness when convinced of the correctness of his reasoning. Oil geology is still an empirical science; it is a follower and not a leader. It is only a decade since it was common for petroleum geologists to declare it was virtually impossible to find oil in the Permian Basin or at least west of the Pecos.

Udden in 1904 in his study of the Terlingua deposits had determined that the concentration of the quicksilver ores of that district had followed the same principles as oil and gas. At that time the quicksilver that had been mined had come from outcropping deposits, most of which proved superficial and were speedily exhausted. The long-sustained, still-lasting, more deep-seated quicksilver production was discovered as a result of Udden's persuading the miners to sink a shaft through a large thickness of barren shale into underlying porous and cavernous limestone in which he deduced that ore was deposited under anticlinal conditions. Schuette showed in recent years that the leading deposits of quicksilver elsewhere in the world also have formed in "closed" structures. The writer extended the Udden theory to the limestone ore-deposits of silver, lead and zinc in the folded Cordillera of northeastern Mexico where it has been accepted by at least two leading mining geologists. In the writer's opinion it will prove to be the most fruitful deduction yet advanced in the discovery and development of new ore deposits in the great "Sierra" limestone of the Lower and Middle Cretaceous which forms the anticlinal and thrust-block mountain ranges of the Sierra Madre Oriental.

Udden's achievements in oil, potash and quicksilver have already added many millions to the wealth of Texas alone and will be responsible for much more in the future. We will turn now to his efforts in pure science during the last twenty years of his life. Outstanding in this respect was his attitude as administrator of the Bureau of Economic Geology. His policy was pure geology first and foremost, stressing the necessity of fundamental research and areal mapping. Its success has been amply demonstrated through continued public support and the constant loyalty of the hundreds of economic geologists engaged in the state. The Bureau has never engaged in propaganda for its own benefit nor has it been the servant of any special interest. This sense of balance and just proportion was an eminent corollary of Udden's broad-mindedness.

In 1916 there appeared the "Review of the Geology of Texas" with the first large-scale geologic map of the state. This is noteworthy if for no other reason than because it became a geologic "best seller," running into three editions, with a total issue of about 50,000 copies. This was due not so much to its merit as to its good fortune. It chanced to appear in time to anticipate the boom in Texas oil which occurred during the war. Its educational value was more than trifling; its map proved to be almost a necessity for the oil promoter within the state and for many without it!

Udden's long connection with the Permian of west Texas began with his field work in the Chinati Mountains in 1903. Although his contributions to the Cretaceous were of no mean order he will be longer and better remembered for his work on the Permian of the Glass Mountains. Concerning this no less an authority than J. P. Smith stated that he had made it the type marine Permian of the world. Yielding to the opinion of his paleontologic colleagues he was induced to draw the Pennsylvanian-Permian contact line perhaps too low but, fortunately, the change was made so late that his own most likely more correct views can be found in reading his paper. The most recent work in the Glass Mountains tends to return more nearly to Udden's original conception. The foundation he laid is pretty certain to endure.

Udden appears to have reached the zenith of his descriptive powers in "A Sketch of the Geology of the Chisos Country, Brewster County, Texas" finished in the spring of 1905, when he was 46 years of age. No doubt his enthusiasm was fired by the fullness of geologic phenomena exhibited in the Trans-Pecos mountains and the welcome relief they must have afforded after years of the monotony of the plains and prairies of the northern Mississippi basin. At least his response to the inspiration of the mountains resulted in a masterpiece which deserves re-printing and a much wider circulation. It ranks with the classic works of Powell, Dutton and Gilbert.

Kindliness, gentleness, a modesty amounting virtually to bashfulness, moral and intellectual honesty, unfailing courtesy and desire to be helpful were outstanding traits of Udden. He was more considerate of others than of himself. He set for himself a higher standard than he exacted. Numerous instances might be given of his sacrificing himself in order to help others. No one was turned away because Udden was tired or busy or had more important affairs. He did not benefit financially by his own discoveries. He was one who, through example and effort, enriched and bettered the world. He maintained until the end that his conspicuous success was the result of chance and not of merit. A model for others to follow, a great geologist, a good man in every sense of the word — he was all of these. Ranking with Sam Houston alone in the measure of his services to the State of Texas Johan August Udden was also one of Nature's noblemen.

C. L. BAKER

Reprinted from Geological Society of America Bulletin, v. 44, 1933, p. 402-406.

APPENDIX III

The Red River Investigations

Following the discovery of oil at Burkburnett in 1918 and in the adjacent Red River valley in 1919, the State of Oklahoma brought suit against Texas asking the Federal government to ascertain and mark the exact boundary between the two states. The problem was to determine the location of the south bank of the meandering Red River at the time it was established as a boundary line in the Adams-Onis Treaty of 1819.

Because some of the lands claimed by both states were oil bearing there were large property values involved. And so in the summer of 1921 the Attorney General requested that the Bureau of Economic Geology and the State Reclamation Department undertake investigations to support the Texas claim.

These investigations were directed toward determining the age of the alluvial lands of the river valley and the physiographic history of the valley lands. Oklahoma officials claimed that at the time of the Adams-Onis Treaty, the Red River flowed at the foot of the Texas bluff, but that by 1921 the river valley was more than a mile in width and the river flowed at the foot of the Oklahoma bluff. They claimed that these disputed valley lands were formerly an island formed when the river opened a new channel along the Oklahoma cut bank. Later, the original channel was completely abandoned by the river and the island was left contiguous with the Texas cut bank. It was a well established principle of law that in such cases the abandoned channel remains the boundary.

Texas officials contended that these disputed river deposits on the south side of the Red River existed in 1819, and that the river had shifted only a little as a result of some land added by accretion on the south bank. E. H. Sellards and R. T. Hill collected sedimentation data in support of their interpretation of the physiographic history of the Red River.[1] Their studies were pioneer contributions to the knowledge of river valley sedimentation and excellent ecological studies by B. C. Tharp supported their conclusions.[2]

[1]E. H. Sellards, B. C. Tharp, and R. T. Hill, "Investigations on the Red River Made in Connection With the Oklahoma-Texas Boundary Suit": Univ. Texas Bull. 2327, 174 p., 1923. Hill's "Summary of Physiographic Investigations" (p. 157-172) is excellent.

[2]B. C. Tharp, "Ecological Investigations in the Red River Valley," *ibid.*, p. 89-155.

The Federal Supreme Court decision was in accordance with the scientific testimony presented by the Texas Attorney General. They ruled that the disputed Red River channel deposits had been built by accretion, in which case the boundary line changes with the shifting river channel. They went further and accepted the claim by Texas that the valley lands existed in roughly their present form back in 1819.

APPENDIX IV

Udden's Correspondence Regarding Bureau Investigations of University Lands

June 22, 1916

Major G. W. Littlefield
Chairman, Board of Regents,
Austin, Texas

Dear Sir:

I have received a letter from Mr. E. J. Mathews, Secretary of the Board of Regents, asking me to write you and make reply to some questions the Board asks, relative to having an examination of the University lands made by the Bureau of Economic Geology and Technology, for the use of the Board.

In reply to Mr. Mathews's first question, "What surveys, if any, have been made heretofore of the University lands?", I can say:—

Early in 1892, W. F. Cummins took notes on the geology of the Llano Estacado and in a report on his work there are found a few notes on Andrews County (Geological Survey of Texas, 3rd Annual Report, 1891, p. 181). This work was of a very general character, determining the principal formations only.

In 1901, Mr. B. F. Hill visited parts of the land belonging to the University in El Paso County, for the University of Texas Mineral Survey, and his reports on the general features of these lands were published by that Survey in 1902. Some very brief notes on the mineral prospects of the tracts visited were published in Bulletin No. 2 of the University of Texas Mineral Survey, and are found on pages 11, 19, and 20. The field work done for this report extended over El Paso, Culberson, Jeff Davis, Presidio, Terrell, and Brewster counties. Hill's observations on the University lands were merely incidental.

In 1902, Mr. G. B. Richardson was engaged by the University Mineral Survey to make a second reconnaissance report in that part of El Paso and Reeves counties lying north of the T. & P. railroad. Mr. Richardson's work was undertaken for the purpose of ascertaining the distribution of the different geological formations and their nature. The results of this season's work were published in Bulletin No. 9 of the University of Texas Mineral Survey, in 1904. The report gives a general description of the geology of that part of the state, with some notes on the occurrence of coal, oil, sulphur and underground water. None of these notes refer to the University lands, but some have a bearing on any work to be done there.

In 1910, two weeks were spent by the undersigned on a hurried examination of the University lands in Irion, Reagan, Crockett, Upton, Crane, Winkler, Loving, Ward and Pecos counties, for the purpose of taking notes on water resources of these lands. Notes were taken on the geological relations of the waters obtained in various wells, with a view of judging of increasing the water supply. The data collected were transmitted with a report to the President of the University, early the following year. They were necessarily very incomplete.

I may say that while a fair general knowledge of the geological conditions in that part of the state has resulted from these expeditions, from the nature of the circumstances under which they were made, they were hurried reconnaissances which did not cover the mineral resources of the country in enough detail to be of much practical value.

Mr. Mathews further asks: "What preliminary surveys for oil or other minerals developments should be made?" It seems to me that a provident management of the landed estate of the University might in some measure be facilitated by an examination of the lands for the mineral possibilities which may exist. For my part I do not believe that these possibilities are very many. I will briefly estimate them as best I can. A map is appended, to further elucidate the conditions described.

1. In the "Alphabet Block" in El Paso County, we have the only possibilities for the occurrence of metallic mineral such as zinc, lead, or silver. The occurrence of such minerals is usually confined to very circumscribed localities, and even if they should exist on these lands, it is not certain that they would be discovered in one season's work, though I consider geologists exceptionally well equipped for doing effective prospecting work.

2. That block of the University lands located in the north part of Culberson County is at least in part covered by a formation known to contain sulphur, not far away. The general structure of the country is also such that there might be conditions favorable for the existence of oil, most likely at considerable depth. If any sulphur beds exist, they can probably be located and if there should be pronounced structure favorable for the accumulation of oil, this also can probably be located.

3. There is also a small chance of finding oil on the University lands in Crane, Upton, and Pecos counties. I regard the possibilities of finding oil on these University lands as quite remote. But there exists an old axis of uplift extending northeast of the northern end of Brewster County for an unknown distance; probably as much as 100 miles. A limestone formation which covers this old mountain axis has not proved oil-bearing except at a single point in the state — South Bosque, where a little oil has been produced. The rocks underlying the limestone in the old mountain axis are in part the oil-bearing rocks of Oklahoma and north Texas. It is not entirely unreasonable to suppose that at some point along this axis there might be a structure into which oil may have gathered from the lower beds.

4. The country east, northeast and southeast from Pecos County is overlain by a limestone from 50 to 800 feet or more in thickness. This limestone contains certain layers in different parts of the country, which are water-bearing. I think it very likely that a close study of these water-bearing strata on University lands might be helpful to a better understanding of the prospects for water in wells on the University lands, if a sufficiently close study could be made of the ground.

5. The general information we have on the formations in the counties east of the Pecos and west of Reagan County shows that this county is underlain by extensive salt beds; in fact, these almost certainly extend 200 to 300 miles to the north through the state. Some observations made a year ago near Amarillo have proved the presence in that region of potash salts. It hardly seems less probable that they should occur in the south end of the Llano Estacado than farther north in the Panhandle. The existence of potash deposits can, however, be determined only by drilling, and it is a question whether or not a geological examination of the ground could throw much light on the advisability of undertaking such work.

6. In Andrews County there are some undrained shallow lakes. It might be worth while to ascertain to what extent the water in these lakes is concentrated. In some lakes of this kind on the western plains in Nebraska, there occur brines that have been worked profitably on a small scale for the winning of potash.

There is, perhaps, another consideration worth mentioning. One of the usual results of work of this kind is that it encourages an intelligent attitude of the people in that the vicinity where the work is done, relative to development of natural resources in general. This sometimes results in unforeseen developments.

I have enumerated these possibilities in the hope that you may consider them and possibly more readily form an opinion yourself regarding the desirability of having a survey made. As I look at it, work of this kind cannot be expected always to result in any positive conclusions. However, I believe that geological examination of these lands would be of some help in judging the values of the different tracts of land; and enable you perhaps to better judge of the merits of offers that may be made for lease or purchase.

In answer to the question, "What would be the cost of such a survey?", I estimate that the total expense of such a survey of the University lands would not exceed $6,000, and might perhaps not cost more than $5,000.

Mr. Mathews asks, "Could your Bureau make these surveys with additional money for expenses, but without extra salaries? If so, how much?" Yes, we could do so, with not to exceed $3,000 for extra expenses.

I shall be glad to discuss this matter with you more in detail at any time, if you so desire.

Very truly yours,

Director.

JAU:ES

C.C. to Mr. Mathews.

Oct. 25, 1919

Dr. R. E. Vinson,
President, University of Texas,
Austin, Texas.

Dear Sir:

Pursuant to your instructions, I arranged last spring to have a geologic survey made of the University lands for the purpose of investigating the possible chances of finding oil or gas on these lands. Dr. J. W. Beede undertook to make an examination of the lands in Hudspeth County. Dr. E. H. Sellards kindly undertook to make a rapid reconnaissance of all other University lands not obviously affected by the northeastward extension of the Marathon Uplift. Mr. R. A. Liddle was detailed to look into the question of the northward extension of the Marathon Uplift in Pecos, Terrell, Crockett, and some adjoining counties.

The field work extended over the greater part of the spring and summer months, and I have now ready for presentation to you five papers with accompanying charts and maps. These are as follows:—

1. "Report on the Oil and Gas Possibilities on the University of Texas Lands in Hudspeth County, Texas," by J. W. Beede, with map and sections; 11 pages.

2. "Notes on the Geology and Oil Possibilities of the Northern Diablo Plateau in Texas," by J. W. Beede, with map and illustrations; 34 pages.

3. "Reconnaissance Report on State University Lands of Andrews, Crane, Culberson, Dawson, Ector, Gaines, Loving, Martin, Pecos, Schleicher, Ward and Winkler Counties," by E. H. Sellards, with map; 6 pages.

4. "Oil Possibilities on some of the University Lands in Crockett, Upton, Reagan, and Irion Counties," by R. A. Liddle and T. M. Prettyman; 11 pages.

5. "Geology and Mineral Resources of Crockett County, with notes on the stratigraphy, structure, and oil prospects of the central Pecos Valley," by R. A. Liddle and T. M. Prettyman, with maps and illustrations; 105 pages.

The first paper is really a separate of a chapter in the second paper, and the fourth is likewise a separate of a chapter in the fifth, arranged for purpose of convenience.

A brief summary oi the results may be stated as follows:—

1. Dr. Beede has located at least three favorable structures on the University Lands in Hudspeth County; one south of Granite Tank (see map) near the southern boundary of the University lands; another north of the old well 8 miles east of Cerro Alto, close to the New Mexico line; and a third crossing the Buckthorn Draw east of Big Tank. His general study of the region of these lands shows that the rocks to be explored consist largely of limestone without much black shale. This circumstance cannot be considered very encouraging.

2. Concerning the lands in Andrews, Martin, Gaines, Dawson, Loving, Winkler, Ward, Crane, Ector counties it can be said that the surface conditions are such that it is unlikely, with few exceptions, that favorable structures can be determined with any degree of certainty in those tracts. Outcrops are scarce. The lands are all underlain by the salt-bearing and other Permian formations which extend to depths of probably more than 4000 feet. The only promise we have of oil or gas in this formation, so far, is from small developments recently obtained at

Amarillo in the Panhandle, and in some borings in the southeast part of New Mexico, where the same series of rocks underlie. So far, no paying production has been obtained from these strata in America, unless we count the shallowest production in Wichita and Clay counties. No one can doubt the possibility of the existence of oil pools either in the Permian or in the Pennsylvanian in the Pecos valley, but the chances for economic development seem very small, except in the belt of the Marathon Uplift referred to below. The University lands lying southeast from a straight line drawn from Garden City in Glasscock County to Fort Stockton, and beyond southward, are all alike in being covered almost entirely by the Comanchean formation. Dr. Sellards observed a north dip on Blocks 17 and 18 of the University Lands in Pecos County, east of Fort Stockton, south of the Sheffield Road. This is apt to be on the northwest slope of the Marathon Uplift, the axis of which lies about 25 miles away to the southeast.

3. As I have stated in an earlier report, the region in which some lands are located is traversed underground by the northeastward extension of the Marathon Uplift. In the belt of the uplift favorable structures for oil accumulation must exist, and it seemed to me that the most important geologic work for oil development was to determine the course of this belt. As it must intersect the Pecos River, I had Mr. Liddle make a detailed section along this river in Crockett, Pecos and Crane counties. This section shows that the underground structure crosses the river in the vicinity of Sheffield, where it has a width of from 15 to 20 miles. From there it continues northeastwardly, and can be traced into the Barnhart tract of University lands. Deep explorations which have already been made in Pecos County also give evidence of the Marathon Uplift near Sheffield.

It is my opinion that it will most likely be impossible to locate many of the underground structures by geologic methods in this belt of buried folding, and I expect that development must be made by more or less haphazard drilling in the indicated belt. Hence it is of utmost importance that careful study should be made of the materials penetrated in borings to be made in that region, as this will furnish the best guide in intelligently locating new tests. If the University could require that parties taking leases on its lands must furnish samples of cuttings for examination, I am satisfied that this would be a most efficient means to promote the discovery of any oil pools that may exist.

In case the publication of the material gathered during this work should be authorized by the Board of Regents, I have had the reports prepared in suitable form for such purpose and submit them herewith for your consideration.

Respectfully yours,

Director.

JAU:ES

APPENDIX V

The Association of American State Geologists

Various directors of the Bureau of Economic Geology have served as officers of the Association of American State Geologists (A.A.S.G.). This has afforded them the opportunity to play an active role on the national scene. The first association of state geologists was the Mississippi Valley Association of State Geologists which was organized in 1906 by H. Foster Bain, State Geologist of Illinois. Bain's idea was that the state surveys should be less dependent on the U.S.G.S. and should organize to facilitate the interchange of information. The name was changed to the A.A.S.G. in 1908.

The A.A.S.G. has served a lobbying function — at the annual meetings all national and state legislation affecting geological surveying, topographical mapping, stream gaging, and environmental studies has been discussed. It has also served its original function as a clearinghouse for the interchange of geological information between states. Typical topics discussed at the annual meeting at Cleveland in October 1922 included the following:

Cooperation between Federal and State Surveys in Geological Work

Cooperative Work between Geological Surveys of Adjoining States

Discussion of Sedimentation Studies (Dr. J. A. Udden)

At the Washington meeting in 1927, the participants considered the Federal topographic appropriation bills. At this meeting, the director of the U.S.G.S., George Otis, announced his desire to cooperate with the state surveys in lines of work demanding experienced and specialized geologists. Some participants noted that Federal parties had been entering states for geological work without making contacts with the state surveys, resulting in the loss of both information and time for the Federal as well as state surveys. Director Smith declared that in the future the state geologists would be promptly informed of such activities. Another facet of the activities of the A.A.S.G. was demonstrated in the motion that the secretary be instructed to write letters to the heads of the departments of geology of universities and colleges with graduate programs inviting their interest in the research problems of the state surveys and extending the facilities of the surveys.

In 1930 the educational extension work of the Illinois Geological Survey was discussed in the Journal of the A.A.S.G. This survey offered free publications to science teachers in public schools and arranged special field trips for them. It is also noted that the Iowa Geological Survey had prepared an introduction to the geology of Iowa for use in the public schools. Today the Bureau of Economic Geology is quite active in similar public school programs.

The journals of the A.A.S.G. include descriptions of the work being done by the various state surveys. Yearly presentations of the budgets of all the surveys provide data used to educate Texas legislatures to the inadequate funding of the Bureau. As secretary of the A.A.S.G. in 1934, Bureau director Sellards proclaimed that one of the most important activities that the Association could undertake would be the establishment of a more widespread understanding of the research and economic contributions of the state geological surveys. Throughout the life of the Bureau, the various directors have been important contributors to the Association, and their membership has kept the Bureau abreast of the innovative activities of other state geological surveys.

APPENDIX VI

A Proposed Institute of Geologic Research at The University of Texas

The following statement relates to a proposed Institute of Geologic Research at The University of Texas. Such an Institute is designed to supplement and enlarge the research opportunities already afforded at that institution and to provide an adequate center in the southwestern United States for the increase and diffusion of knowledge in the geologic sciences.

Need and Opportunity for Investigations in the Geologic Sciences in Texas and the Southwestern Part of the United States

The geologic sciences are devoted to the history of the earth, its resources and inhabitants. For the increase of knowledge in these sciences Texas affords exceptional opportunities. By reason of its location adjacent to the Gulf Coast on the one side and the Rocky Mountain region on the other the state presents great diversity of geologic conditions and affords great range of geologic observations. The climatic conditions favor outdoor work. The rocks are well exposed and access to field problems relatively easy. The results obtained in this science may have either a cultural or an economic value and fortunately many of the results have both.

The geologic sciences have heretofore found their largest development in the northeastern part of the United States and it is there, as in New York and Washington, that the largest museums and collections of fossils are now found. This early development has come about in part by the fact of early settlement in that region but more particularly by the greater concentration of wealth in the northeastern states. The Texas region, including under this term Texas and the adjacent territory in the United States and Mexico, is in a way set apart from this northeastern region, the distance being so great that the results of investigations made there are not readily diffused to the people of the Texas region.

The opportunities of development in the geologic sciences are distinctly better in the Texas region than in the New England or adjacent territory. The Texas region has the advantage of greater variety of geologic phenomena, a better climate for outdoor work, better exposures of the rocks, and vast and varied surface features. Texas is the largest and wealthiest unit of the southwestern part of this continent and may rightly expect to lead in the development of geology. It is the purpose of this undertaking to establish here a center for the increase and diffusion of knowledge in the geologic sciences that shall be second to none.

Progress Made Towards Providing for Such Investigations

In a small way some progress has been made towards providing for such an Institute. In 1928, with consent of President Benedict and the Regents of the University, some plans were made for enlarged geologic investigations in this part of the country. As a result of the plan initiated at that time four small funds for geologic purposes have been established as follows:

1. Geologic mapping fund, contributed by the American Association of Petroleum Geologists, the West Texas, Fort Worth, and North Texas Geological Societies and some individuals. Accepted by the Regents February 21, 1929. Applied in drafting, printing, and circulating county geologic maps.
2. Paul Franklin Morse Memorial Publication Fund, contributed by Professor and Mrs. W. C. Morse of State College, Mississippi, the San Antonio, Houston, Dallas, West Texas, and Panhandle Geological Societies, and some individuals. Accepted by the Regents July 16, 1930. Applied in printing publication on the Geology of the Texas Coastal Plain West of Brazos River, 1930, and a volume under the title of Contributions to Geology, 1930.
3. Fort Worth Geological Society fund, contributed by the Fort Worth Geological Society. Accepted by the Regents March 7, 1930. Applied in printing Geology of the Glass Mountains of Texas, Part 1, 1930.
4. Johan August Udden Fund, contributed by the Udden family, the Dallas, North Texas, San Antonio, and West Texas Geological Societies, and about 108 individuals. Accepted by the Regents May 30, 1932. Applied in publishing a volume on Contributions to Geology, 1932, as memorial to Dr. J. A. Udden.

No one of these funds is large enough to serve as endowment, and hence all are devoted to publication and applied as revolving funds, accounting both as to expenditure and income being made by the Auditor of the University.

As revolving funds these contributions, although necessarily small, will be continued indefinitely to the purpose of disseminating geologic knowledge in Texas. The plan here proposed is for a fund that will serve as an endowment, the income from which will serve as long as the University continues to carry on investigations with publication of results.

Plan of Operation of the Proposed Geologic Institute

It seems evident that any plan to increase geologic knowledge in the Texas region should include diffusion of non-technical geologic knowledge as well as increase of knowledge in these sciences.

Non-technical Information

Since this plan contemplates a center in the Southwest for the increase and diffusion of knowledge the set-up should include means of disseminating non-technical as well as technical information. It would possibly require some experimenting to determine just how this could best be accomplished. Visits by those interested and inquiries by letter would naturally be a part of the plan. Possibly circulars, either mimeographed or printed, would be helpful. Some serial publication, such as an annual report, would be helpful. Under this heading would come also the aid that can be extended to communities to encourage the preservation of natural scenic features. In matters of this kind the work of the Institute should be to advise and encourage preservation and markings of objects of geologic interest.

To increase knowledge in the geologic sciences is an important part of this plan. To carry out this purpose requires both investigation and publication.

The publications would preferably be in the form of memoirs which should be published according to the best standards of printing, engraving, and illustrating fossils, minerals, or other objects described. This involves considerable expense, but publication of the highest grade is necessary to the standing and reputation of the Institute.

The investigations to be carried on are varied and any attempt to enumerate lines of investigation would fall short of being complete. In general it is contemplated that any investigation that will extend the bounds of geologic knowledge may be undertaken. The following are illustrative.

Fossils of the Vertebrate Animals

The Texas formations contain the fossil remains of vertebrate animals which are in some respects the most notable to be found in any part of the world. As yet this rich field has been almost wholly neglected by the state, and the fine material is gradually going to the collections of other states or foreign countries, or, unfortunately, is being lost when uncovered or otherwise brought to light.

In the not distant past such animals as the elephant, its associate in size and kind, the mastodon, as well as a variety of scarcely less striking creatures, inhabited the plains where now are cities and farms. In the more remote past, hordes of other and widely different animals dwelt in the greatly different lands and landscapes of that early time. To preserve and make known what remains of the records of this early life is a duty of this generation. That animal life in such varying succession has occupied the land that is now Texas cannot fail to appeal to the imagination of those interested in the history of this state.

Fossils of Other Animals and Shells

Aside from the vertebrate animals which are so well preserved in some of the formations, there is need for more complete studies of the invertebrates. The marine invertebrate faunas of Texas are wonderfully preserved through a long series of formations from early geologic time to the present. Collections of these fossils have been made to some extent in the educational institutions. However, very large and rich invertebrate faunas remain to be collected, described, properly preserved, and applied in geologic and biologic studies.

Minerals

Similar conditions exist in respect to the minerals. In the busy commercial life of Texas attention has been given to the minerals and mineral products of direct economic value. However, a wealth of material awaits the student of the minerals of the state as a whole.

Regional Geologic Studies

Other geologic studies of varied nature will naturally be undertaken. In some instances these may relate to climates of the past ages, or again to the time and extent of the great floodings of the region by ocean which has occurred several times. Problems of earth history come under this heading. As already stated, any problem that seems likely to extend the bounds of geologic knowledge is to be considered.

To what extent a Geologic Institute at The University of Texas should have a staff devoting their time wholly to this work probably could only be determined with the experience of years. An ideal arrangement, from the standpoint of maximum results at least cost, would be a small full-time staff supplemented by cooperative relation with all geologic activities in the Texas region. The Institute would thus serve as a clearinghouse, as it could well do, for the geologic sciences.

Administration of the Funds of the Institute

It is probable that any funds provided for the purpose of a Geologic Institute would be in the form of a trust fund, administered by the Regents of the University or other similar corporate body. The scientific work of the Institute, likewise, would be under supervision of a similar body. The geologic work of an economic nature now being carried on in Texas by the Bureau of Economic Geology of the University and by the United States Geological Survey would not be duplicated or interfered with by the Institute but would be supplemented, more particularly in the field of pure science and fundamental research.

Relationship to Other Similar Scientific Institutions

The Smithsonian Institution at Washington, D. C., established to serve all branches of science, has some elements of similarity to this proposed Institute for the Geologic Sciences. The original contribution upon which the Smithsonian Institution was established was made by James Smithson, an Englishman who had never visited the United States. Smithson's original bequest to the United States of America has since been largely added to by others in this country. The purpose of the Smithsonian Institution, as expressed by Smithson, was to aid in the "increase and diffusion of knowledge among men."

The University of Texas has in the McDonald Observatory an institution for the astronomical sciences comparable to that here proposed for the geologic sciences. The purpose to be served by this fund, as stated by the donor, is that of "aiding in erecting and equipping an Astronomical Observatory to be kept and used in connection with and as a part of the University (of Texas) for the study of Astronomical Science."

The set-up for geologic investigations necessarily differs from that for astronomical investigations. The geologic sciences are in a degree localized, at least to the extent of relating to the earth. The astronomical sciences use the earth as a footstool, their study being almost entirely of other bodies. For astronomical observations a favorable site is important, such as the Davis Mountains in which the McDonald Observatory has been located. Some buildings are likewise required, including an observatory. The entire income from a fund for the geologic sciences aside from essential equipment can be applied to productive results in these sciences.

The Endowment Necessary to Establish A Geologic Institute

Funds of any amount, however small, can be used advantageously in promoting the geologic sciences at The University of Texas. That this is true has been shown by the use made of the several small funds already received. These small funds, however, are applied in full, used for publication only, and administered as revolving funds. To make extended investigations an income-bearing endowment fund is necessary.

For funds of this kind where permanency is a prime consideration, naturally conservative and hence usually relatively low interest-bearing investments prevail. All investments of the McDonald Observatory fund, by the terms of the bequest, "are to be made in such bonds and securities as are prescribed by law for the investment of the State Common School Fund." That fund was given in trust to the Regents of the University with the injunction that they "shall proceed at such time and manner as in their judgment may seem best to execute the trust and they shall have full power and authority to administer and handle the same for the purpose of carrying out its purpose and object, and in so doing they may apply all or any part of the income from the bequest and all or any part of the corpus of the bequest as they may deem proper to use towards such erecting and equipping the Observatory, it being my intention that the Regents shall have full power and authority to handle, use and appropriate this bequest, corpus and all accrued income in such manner as to them may seem best in order to carry out the object of the bequest, the only limitation on their authority and power being that the bequest is intended solely for the use and benefit of an Astronomical Observatory in one or all the ways hereinbefore mentioned."

While small funds can be made to produce useful results, an endowment, the income of which from conservative investment shall be sufficient to carry on fully the program outlined on pages 3, 4, and 5, needs to be of the order of a half million dollars. This, to be sure, is approximately that which served as the foundation for the Smithsonian Institution which was designed for all of the sciences. However, conditions a century ago were very different to those of the present time. Moreover, the original Smithson bequest has since been greatly added to by other gifts mostly by residents of the United States, as well as by Federal appropriations. The purpose of this proposed Institute is, after all, a step towards accomplishing Smithson's great desire for the "increase and diffusion of knowledge among men."

Permanency of the Institute

Actual permanency, perhaps, cannot be secured for anything that is of the earth. Relative permanency, however, can in a measure be secured and safeguarded. With respect to endowments, two principal factors are involved, first, security of the investments on which income depends and, secondly, permanency of the institution or project that is being provided for. The institution to which it is proposed that this Institute shall be attached, The University of Texas, has an unexcelled expectation of permanency. It is provided for in the State constitution and is an integral part of the State government, has an extensive physical plant, has a small, but not negligible, endowment accrued from its own lands, the good will of successive generations of students, and a purpose for existence, namely, the education of youth, that will continue so long as the present civilization continues.

The Institute itself is to deal with earth science problems. These sciences originated in a very limited way with the Greeks, were subsequently lost, and again revived with the quickened civilization of the Fifteenth Century. The pronounced development of these sciences in the modern sense has come about within the past 100 years. To what the science will lead in the course of a few more centuries no one can tell, but it would be an optimistic worker who would imagine that all of the geologic problems would be solved within a few generations. On the contrary, it is to be expected that these sciences will progress as do other sciences and that the advance towards the solution of the problems now before us will but open the vista to others that even more directly concern the welfare of the human race. The Institute, if established with secure investment of funds, will be as permanent through the centuries as is the University of which it becomes a part.

APPENDIX VII

Some Events in the Geologic History of Texas

by

E. H. Sellards

The geology of Texas makes an important chapter in the geologic history of the North American continent. No one of the great geologic time divisions is unrepresented in the state, and the sediments, although by no means continuously deposited, nevertheless present an approach to a completeness of the record that is gratifying. Within the confines of the state is found great diversity of deposits and of geologic conditions, and one who would know the history of this region must appreciate the reality of great changes, through geologic time, in the distribution of land and sea and in the succession of faunas and floras.

In considering the history of the region it will be helpful to view in succession conditions attending each of the several great geologic time divisions. Of these divisions the most distant in time and the most nearly concealed from view is the pre-Cambrian, the rocks of which are but sparingly exposed, and yet such exposures as occur record plainly pre-Cambrian conditions and events in this part of the world. At three localities in the state, the Llano region in central Texas, and the Van Horn and El Paso regions of west Texas, pre-Cambrian rocks lie at the surface. The record which they afford is supplemented by several other localities or regions where information on the pre-Cambrian has been obtained from wells. The drill has worked wonders in extending geologic knowledge of the older formations. By it, mountain systems, whose existence was unsuspected from flat-lying surface strata, have been proven to exist underground. It was in this way that the Red River mountain chain of pre-Cambrian and Paleozoic rocks, now known through an extent of 200 miles, was revealed underlying Cretaceous and Permian strata. The Amarillo Mountains, on whose slopes is a rich storehouse of oil and gas, have been explored by the drill in their extent entirely across the Panhandle of Texas. Middle Ordovician sediments, otherwise unknown in central Texas, have been revealed by drilling in the southern Permian basin at a depth in excess of 8,000 feet.

A brief look at the history of pre-Cambrian in Texas will show that these old formations hold never failing interest. Of the three regions of surface exposure the largest is found in the erosion basin that forms the top of the great Llano dome or Llano uplift of central Texas. Here sediments accumulated in pre-Cambrian time to a great, although undetermined, thickness. These sediments are chiefly of clastic materials, sands, clays, and silts, now, by alteration, quartzites, schists, and slates. However, limestones, now marbles, are present as thin bands. The pre-Cambrian history of this region is complicated. Before Paleozoic time, the sediments had been intruded by great masses of igneous rocks, chiefly granitic, had been uplifted, and strongly deformed with folds trending northwest-southeast, and these folds and igneous masses in turn beveled by extrusive erosion.

The pre-Cambrian rocks found in the Van Horn uplift of Trans-Pecos Texas had likewise a complicated pre-Cambrian history. In this region two great periods of deformation are recognizable previous to Cambrian time. The oldest deposits of this region, a great body of clastic sediments, known as the Carrizo Mountain formation, were deformed by an early pre-Cambrian mountain-making movement in which the trend is northeast-southwest. A later group of pre-Cambrian sediments, consisting of fine sandstone and massive limestones or marbles, the Millican formation, was affected by a deformation which established northwest-southeast trends. The thickness of both the Carrizo Mountain and Millican formations is undetermined. Both contain some, although not large, amounts of igneous rock.

The third locality of surface exposures of pre-Cambrian in the state is in the Franklin Mountains of the El Paso region. The sediments of this region, chiefly quartzites of the Lanoria formation, although intruded by granites remained relatively undisturbed, not only through pre-Cambrian time but through Paleozoic time as well and into the Cenozoic, when with the overlying formations they were faulted and sharply tilted westward. The Franklin Mountain pre-Cambrian is exceptional in that it long escaped deformation. It is also exceptional, in the Texas region, in that it received within pre-Cambrian time a great outpouring of lava. Although subjected to erosion previous to receiving a protecting covering of Cambrian sediments, there still remains 1,500 feet or more of this pre-Cambrian lava sheet lying between the Cambrian above and the Lanoria quartzite below. The rock of this lava flow contains numerous feldspar crystals and is a slightly pinkish, rhyolite porphyry of pleasing color. No fossils have been obtained from the Texas pre-Cambrian.

The Paleozoic sediments in Texas are so widely distributed, so diverse in character, and so abundantly supplied with fossils that they do not lend themselves easily to a brief description. To understand fully the Texas Paleozoic one must be prepared to visualize a vastly changed distribution of land and sea to that which now exists. At the present time the highlands of the state are in its western part, in the Trans-Pecos and High Plains regions. From the high plains the land slopes gulfwards, and the streams, originating either in the interior of the state or beyond the state boundaries, reach the Gulf of Mexico after threading their way across the nearly level coastal plains. Throughout the long Paleozoic era exactly reverse conditions prevailed. The present coastal plains region, or a part of it, was the highland belt of the state, the central and western part being the lowlands, more or less occupied by fluctuating intermittent seas. The drainage instead of being to the southeast, as now, was in the reverse direction, west or northwest. Throughout all of Paleozoic time the streams were carrying sediments westward. Suggestive of the futility of some human efforts is the fact that throughout a part of the Mesozoic and all of Cenozoic time the streams have been carrying these same sediments eastward.

With the setting of early Paleozoic conditions in mind, one may follow with ease the succession of events of Paleozoic time. At the dawn of the Paleozoic the highlands in the Texas region, as previously explained, lay in the present coastal plains, the ancient land mass being known as Llanoria. Earliest Paleozoic, lower and middle Cambrian, is unrepresented by deposits in Texas, and hence it can only be assumed that the Texas region was relatively emergent and largely or entirely a land area, as was also much of the North American continent at that time. However, as the Cambrian approached a close, or at least when the middle of the period had been reached or passed, a shallow sea, a part of the great sea that at that time so extensively flooded the entire continent, spread across the lowlands of central Texas. The Llano region, which had experienced pre-Cambrian deposition, igneous intrusion, uplift and erosion, was again submerged. The Cambrian sea, encroaching upon the somewhat uneven land surface of that time, accumulated first of all deposits of sand which were washed clear by the waves and in which were included fragments of the protruding pre-Cambrian rocks. Following deposition of the sands, in a clearer sea, marine life abounded and organic limestones accumulated. This late Cambrian sea submerged not only central Texas but extended westward across the Trans-Pecos region. The earliest recorded events of the Paleozoic era are therefore associated with the extensive submergence of late Cambrian time which established a great interior sea across the central part of the Texas region.

Once established, the Paleozoic seas in central Texas, although shifted by warpings of the earth's crust, and at times temporarily entirely excluded from the region, in the main persisted through most of Paleozoic time. As shown by these seas, the Paleozoic era, from Cambrian to Mississippian, is characterized by relative stability. The earth's surface underwent changes in elevation, but during this great expanse of time no pronounced upheaval or mountain-making period is recorded in the Texas region, and the history of central Texas for this large part of Paleozoic time is concerned with these fluctuating mostly shallow seas and mainly relatively low bordering lands.

The Cambrian sea was followed in central Texas after some appreciable fluctuation in level with an equal or more extensive sea of Lower Ordovician time. Drainage coming in from the highlands at the east resulted in the deposition of near shore clastic sediments, covered except in a few localities by later deposits. Westward from the highlands in clear seas in central and Trans-Pecos Texas extensive limestones and dolomites accumulated during lower Ordovician time.

Middle Ordovician must be set aside as the great question mark of Texas Paleozoic geology. What part of the central lowland region of the state was inundated during middle Ordovician time? Marine deposits of this age are revealed by the Marathon and Solitario uplifts of Trans-Pecos Texas. Their absence elsewhere seemed to point to land conditions at this time in central Texas until wells, penetrating to a depth in excess of 8,000 feet, proved that marine middle Ordovician sediments are present in the southern Permian basin. But even with this important discovery, it still is probably true that a part only of central Texas was covered by the Middle Ordovician seas. The known extensive absence of Middle Ordovician deposits, while possibly due in some measure to subsequent erosion, seems rather to indicate relative emergence at this time, resulting in restricted seas in central and Trans-Pecos Texas. Upper Ordovician, likewise, seems to have been a time of relative emergence, for while marine deposits of this age are present in Trans-Pecos, none have been found in central Texas.

It is necessary in this paper to pass the Paleozoic systems rapidly in view. Of the Silurian period, it may be said only that while deposits of middle Silurian time are found in extreme west Texas, elsewhere in the state these deposits are unknown, except, possibly, under the Cretaceous covering adjacent to the Red River in northeast Texas. The absence of these deposits over wide areas in the state is interpreted as indicating a continuance, with some variations, of the generally emergent or land condition of much of the Texas region.

A like record is to be made regarding the Devonian. Marine deposits of the Devonian system are known in Trans-Pecos Texas and are probably present under the Cretaceous covering of northeast Texas but elsewhere in the state, so far as known, are wanting. A relatively emergent condition in central Texas is therefore assumed to have continued through Devonian time.

The Mississippian period affords once more positive evidence of the invasion of the lowland region of central Texas by shallow seas. These seas were fluctuating and of limited extent. Of earliest Mississippian no record whatever is found, but later a clear sea in which organic limestones accumulated occupied a part at least of central Texas and once more lapped against or spread across the Llano region. No more than thin deposits remain as the record of this sea. In late Mississippian time a sea in which fine muds and some limestones were deposited spread across the same region and were present likewise across the region of the present Franklin Mountains and Diablo Plateau of extreme west Texas.

With all of the fluctuations in level here indicated, the entire Texas region for the long interval of time represented by these several periods, Cambrian, Ordovician, Silurian, Devonian, and Mississippian, had been relatively quiescent, as far at least as the available record shows. Orogenies may have occurred in the now concealed coastal plains region, but if so their effect was so far reduced in the central Texas region as to appear only as disconformities, overlaps, and apparently relatively slight regional elevation and depression.

The crust of Mother Earth, never entirely quiescent, apparently can not remain at ease indefinitely, and at or near the close of Mississippian or the beginning of Pennsylvanian time occurred unmistakably great orogenies in the coastal plains of Texas. This region is now hopelessly concealed from view by a Cretaceous and Cenozoic covering, but the reality of mountain making at the close of Mississippian time is eloquently witnessed by the great body of sediments carried westward from that region in early Pennsylvanian time. A great thickness of the sediments which accumulated may be seen in the Marathon region of Texas and in the Ouachita region in Oklahoma adjacent to Texas. In the intervening region between these exposures along the inner margin of the coastal plain these sediments are largely covered by later deposits and are known only from drill records. In the early Pennsylvanian, likewise, the seas spread extensively across central and Trans-Pecos Texas, and in these seas were accumulated the limestones and shales of the early Pennsylvanian Bend series.

The orogeny of late Mississippian time was followed, after a more or less quiescent interval, by possibly equally pronounced disturbances occurring in or near mid-Pennsylvanian time. This second outbreak of diastrophism affected, as did the first, the land mass of the coastal plains. Its center of influence was apparently farther to the west than was that of the earlier disturbance, since it obliterated some of the earlier basins of deposition adjacent to the land mass and established others farther west. Its effect in the central Texas region was likewise pronounced. The Llano region was folded and uplifted and subsequently severely eroded. The buried Red River and Amarillo Mountains are believed to have been chiefly made at this time. The geosyncline of early Pennsylvanian time adjacent to the Llanoria land mass was so far elevated as to become subject to erosion, and the Strawn and Kerr basins farther west were established.

Once again before the close of Pennsylvanian time, or near its close, extensive orogeny including folding and overthrusting manifested itself particularly in the Marathon region of Trans-Pecos Texas. Thus in contrast to the long quiescence of earlier Paleozoic time, three great periods of orogeny occurred within or almost within the period of time represented by the Pennsylvanian system.

Notwithstanding these orogenies the seas spread widely in Texas during the Pennsylvanian particularly in late Pennsylvanian, at which time more than half of the state, including all of central and much or all of west Texas, was submerged.

The Permian in Texas is difficult to characterize. In the main it was a time of increased, probably gradual, emergence so that the extensive seas inherited from Pennsylvanian time gave place at the end of the period to drylands. Facies of deposition in these basin seas which were undergoing dessication [desiccation] are extremely varied. Forming contemporaneously or essentially so in different basins, or to some extent in different parts of the same basin, were limestones, shales, sands, extensive reefs, anhydrite salt, and "red beds." Frequent alteration in the deposition of salt, anhydrite, or red beds so complicates the problem that the correlation of strata over an extensive area with their intergrading facies becomes the *bête noire* of the geologist. However, the success already obtained by a host of federal, state, and oil company geologists in correlating these Permian beds on the surface and, by means of well records, underground justifies quotation of that modern expression, "Who's afraid of the big, bad wolf." In other words, the Permian sediments of Texas, with all their varying facies, will in time be minutely subdivided and correlated from basin to basin. The wealth of oil which these formations contain is the principal driving force in these detailed studies, as it is also for many other formations. The Permian contains in addition valuable potash deposits.

The gradually emerging condition of the land, which resulted in the unusual depositional facies of the Permian seas, continued until these seas were obliterated and their basins a dry land plain.

The Mesozoic era marks a turning point in the history of the Texas region. The continental uplift at the end of the Paleozoic had excluded the sea from all of central and west Texas and so far as known from all parts of the state. This emergent condition continued through the first of the Mesozoic periods, the

Triassic, and in the main through the second period of this era, the Jurassic. Of the Triassic, continental and stream-laid deposits only are known in Texas. However, near the close of Jurassic time, a sea reaching northward from Mexico came into west Texas in the region of the Malone Mountains.

The encroachment of the Jurassic sea was a harbinger of the oncoming floods, for in Cretaceous time gradual submergence carried the entire Texas region below sea level, and the condition of complete emergence which closed Paleozoic time, gave place in the latter part of the Mesozoic era to that of complete submergence. This great change in relative elevation, which affected not the Texas region alone, but a large part of the continent, did not take place suddenly but on the contrary, very gradually. Beginning in Jurassic time, it reached the maximum flood stage for the Texas region near mid-Cretaceous time. Its retreat, likewise, was gradual and was characterized by fluctuations. By the close of the Cretaceous period land conditions had been re-established in the Texas region much as at present except that the extent to which the coastal plain was lifted above water is undetermined.

It was with the approach of this great continent-wide submergence that the drainage conditions previously referred to were reversed in Texas. The highlands which had occupied the site of the present coastal plains during Paleozoic time became lowlands and later were submerged. The westward-flowing rivers of Paleozoic time ceased and were replaced in time by eastward-flowing streams which emptied into the encroaching seas. Upon retreat of the seas approaching the close of Cretaceous time, the streams established themselves as eastward- or southeastward-flowing drainage systems.

Approximately marking the close of the Mesozoic era was a great world-wide orogenic movement known as the Laramide mountain-making period. The Laramide orogeny affected particularly west Texas but was there much less pronounced than in either the Rocky Mountain region to the north or to the south in Mexico. However, with the close of the Mesozoic era was fully established the present relative prosition [position] of highland and lowland in Texas, the highlands being in the west and the lowlands adjacent to the Gulf coast, with the drainage prevailingly eastward.

At the beginning of the Cenozoic era much of the present coastal plain was submerged. The history of this era is that of gradual retreat of the sea from the partly submerged coastal plain, not, however, without decided fluctuations. Advance and retreat of the sea seems to have occurred with almost pulsating regularity, but with retreat in excess of advance until through these successive stages the present coast line was established. During this era, possibly in its middle part, great lava flows occurred in Trans-Pecos Texas. Extensive block faulting likewise affected the west Texas region. These faults trend in the main northwest-southeast to north-south. To their influence is to be credited such outstanding structural features as the Franklin Mountains of the El Paso region and the great Salt Flat valley or graben of the Van Horn region.

This survey, restricted as it has been, nevertheless indicates a succession of events in Texas which are a part of the greater history of the continent as a whole. The limitations have been such that it has been necessary to pass almost unmentioned the records afforded by the fossils of the Texas region. Space does not permit speaking of the invertebrates which abound from the Cambrian to the

Pleistocene, nor of the Permian and Triassic reptiles among which are such bizarre forms as are not elsewhere seen. Fossil plants are present in Pennsylvanian, Permian, and some later formations. An important record has been left, also, of the mammalian faunas which inhabited the Texas plains during the latter part of the Cenozoic era.

In the orogenies of the Texas region two trends are dominant, northeast-southwest and the opposite direction, northwest-southeast. Northeast trends are notably pronounced in the orogenies of Pennsylvanian time but are found attending pre-Cambrian and other disturbances. Southeast trends, likewise, occur repeatedly with successive orogenies. This trend, varying to almost or quite north-south is notably dominant in the Cordilleran orogeny.